SOLIDWORKS 2016
BLACK BOOK

By
Gaurav Verma
Matt Weber
(CADCAMCAE Works)

Edited by
Kristen

ℋCW

ISBN # 978-0-9950974-0-7

NOTICE TO THE READER

DEDICATION

To teachers, who make it possible to disseminate knowledge
to enlighten the young and curious minds
of our future generations

To students, who are the future of the world

THANKS

To my friends and colleagues

To my family for their love and support

Training and Consultant Services

At CADCAMCAEWORKS, we provides effective and affordable one to one online training on various software packages in Computer Aided Design(CAD), Computer Aided Manufacturing(CAM), Computer Aided Engineering (CAE), Computer programming languages(C/C++, Java, .NET, Android, Javascript, HTML and so on). The training is delivered through remote access to your system and voice chat via Internet at any time, any place, and at any pace to individuals, groups, students of colleges/universities, and CAD/CAM/CAE training centers. The main features of this program are:

Training as per your need

Highly experienced Engineers and Technician conduct the classes on the software applications used in the industries. The methodology adopted to teach the software is totally practical based, so that the learner can adapt to the design and development industries in almost no time. The efforts are to make the training process cost effective and time saving while you have the comfort of your time and place, thereby relieving you from the hassles of traveling to training centers or rearranging your time table.

Current Software Packages on which we provide basic and advanced training are:

CAD/CAM/CAE: CATIA, Creo Parametric, Creo Direct, SolidWorks, Autodesk Inventor, Solid Edge, UG NX, AutoCAD, AutoCAD LT, EdgeCAM, MasterCAM, SolidCAM, DelCAM, BOBCAM, UG NX Manufacturing, UG Mold Wizard, UG Progressive Die, UG Die Design, SolidWorks Mold, Creo Manufacturing, Creo Expert Machinist, NX Nastran, Hypermesh, SolidWorks Simulation, Autodesk Simulation Mechanical, Creo Simulate, Gambit, ANSYS and many others.

Computer Programming Languages: C++, VB.NET, HTML, Android, Javascript and so on.

Game Designing: Unity.

Civil Engineering: AutoCAD MEP, Revit Structure, Revit Architecture, AutoCAD Map 3D and so on.

We also provide consultant services for Design and development on the above mentioned software packages

For more information you can mail us at:
cadcamcaeworks@gmail.com

TABLE OF CONTENTS

Analysis Express

Mold Tools

Sheet Metal Introduction

Weldments

Preface

SolidWorks 2016 is a parametric, feature-based solid modeling tool that not only unites the three-dimensional (3D) parametric features with two-dimensional (2D) tools, but also addresses every design-through-manufacturing process. The continuous enhancements in the software has made it a complete PLM software. The software is capable of performing analysis with an ease. Its compatibility with CAM software is remarkable. Based mainly on the user feedback, this solid modeling tool is remarkably user-friendly and it allows you to be productive from day one.

The **SolidWorks 2016 Black Book** is third edition of our series on SolidWorks. With lots of additions and thorough review, we present a book to help professionals as well as learners in creating some of the most complex solid models. The book follows a step by step methodology. In this book, we have tried to give real-world examples with real challenges in designing. We have tried to reduce the gap between university use of SolidWorks and industrial use of SolidWorks. In this edition of book, we have included self-assessment questions at the end of each chapter. The book covers almost all the information required by a learner to master the SolidWorks. The book starts with sketching and ends at advanced topics like Mold Design, Sheetmetal, and Weldment. Some of the salient features of this book are :

In-Depth explanation of concepts
Every new topic of this book starts with the explanation of the basic concepts. In this way, the user becomes capable of relating the things with real world.

Topics Covered
Every chapter starts with a list of topics being covered in that chapter. In this way, the user can easy find the topic of his/her interest easily.

Instruction through illustration
The instructions to perform any action are provided by maximum number of illustrations so that the user can perform the actions discussed in the book easily and effectively. There are about 1000 illustrations that make the learning process effective.

Tutorial point of view
At the end of concept's explanation, the tutorial make the understanding of users firm and long lasting. Almost each chapter of the book has tutorials that are real world projects.

Project
Free projects and exercises are provided to students for practicing.

For Faculty
If you are a faculty member, then you can ask for video tutorials on any of the topic, exercise, tutorial, or concept.

Formatting Conventions Used in the Text
All the key terms like name of button, tool, drop-down etc. are kept bold.

Free Resources
Link to the resources used in this book are provided to the users via email. To get the resources, mail us at *cadcamcaeworks@gmail.com* with your contact information. With your contact record with us, you will be provided latest updates and informations regarding various technologies. The format to write us mail for resources is as follows:

Subject of E-mail as *Application for resources of _____book*.
Also, given your information like
Name:
Course pursuing/Profession:
Contact Address:
E-mail ID:

Note: We respect your privacy and value it. If you do not want to give your personal informations then you can ask for resources without giving your information.

For Any query or suggestion
If you have any query or suggestion, please let us know by mailing us on *cadcamcaeworks@gmail.com*. Your valuable constructive suggestions will be incorporated in our books and your name will be addressed in special thanks area of our books on your confirmation.

About Authors

The author of this book, Matt Weber, has written many books on CAD/CAM/CAE available already in market. **SolidWorks Simulation Black Book** is one of the most selling book in SolidWorks Simulation field. The author has hands on experience on almost all the CAD/CAM/CAE packages. If you have any query/doubt in any CAD/CAM/CAE package, then you can contact the author by writing at cadcamcaeworks@gmail.com

The author of this book, Gaurav Verma, has written and assisted in more than 10 titles in CAD/CAM/CAE which are already available in market. He has authored **AutoCAD Electrical 2015 Black Book** which is available in both **English** and **Russian** language. He has provided consultant services to many industries in US, Greece, Canada, and UK.

Starting with SolidWorks

Chapter 1

Topics Covered

The major topics covered in this chapter are:

- *Installing SolidWorks 2016.*
- *Starting SolidWorks 2016.*
- *Starting a new document.*
- *Terminology used in SolidWorks.*
- *Opening a document.*
- *Closing documents.*
- *Basic Settings for SolidWorks*
- *Workflow in Industries using the SolidWorks*

INSTALLING SOLIDWORKS 2016

- If you are installing SolidWorks using the CD/DVD provided by Dassault Systemes then go to the folder containing **setup.exe** file and then right click on **setup.exe** in the folder. A shortcut menu is displayed on the screen; refer to Figure-1.

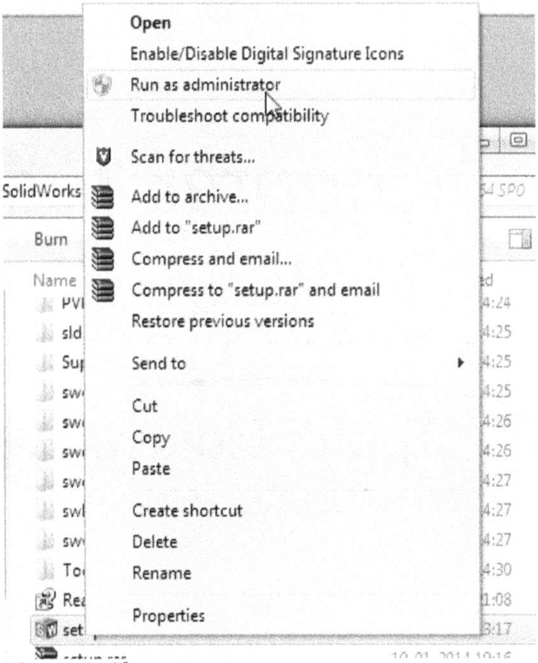

Figure-1. Shorcut menu

- Select the **Run as Administrator** option from the menu displayed; refer to Figure 1.

- Select the **Yes** button from the dialog box displayed. The **SolidWorks 2016 Installation Manager** will be displayed. Follow the instructions given in the dialog box. Note that you must have the **Serial Number** with you to install the application. To get more about installation, double click on the **Read Me** documentation file displayed above the **setup.exe** file in the Setup folder.

- If you have downloaded the software from Internet, then you are required to browse in the **SolidWorks Download** folder in the **Documents** folder. Open the folder of latest version of software and then run **setup.exe**. Rest of the procedure is same.

STARTING SOLIDWORKS 2016

- To start SolidWorks from **Start** menu, click on the **Start** button in the Taskbar at the bottom left corner, click on the **All Programs** folder and then on the **SolidWorks 2016** folder. In this folder, select the SolidWorks 2016 icon; refer to Figure-2

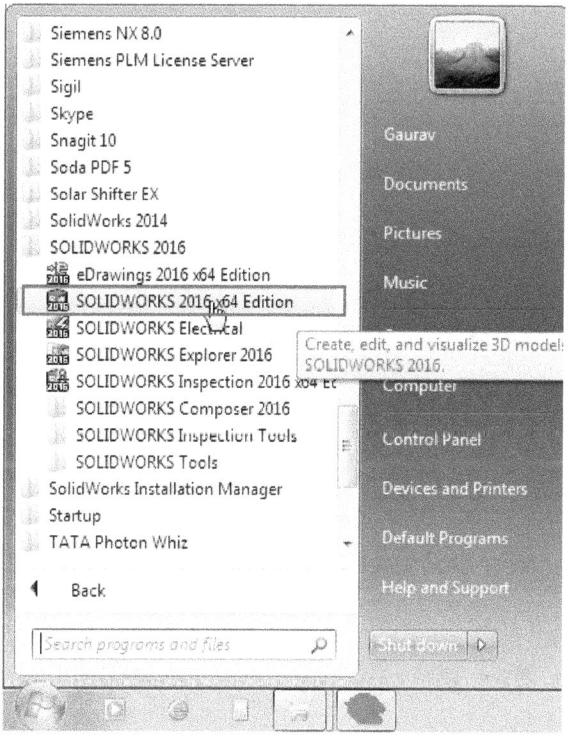

Figure-2. Start menu

- While installing the software, if you have selected the check box to create a desktop icon, then you can double click on that icon to run the software.

- If you have not selected the check box to create the desktop icon but want to create the icon on desktop, then right click on the **SolidWorks 2016** icon in the Start menu and select the **Send To-> Desktop (Create icon)** option from the shortcut menu displayed.

After you perform the above steps, the SolidWorks 2016 application window will be displayed; refer to Figure-3.

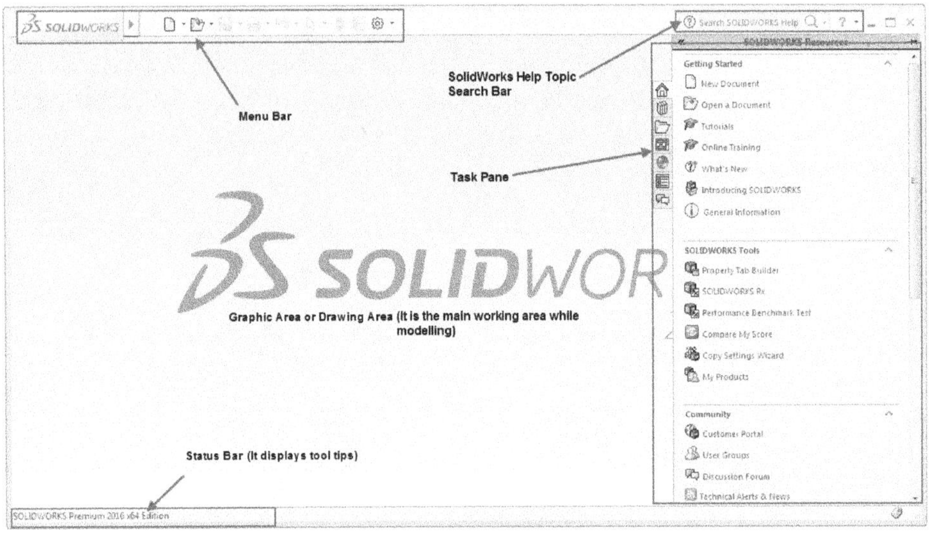

Figure-3. SolidWorks 2016 application window

STARTING A NEW DOCUMENT

You can start SolidWorks by four ways:

1. Click on the **New Document** link in the Task Pane; refer to Figure-3.
Or
2. Click on the **New** button in the **Menu Bar** .
Or
3. Move the cursor on the left arrow near the **SolidWorks icon**; refer to Figure-4 and then click on the **File** menu button and click on the **New** button; refer to Figure-4.
Or
4. Press **CTRL** and **N** together from the Keyboard.

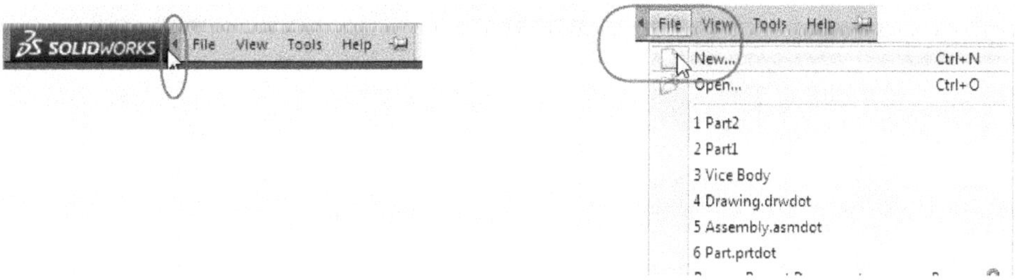

Figure-4. File menu

- After performing any of the above steps, the **New SolidWorks Document** dialog box will be displayed as shown in Figure-5.

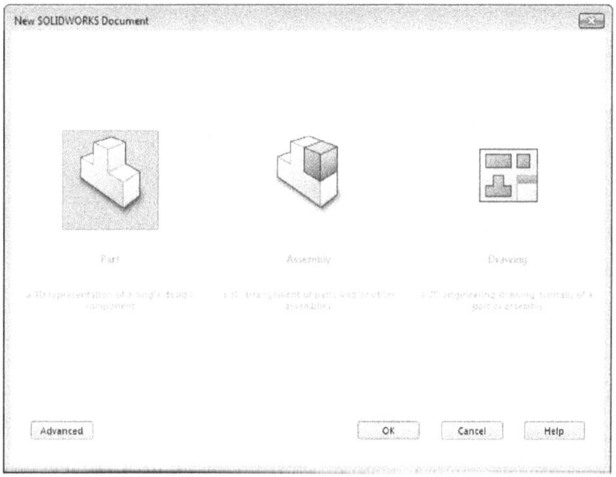

Figure-5. New solidworks document

- There are three buttons available in this dialog box; **Part, Assembly** and **Drawing**.

The **Part** button is used to create Solid or Surface models.

The **Assembly** button is used to create Assemblies.

The **Drawing** button is used to create drawings from the solid/surface models or assemblies.

You will learn more about solids, surfaces, assemblies, and drawings later in the book.

Note that the building block of CAD is solid models. In SolidWorks, solid models are created by using the tools available in the **Part** mode. You can start with the **Part** mode by selecting the **Part** button.

• Double-click on the **Part** button to start the Solid/ surface modeling environment of SolidWorks. On doing so, the application interface will be displayed as shown in Figure-6.

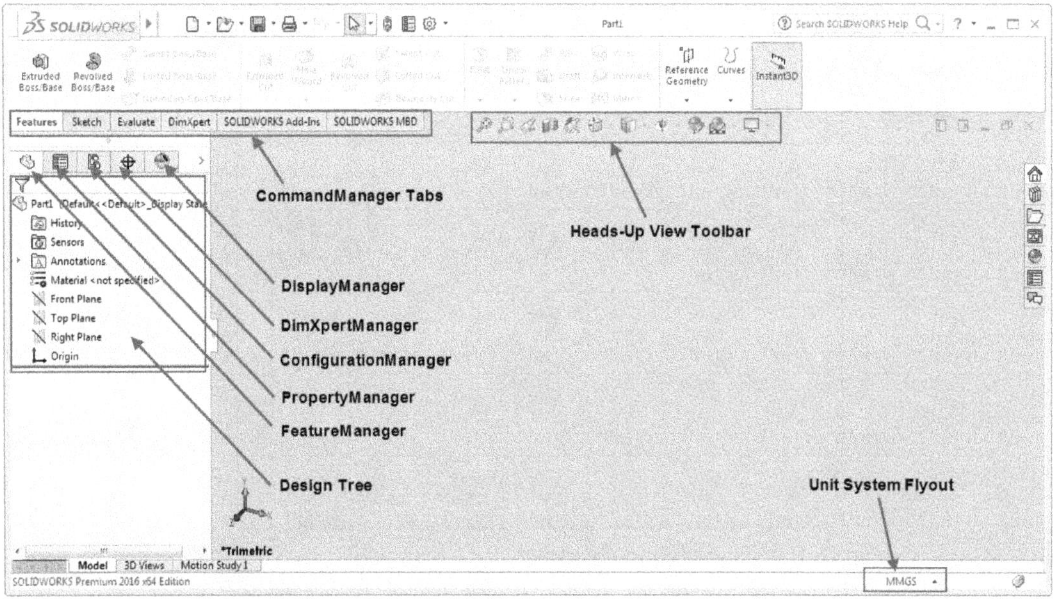

Figure-6. Application interface

The tools available in the **Part, Assembly** and **Drawing** mode are compiled in the form of CommandManagers. Various CommandManagers available in SolidWorks are discussed next.

Part Mode CommandManagers

A number of **CommandManagers** can be invoked in the **Part** mode. These **CommandManagers** with their functioning are described next.

Sketch CommandManager

The tools available in this **CommandManager** are used to draw sketches for creating solid/surface models. This **CommandManager** is also used to add relations and smart dimensions to the sketched entities. The **Sketch CommandManager** is shown in Figure-7.

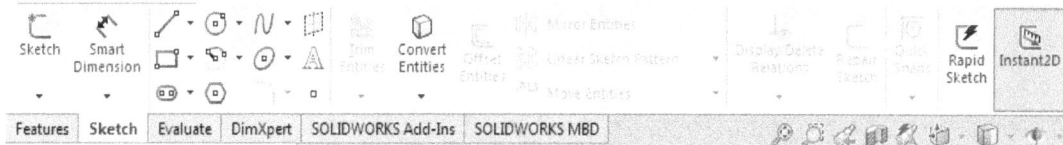

Figure-7. Sketch CommandManager

Features CommandManager

This **CommandManager** provides all modeling tools that are used for feature-based solid modeling. The **Features CommandManager** is shown in Figure-8.

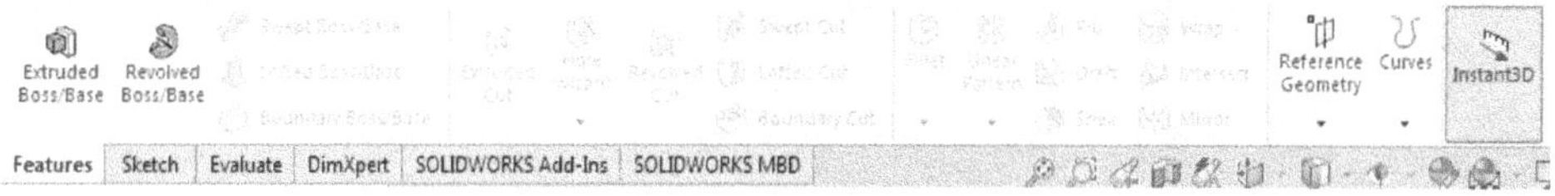

Figure-8. Features CommandManager

DimXpert CommandManager

This **CommandManager** is used to add dimensions and tolerances to the features of a part. The **DimXpert CommandManager** is shown in Figure-9.

Figure-9. DimXpert CommandManager

Sheet Metal CommandManager

The tools in this **CommandManager** are used to create the sheet metal parts. The **Sheet Metal CommandManager** shown in Figure-10. If this **CommandManager** is not added in the **Ribbon**, then right-click on any of the CommandManager tab and select the **Sheet Metal** option from the menu; refer to Figure-11.

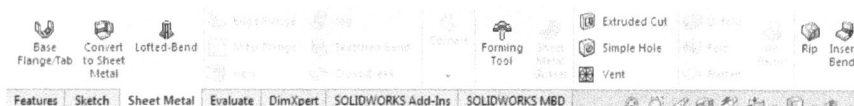

Figure-10. Sheet Metal CommandManager

Figure-11. Adding hidden tabs in Ribbon

Mold Tools CommandManager

The tools in this **CommandManager** are used to design a mold and extract its core and cavity. The **Mold Tools CommandManager** is shown in Figure-12.

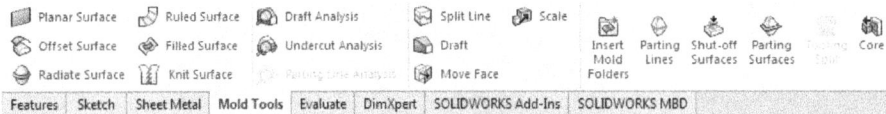

Figure-12. Mold Tools CommandManager

Evaluate CommandManager

This **CommandManager** is used to measure entities, perform analysis and so on. The **Evaluate CommandManager** is shown in Figure-13.

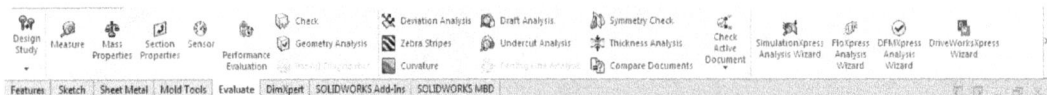

Figure-13. Evaluate CommandManager

Surfaces CommandManager

This **CommandManager** is used to create complicated surface features. The **Surfaces CommandManager** is shown in Figure-14.

Figure-14. Surfaces CommandManager

Direct Editing CommandManager

This **CommandManager** consists of tools (Figure-15) that are used for editing a feature.

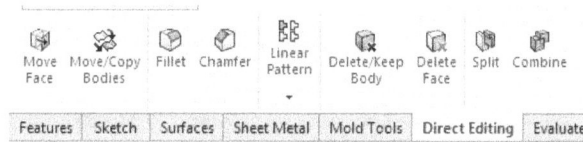

Figure-15. Direct Editing CommandManager

Data Migration CommandManager

This **CommandManager** consist of tools (Figure-16) that are used to work with the models created in other packages or in different environments.

Figure-16. Data Migration CommandManager

Weldments CommandManager

This **CommandManager** is used to create welding joints in the model and assembly. The **Weldments CommandManager** is shown in Figure-17.

Figure-17. Weldments CommandManager

Assembly Mode CommandManagers

The **CommandManagers** in the **Assembly** mode are used to assemble the components. The **CommandManagers** in the **Assembly** mode are discussed next.

Assembly CommandManager

This **CommandManager** is used to insert a component and apply various types of mates to the assembly. The **Assembly CommandManager** is shown in Figure-18.

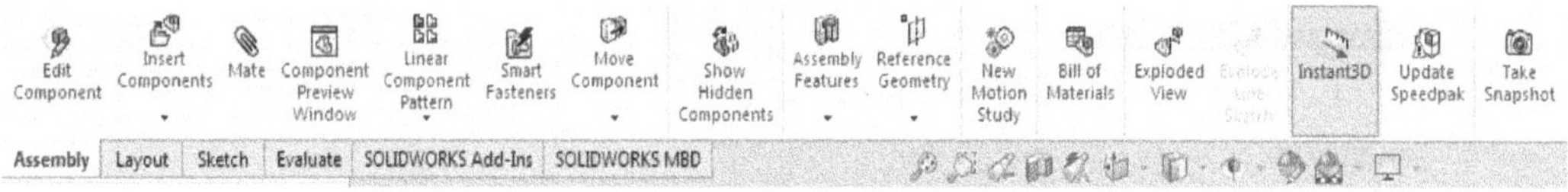

Figure-18. Assembly CommandManager

Layout CommandManager

The tools in this **CommandManager** (Figure-19) are used to create and edit blocks.

Figure-19. Layout CommandManager

Drawing Mode CommandManagers

You can invoke a number of **CommandManagers** in the **Drawing** mode. The **CommandManagers** that are extensively used during the designing process in this mode are discussed next.

View Layout CommandManager

This **CommandManager** is used to generate the drawing views of an existing model or an assembly. The **View Layout CommandManager** is shown in Figure-20.

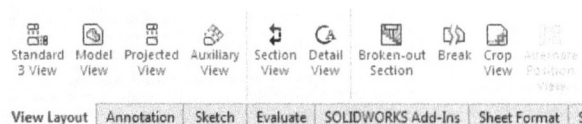

Figure-20. View Layout CommandManager

Annotation CommandManager

The **Annotation CommandManager** is used to generate the model items and to add notes, balloons, geometric tolerance, surface finish symbols, and so on to the drawing views. The **Annotation CommandManager** is shown in Figure-21.

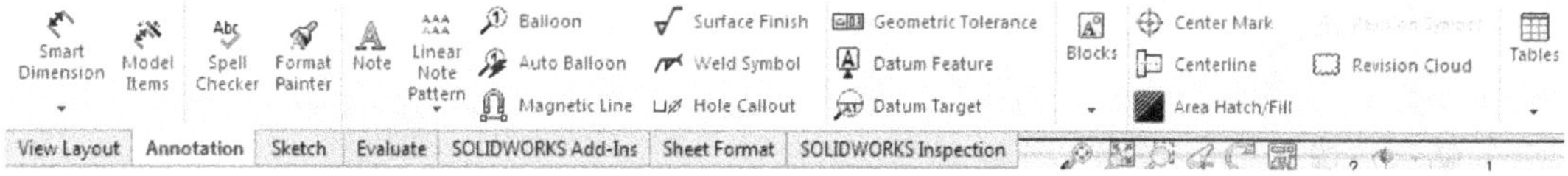

Figure-21. Annotation CommandManager

The commands available in these CommandManagers will be discussed one by one later in this book.

OPENING A DOCUMENT

Like creating new documents, there are many ways to open documents, some of them are discussed next.

1. Click on the **Open a Document** link in the Task Pane; refer to Figure-3.

2. Click on the **Open** button in the **Menu Bar** .

3. Move the cursor on the left arrow near the **SolidWorks icon**; refer to Figure-4 and then click on the **File** menu button and click on the **Open** button; refer to Figure-4.

4. Press **CTRL** and **O** together from the Keyboard.

5. After performing any of the above step, the **Open** dialog box will be displayed; refer to Figure-22.

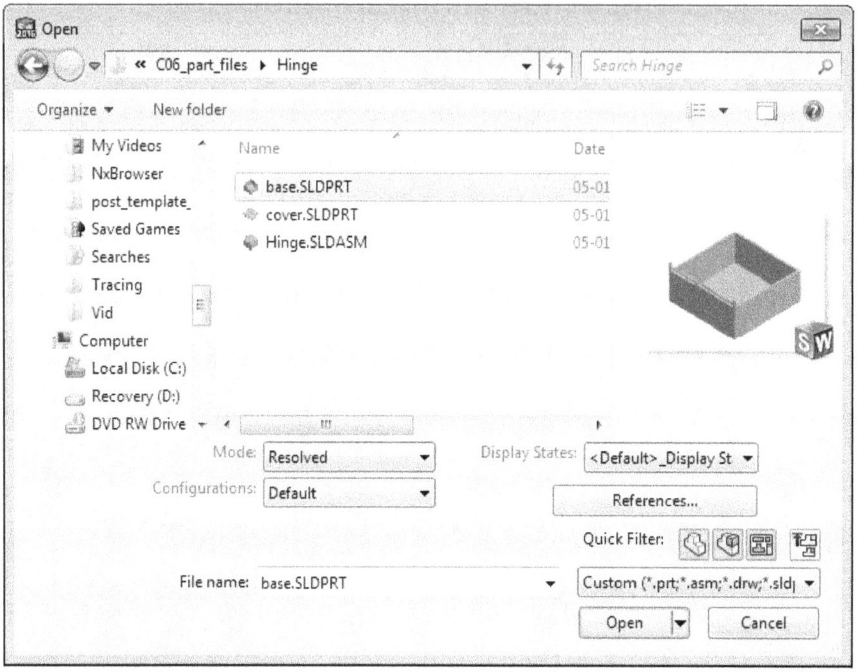

Figure-22. Open dialog box

6. Select the file type of your file from the **File Type** fly-out `Custom (*.prt;*.asm;*.drw;*.sld| ▼)` in the bottom right corner of the dialog box.

7. Browse to the folder in which you have saved the file and then double-click on it to open.

Most of the time people close the application to close the current file and then restart the application to start working again. In SolidWorks, you can close the current file while the application remains open. The steps to do so are given next.

CLOSING A DOCUMENT

To close a document, there are two ways displayed in Figure-23.

- Open the **File** menu and then click on the **Close** option from it.
- Click on the **Close** button ✕ at the top-right of the current viewport.(Viewport is the area in which the model is displayed.)

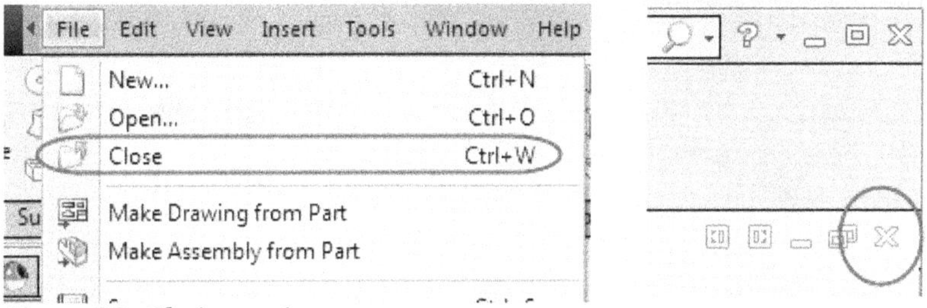

Figure-23. Closing options

- If you have done some editing in the document then the dialog box will be displayed, prompting you to save the document; refer to Figure-24.

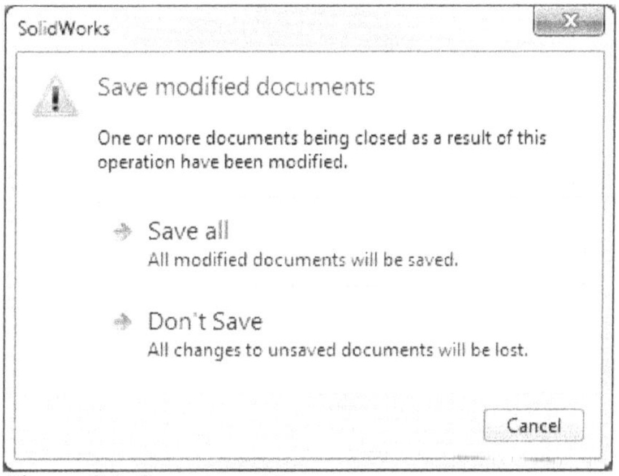

Figure-24. Save prompt

- Click on the **Save all** button to save the changes or click on the **Don't Save** button to reject the changes.

Till this point you know, starting SolidWorks; Creating, Opening, and Closing documents; and you have idea about the interface of SolidWorks. Now, we will discuss about some basic settings that are required for easy working with SolidWorks.

BASIC SETTINGS OF SOLIDWORKS

All the settings of SolidWorks are compiled in the **Options** dialog box. The steps to change the settings for SolidWorks are given next.

- Click on the **Options** option in the **Tools** menu or click on the **Options** button ⚙ from the Menu Bar.

- On performing the above step, the **System Options** dialog box will be displayed as shown in Figure-25. Note that if you have a document opened then the **Document Properties** tab is also added with the **System Options** tab. To get the detail about each and every option, you need to refer to SolidWorks Help Documentation. In this section, we will discuss about some of the important options that are generally required.

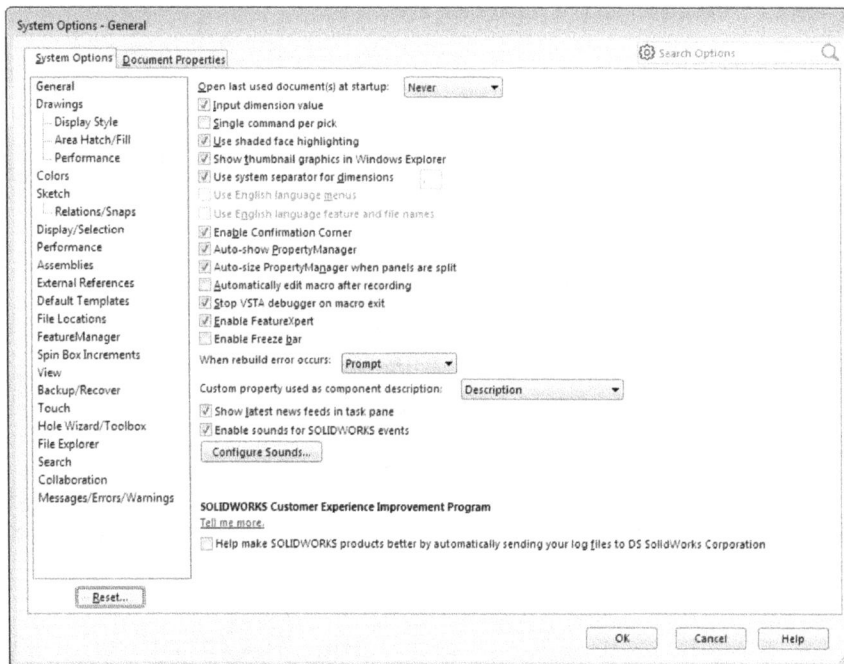

Figure-25. Partial System Options dialog box

- Click on the **Sketch** option in the left of the dialog box, select the **Enable on screen numeric input on entity creation** check box from the right to enter dimensions while creating the sketch.
- Click on **Relations/Snaps** in the left of the dialog box and then select the **Enable snapping** check box to enable auto snapping to the key points.

- Click on the **Documents Properties** tab if you have any document opened in the viewport. Click on the **Units** option from the left. The **Options** dialog box will be displayed as shown in Figure-26.

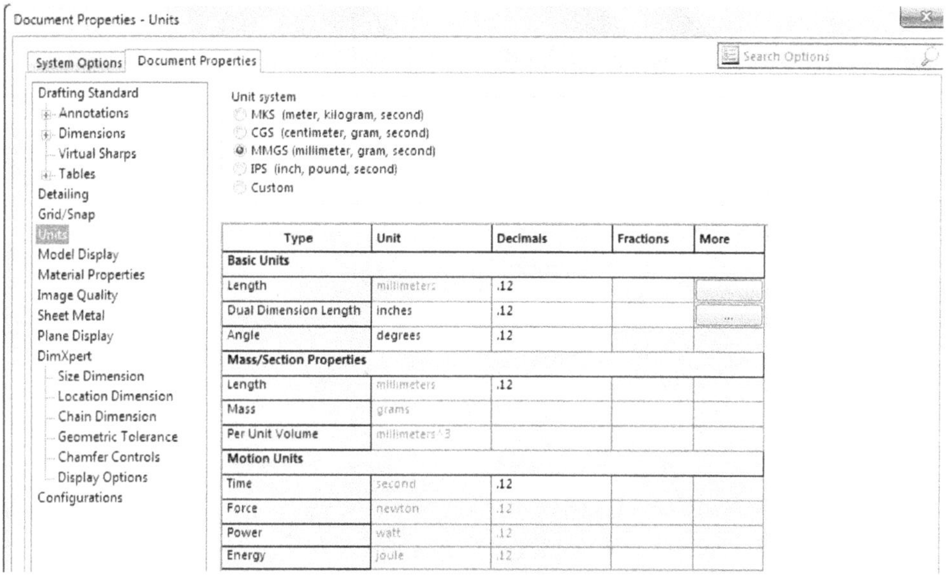

Figure-26. Document properties tab

- Select the desired radio button from the right to set the unit system for the current document.
- Click on the **OK** button from the bottom of the dialog box to save the settings.

You can also change the units of document by selecting the desired option from the list displayed on clicking on the **Unit system** flyout at the bottom of the viewport; refer to Figure-27.

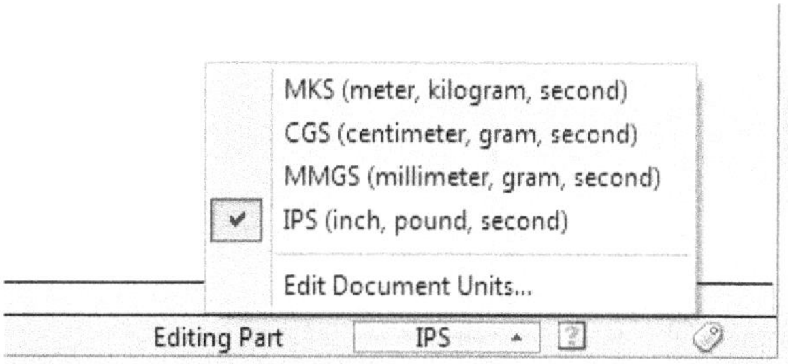

Figure-27. Unit system flyout

WORKFLOW IN SOLIDWORKS

The first step in SolidWorks in to create a sketch. After creating sketch of the desired feature, we create solid or surface model from that sketch. After doing the desired operations on the solid/surface model, we go for assembly or analyses. After, we are satisfied with the assembly/ analyses, we create the engineering drawings from the model to allow manufacturing of the model into a real world object.

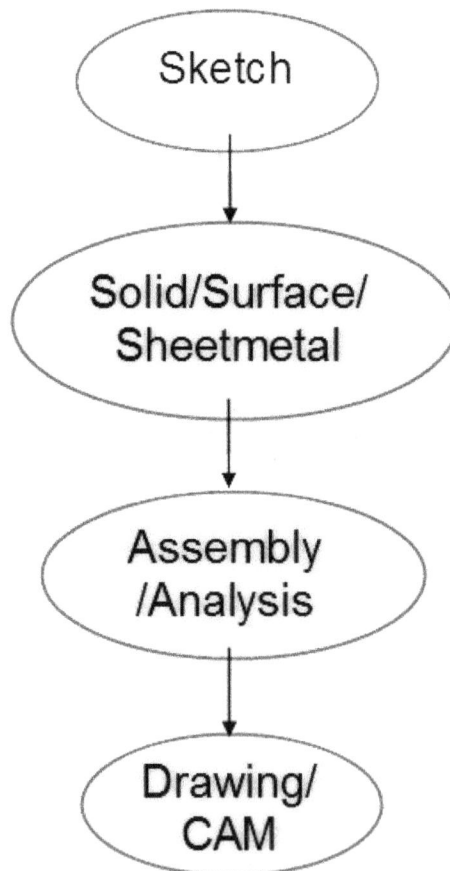

```
        ┌─────────────┐
        │   Sketch    │
        └──────┬──────┘
               │
               ▼
    ┌──────────────────────┐
    │   Solid/Surface/     │
    │     Sheetmetal       │
    └──────────┬───────────┘
               │
               ▼
    ┌──────────────────────┐
    │     Assembly         │
    │     /Analysis        │
    └──────────┬───────────┘
               │
               ▼
        ┌─────────────┐
        │  Drawing/   │
        │    CAM      │
        └─────────────┘
```

SELF ASSESSMENT

Q1. If you have downloaded the SolidWorks Setup files from Internet then the files will be available in Downloads folder of Windows by default. (T/F)

Q2. We cannot create the desktop icon of SolidWorks 2016 if we have not opted for it while installing. (T/F)

Q3. Status Bar in SolidWorks window also displays the tips related to current tool. (T/F)

Q4. Which of the following operation results in display of **New SolidWorks Document** dialog box?
(a)   Press CTRL+N
(b)   Click on **New** button from Menu Bar
(c)   Click on **New Document** link button from Task Pane.
(d)   All of the above

Q5. Which of the following is not an option in the New SolidWorks Document dialog box?
(a)   Part
(b)   Assembly
(c)   Drawing
(d)   Sketch

Q6. The _____ CommandManager provides all modeling tools that are used for feature-based solid modeling.

Q7. The _____ CommandManager is used to add dimensions and tolerances to the features of a part.

Q8. The _____ CommandManager is used to insert a component and apply various types of mates to the assembly.

Q9. Write down the steps to close the current document in SolidWorks.

Q10. How can we change the unit system of current SolidWorks document?

FOR STUDENT NOTES

Sketching
Basic to Advanced

Chapter 2

Topics Covered

The major topics covered in this chapter are:

- *Basics for Sketching.*
- *Entity creation tools.*
- *Entity Editing tools.*
- *Dimensioning and Constraining.*
- *3D Sketching.*
- *Printing and exporting sketch.*

BASICS FOR SKETCHING

In Engineering, sketching do not mean sketches of birds or animals. It means sketches that are based on real dimensions of real-world objects. In this chapter, we will be working with geometric entities like; line, circle, arc, ellipse and so on. But this time, we will be using the software tools in place of pencil, scale and other geometry tools. Note that the sketching environment is the building base of 3D Models so you should be proficient in sketching.

To start with Sketching, we must have a good understanding of planes in SolidWorks. Next figure (Figure-1) shows the names of planes and their respective faces.

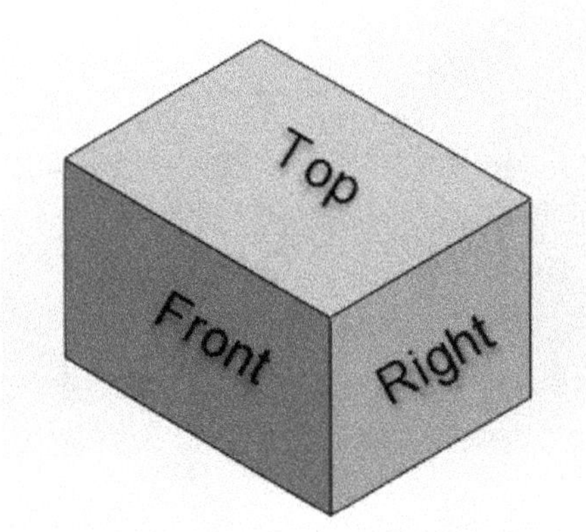

Figure-1. Planes

In SolidWorks, the planes are displayed in the same orientation as shown in the above figure. To check the planes of SolidWorks, click on the planes (Front Plane, Top Plane, Right Plane) in the **FeatureManager Design Tree**; refer to Figure-2.

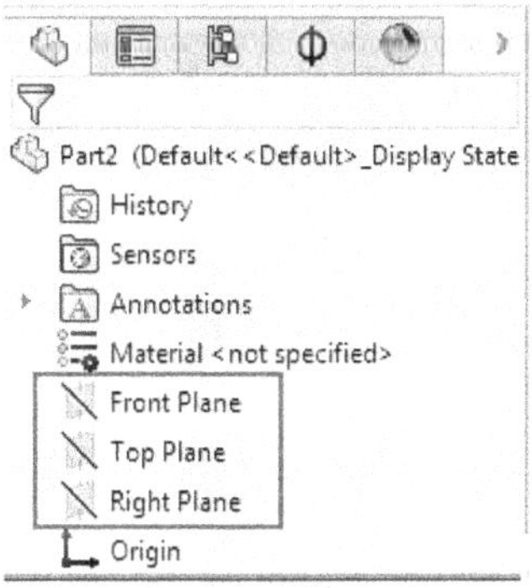

Figure-2. Planes in Property-Manager

- To show these planes, select them one by one while holding the CTRL key and right-click. A shortcut menu will be displayed; refer to Figure-3.

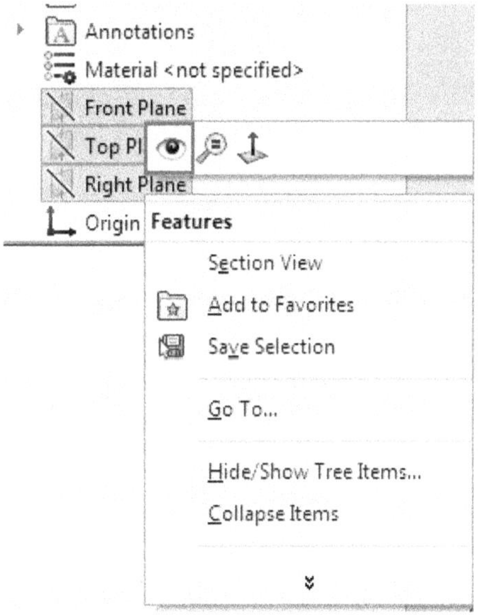

Figure-3. Shortcut menu on right clicking on planes

- Select the **Show** button 👁 from the shortcut menu; the planes will be displayed. To hide them again, you are required to do the above procedure again.

The **Heads-up View Toolbar** contains tools to change the view and orient the modeling area; refer to Figure-4.

Figure-4. Heads-up View toolbar

The functions of the tools displayed in above figure are discussed next.

Zoom to Fit : This tool is used to display all the objects created in the viewport. To use this tool, click on the tool once. The objects will automatically fit in the current viewport.

Zoom to Area : This tool is used to display a specific area in the viewport zoomed to the full extent. To use this tool, click on it. The cursor will change to a zoom box selection cursor and you are asked to create a boundary box surrounding the entities you want to zoom in. Click to specify the starting point of the zoom box and then drag to the point till where you want to complete the zoom box. The area in the box will zoom automatically. Figure-5 shows a zoom box drawn to zoom.

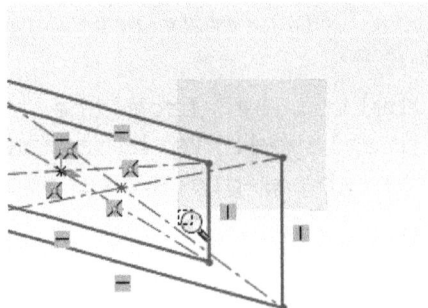

Figure-5. Zoom box

Previous View : This tool is used to zoom to the previous level. To use this tool, click on it once. The viewport will be displayed at the previous level.

Section View : This tool is used to display section of a solid model. (Section is created when you cut a solid from a plane. It is mainly used to see the inside of the model.)

Dynamic Annotation View : This tool is used to display or hide the annotations applied to the model in Part/Assembly environment. You will learn more about this tool in Solid Modeling.

View Orientation : This tool is used to change the view orientation of the model. When you click on this button, a toolbox will be displayed as shown in Figure-6. The buttons perform the action mentioned in the figure.

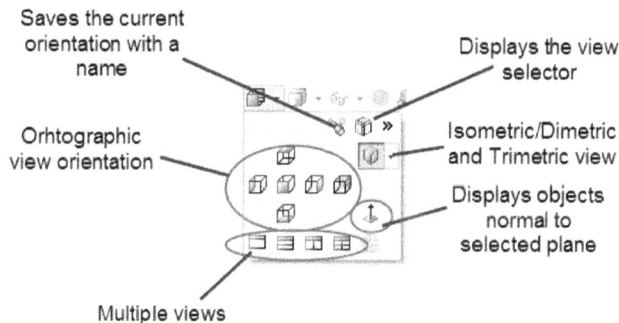

Figure-6. View orientation toolbox

Display Style : These tools are used to display the model is different styles, like shaded, hidden, no hidden and so on.

Hide/Show Items : When you click on this button, a tool box will be displayed with various toggle buttons. These buttons allow to display or hide key feature like, geometric relations, center lines, annotations, grids, and so on. To enable or disable the view of a key feature, select the respective button.

Next three buttons will be discussed later in the book. Now, we are ready to start with sketch.

Start SolidWorks, create a new Part document and then click on the **Sketch** tab. The tools for sketching are displayed in the **Sketch CommandManager**.

STARTING SKETCH

- Click on the **Sketch** button at the left in the **Sketch** CommandManager. Three main planes are displayed with their name in the viewport; refer to Figure-7.

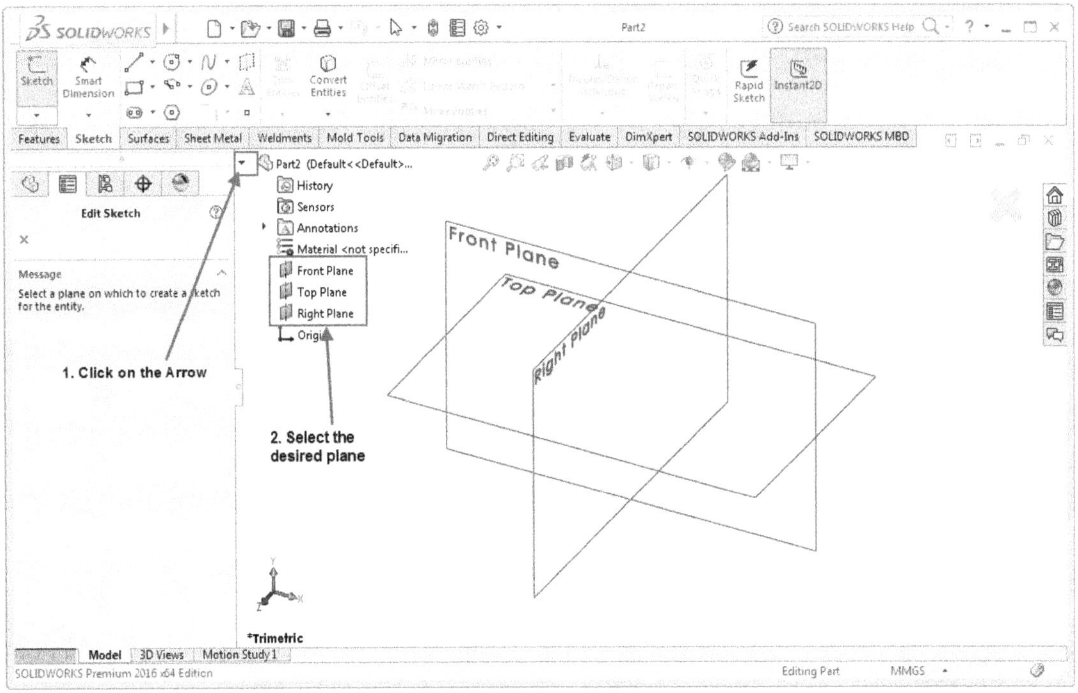

Figure-7. Sketching on default planes

- Click on the desired plane from the viewport or click on the arrow displayed in the top-left corner of the viewport and select the name of the plane from the list of features; refer to Figure-7.
- On clicking on a plane, the selected plane will become parallel to the screen. Now, we are ready to draw sketch on the plane. The **Sketch CommandManager** is divided into

sections depending on the functions of the tools. Figure-8 shows the **Sketch CommandManager** and the division of tools.

Figure-8. Sketch CommandManager

First, we will start with the sketch creation tools and then one by one we will discuss about other tools.

SKETCH CREATION TOOLS

The standard tools to draw sketch entities are available in this section of **Ribbon**. These tools are discussed next.

Line

There are three tools in the **Line** drop-down to create different type of lines; **Line, Centerline** and **Midpoint Line**; refer to Figure-9.

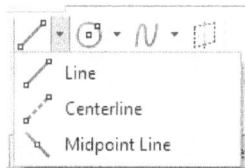

Figure-9. Line drop-down

Line

We use this tool to create every type of lines required in creating sketch. The procedure to create line is explained in the next given steps:

- Click on the **Line** tool from the drop-down. The **Line PropertyManager** will be displayed; refer to Figure-10.

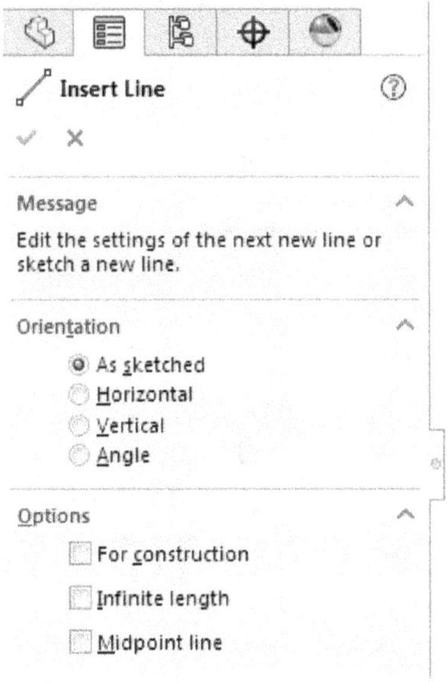

Figure-10. Line PropertyManager

- The radio buttons in the **Orientation** rollout are used to set the orientation of the line before drawing it. There are four radio buttons available in this rollout; **As sketched**, **Horizontal**, **Vertical**, and **Angle**. The **As sketched** radio button is selected by default. So, you do not need to define the orientation of line, you can just start creating line. Click in the drawing area to create the line.

Select the **Horizontal** radio button, if you want to create horizontal line. On selecting this radio button, the **Parameters** rollout will be displayed at the bottom of the rollout. Specify the length of the line in the edit box displayed in the **Parameters** rollout and click to specify start point of line to create the line with specified length.

Select the **Vertical** radio button, if you want to create the vertical line. This works in the same way as the horizontal option works.

Select the **Angle** radio button, if you want to create the lines at the specified angle. On selecting this radio button, two edit boxes will become available. Specify the length of the line and angle of the line in the respective edit boxes; refer to Figure-11.

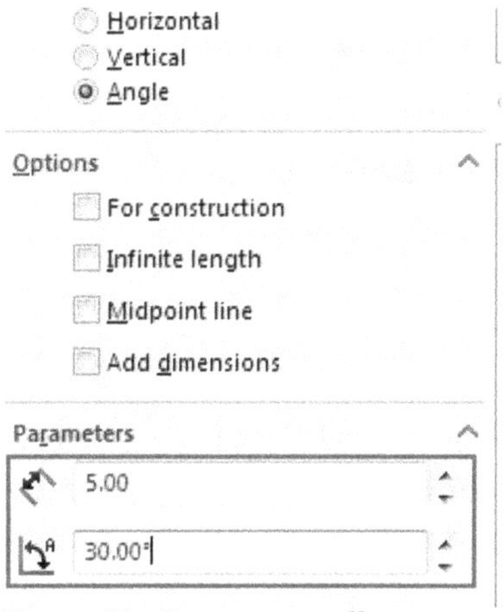

Figure-11. Parameters rollout

- The check boxes in the **Options** rollout are used to modify the properties of line while creating it. On selecting the **For construction** check box, you will create a line in construction mode. On selecting the **Infinite length** check box, you can create a line of infinite length. The **Midpoint line** check box was added in SolidWorks 2015. On selecting this check box, you can create a line with the help of mid point and end point. The **Add dimensions** check box is used to add dimensions while creating the line.
- Most of the time, we use the **As sketched** radio button to create lines in sketch. Select the **As sketched** radio button. If you have earlier created line by using any other radio button then click on the OK button from the **PropertyManager** and select the As sketched radio button.
- When you move the cursor in the viewport. By default the cursor snaps to the key points like horizontal/ vertical to coordinate system, coincident to the coordinate system and so on. Click to specify the start point on the screen. Figure-12 shows the creation of line.

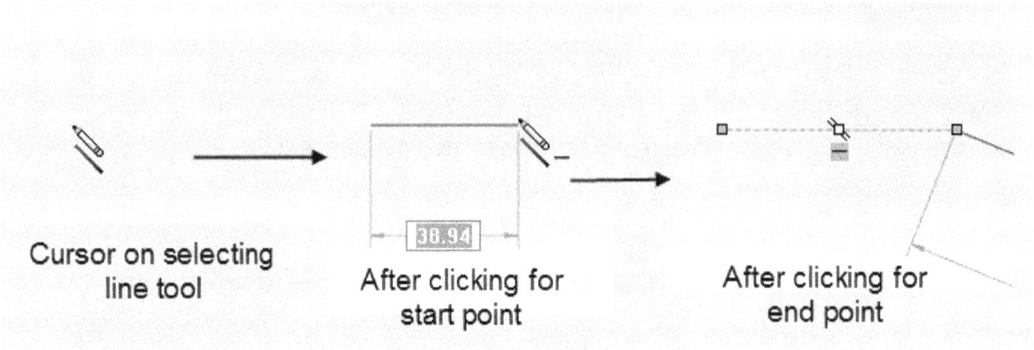

Figure-12. Line creation

- Press ESC button from the keyboard to exit the tool.

Now, stop reading the book and first practice on the tool by using all the options one by one. From now onwards, you should practice on the tool as soon as it has been discussed because we are not in theory business!!

Centerline

The **Centerline** tool is used to create center line in the viewport. The procedure to create a center line is the same as line creation. Click on this tool, specify the start point of the center line and then specify the end point of the center line. Hence, the center line is created.

Note that while drawing a line using the **Line** tool, if you select the **For construction** check box; then the line is created similar to center line but it will not act as centerline when you will be using **Revolve** tool later.

Midpoint Line

The **Midpoint Line** tool was added in SolidWorks 2015 on the demand of users. Generally, we draw a line in SolidWorks by using the start point and end point but now we have the flexibility to draw a line by using the mid point and end point of the line. The method to use this tool is given next.

- Click on the **Midpoint Line** tool from the **Line** drop-down. The **Insert Line PropertyManager** will be displayed as shown in Figure-13.

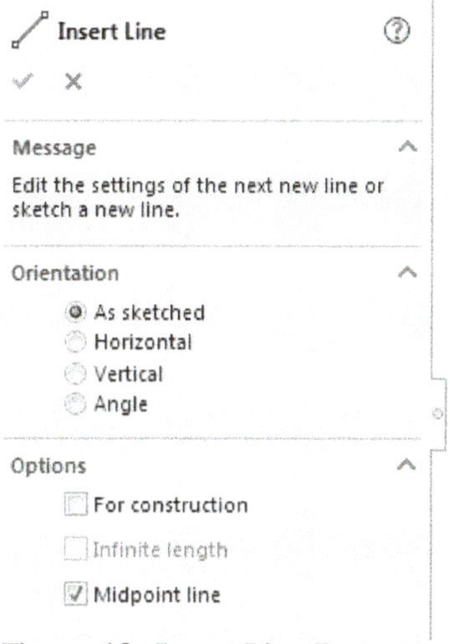

Figure-13. Insert Line Property Manager

- Click to specify the mid point of the line. The end point of line will get attached to the cursor.
- Move the cursor in desired direction and click to specify the end point of line. You can also enter the value in the edit box displayed with the cursor while specifying end point; refer to Figure-14.

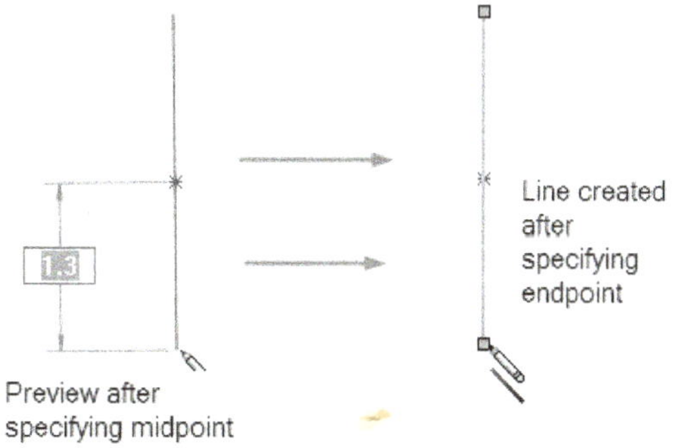

Figure-14. Midpoint line creation

Rectangle

There are five tools in the **Rectangles** drop-down; **Corner Rectangle, Center Rectangle, 3 Point Corner Rectangle, 3 Point Center Rectangle**, and **Parallelogram**; refer to Figure-15.

Figure-15. Rectangles drop-down

Procedures to use these tools are discussed next.

Corner Rectangle

• Click on the **Corner Rectangle** tool from the drop-down. The **Rectangle PropertyManager** will be displayed as shown in Figure-16.

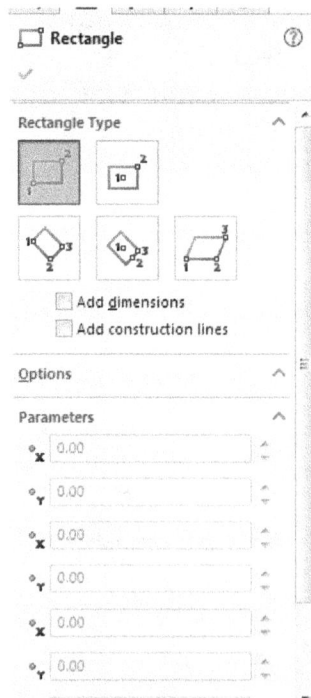

Figure-16. Rectangle PropertyManager

- Click in the viewport to specify the first point. The current dimensions will be displayed along the lines of rectangle. Specify the desired value and press the **Tab** key from the Keyboard to switch to the other dimension.
- Specify the other dimension and then press **Enter** from the Keyboard. The rectangle will be created with the specified dimensions.
- Select the **Add dimensions** check box if you want to apply the dimensions while creating the rectangle.
- You can add the construction lines for rectangle being created by selecting the **Add construction lines** check box. There are two ways by which you can add the construction lines in a rectangle; refer to Figure-17.

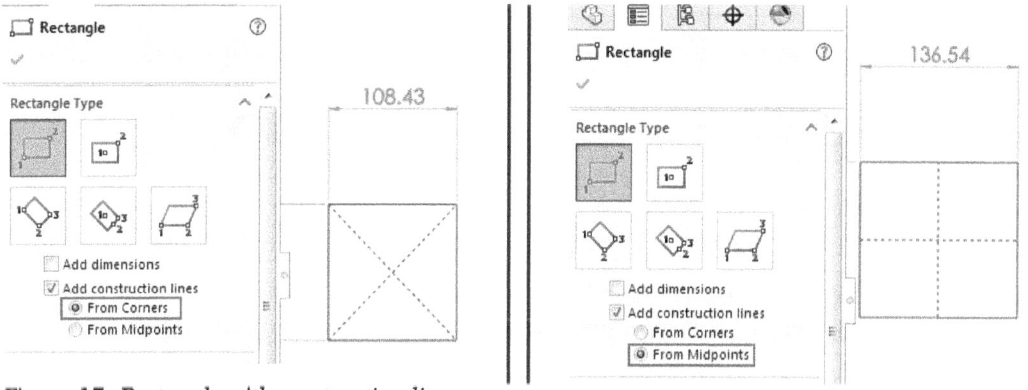

Figure-17. Rectangle with construction lines

- You can switch to the other types of rectangle by using the five buttons available in the top section of the **PropertyManager**.

Center Rectangle

- Click on the **Center Rectangle** tool from the drop-down. The **Rectangle PropertyManager** will be displayed similar to the one displayed earlier. In this **PropertyManager**, the **Center Rectangle** button is selected by default.
- Click to specify the center point of the rectangle.
- Specify the corner point of the rectangle. You can specify any of the corner point by moving the cursor in desired direction.

3 Point Corner Rectangle

- Click on the **3 Point Corner Rectangle** tool from the drop-down. The **Rectangle PropertyManager** will be displayed similar to the one displayed earlier. In this **PropertyManager,** the **3 Point Corner Rectangle** button is selected by default.
- Click to specify the starting point of the rectangle.
- Click to specify the end point of the base line.
- Click to specify the end point for the height. Refer to Figure-18 for procedure.

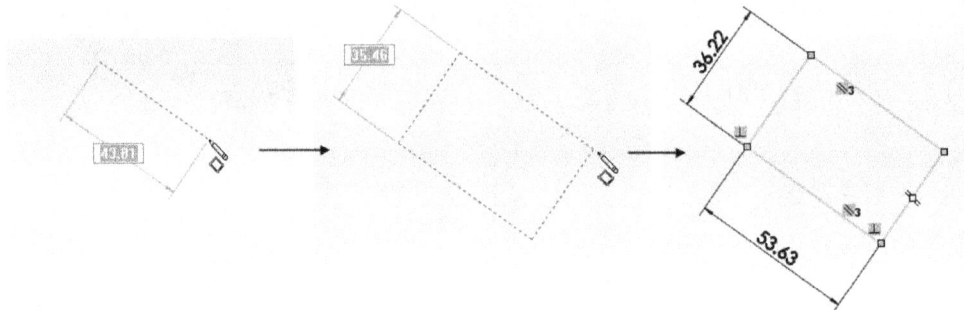

Figure-18. 3Point Rectangle Creation

3 Point Center Rectangle

- Click on the **3 Point Center Rectangle** tool from the drop-down. The **Rectangle PropertyManager** will be displayed similar to the one displayed earlier. In this **PropertyManager,** the **3 Point Center Rectangle** button is selected by default.
- Click to specify the center point of the rectangle.
- Click to specify the half length of the base line
- Click to specify the half length of vertical line. Refer to Figure-19 for the procedure.

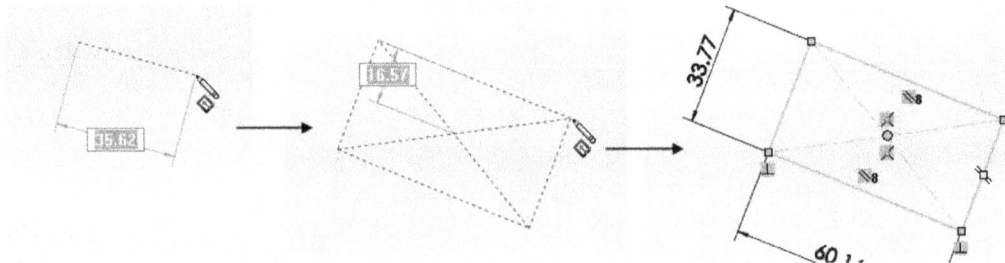

Figure-19. 3Point Center Rectangle Creation

Parallelogram

- Click on the **Parallelogram** tool from the drop-down. The **Rectangle PropertyManager** will be displayed similar to the one displayed earlier. In this **PropertyManager**, the **Parallelogram** button is selected by default.
- Click to specify the start point of the base line.
- Click to specify the end point of the base line.
- Click to specify the end point of the line defining height of the parallelogram. While defining this line, you can move the cursor in left/right and vertical direction.

Slot

There are four tools in the **Slot** drop-down; **Straight Slot, Center Straight Slot, 3 Point Arc Slot, Centerpoint Arc Slot**; refer to Figure-20.

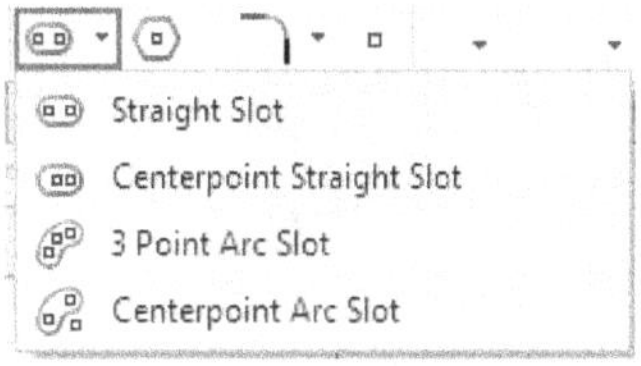

Figure-20. Slots drop-down

The tools in this drop-down are explained next.

Straight Slot

- Click on the **Straight Slot** tool from the drop-down. The **Slot PropertyManager** will be displayed as shown in Figure-21.

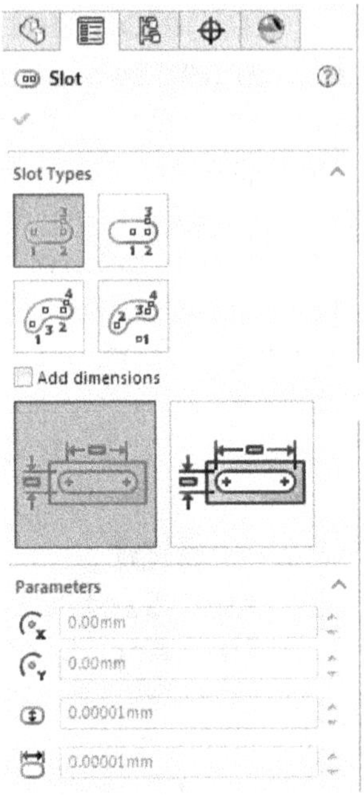

Figure-21. Slot PropertyManager

- Click on the **Center to Center** button or **Overall Length** button to set the length dimension style for slot. By default, the **Center to Center** button is selected in **PropertyManager**.
- Click in the viewport to specify the center of the first semi-circle of slot.
- Click to specify the center to center distance of the two end semi-circles.
- Click to specify the width of the slot. Figure-22 shows the procedure of creating straight slot.

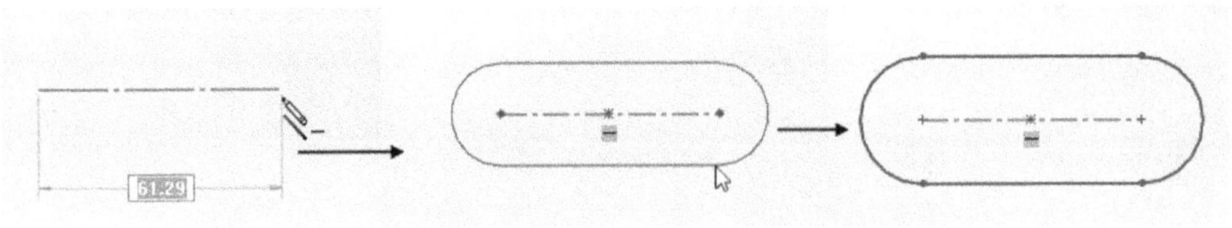

Figure-22. Straight Slot Creation

Centerpoint Straight Slot

- Click on the **Centerpoint Straight Slot** tool from the drop-down. The **Slot PropertyManager** will be displayed as discussed earlier. In this **PropertyManager**, the **Centerpoint Straight Slot** button is selected by default in the top section.

- Click to specify the center of the slot.
- Click to specify the half length of the center to center distance of the slot.
- Click to specify the length of the slot. Figure-23 shows the procedure of creating the centerpoint straight slot.

Figure-23. Centerpoint Straight Slot Creation

3 Point Arc Slot

- Click on the **3 Point Arc Slot** tool from the drop-down. The **Slot PropertyManager** will be displayed as earlier. In this **PropertyManager** the **3 Point Arc Slot** button is selected by default in the top section.
- Click to specify the start point of the center arc.
- Click to specify the end point of the center arc.
- Click to specify the radius of the slot.
- Click to specify the width of the slot.

Centerpoint Arc Slot

- Click on the **Centerpoint Arc Slot** tool from the drop-down. The **Slot PropertyManager** will be displayed as earlier. In this **PropertyManager** the **Centerpoint Arc Slot** button is selected by default in the top section.
- Click to specify the center of the slot center arc.
- Click to specify the start point of the center arc.
- Click to specify the end point of the slot.
- Click to specify the width of the slot.

Circle

There are two tools in this drop-down; **Circle** and **Perimeter Circle**; refer to Figure-24.

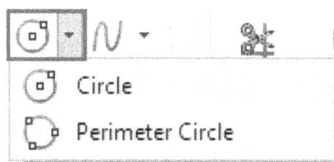

Figure-24. Circle drop-down

The tools in this drop-down are explained next.

Circle

- Click on the **Circle** tool from the drop-down. The **Circle PropertyManager** will be displayed; refer to Figure-25.

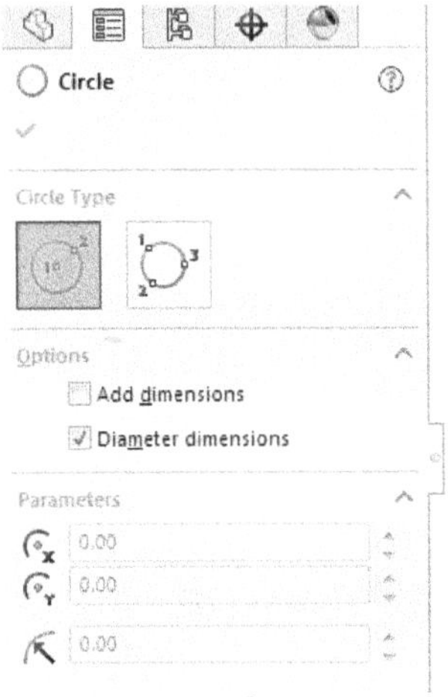

Figure-25. Circle Property-Manager

- Click to specify the center point of the circle.
- Click to specify the diameter of the circle. If you want to specify the radius of the circle then clear the **Diameter dimensions** check box from the **PropertyManager** after selecting the **Circle** tool.

Perimeter Circle

- Click on the **Perimeter Circle** tool from the drop-down. The **Circle PropertyManager** will be displayed as earlier. In this **PropertyManager**, the **Perimeter Circle** button is selected by default.
- Click one by one at three locations to specify three perimeter points through which the circle should pass.

Arc

There are three tools in this drop-down; **Centerpoint Arc, Tangent Arc, 3 Point Arc**; refer to Figure-26.

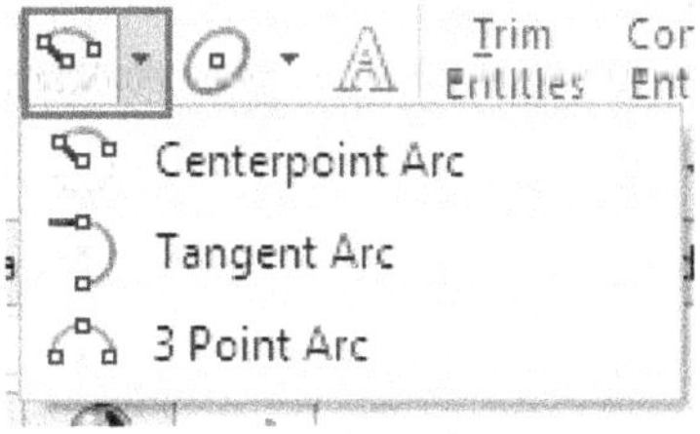

Figure-26. Arc drop-down

The tools in this drop-down are explained next.

Centerpoint Arc

- Click on the **Centerpoint Arc** tool from the drop-down. The **Arc PropertyManager** will be displayed as shown in Figure-27.
- Click to specify the center point of the arc.
- Click to specify the start point of the arc.
- Click to specify the end point of the arc. Refer to Figure-28 for the creation of the arc.

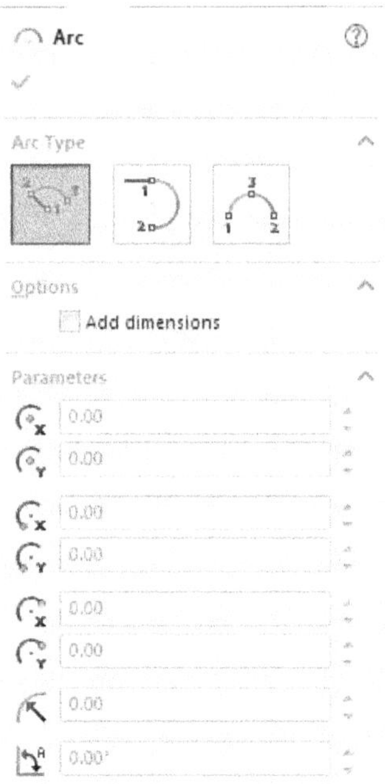

Figure-27. Arc PropertyManager

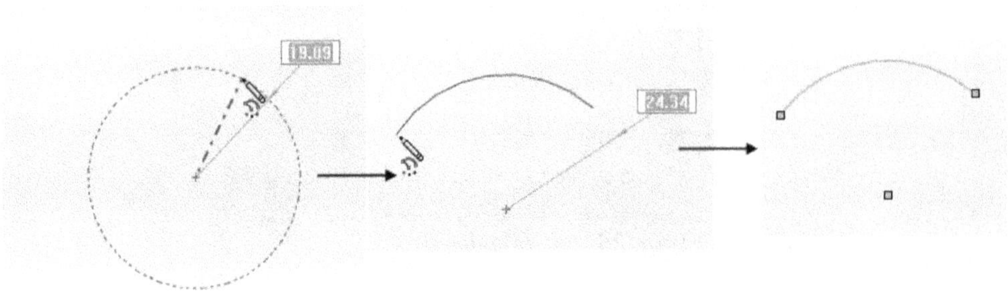

Figure-28. Arc Creation

Tangent Arc

- Click on the **Tangent Arc** tool from the drop-down. The **Arc PropertyManager** will be displayed as earlier. In this **PropertyManager**, the **Tangent Arc** button is selected by default.
- Select end point of the entity to which the arc should be tangent.
- Click to specify the end point of the arc. Note that the tangent constraint is automatically applied at the arc.

3 Point Arc

- Click on the **3 Point Arc** tool from the drop-down. The **Arc PropertyManager** will be displayed as earlier. In this **PropertyManager**, the **3 Point Arc** button is selected by default.
- Click to specify the start point of the arc.
- Click to specify the end point of the arc.
- Click to specify a point on the arc to set the radius of the arc.

Polygon ⊕

The **Polygon** tool is used to create polygons of desired number of sides. The procedure to create polygon is explained next.

- Click on the **Polygon** tool from the **Ribbon**. The **Polygon PropertyManager** will be displayed as shown in Figure-29.

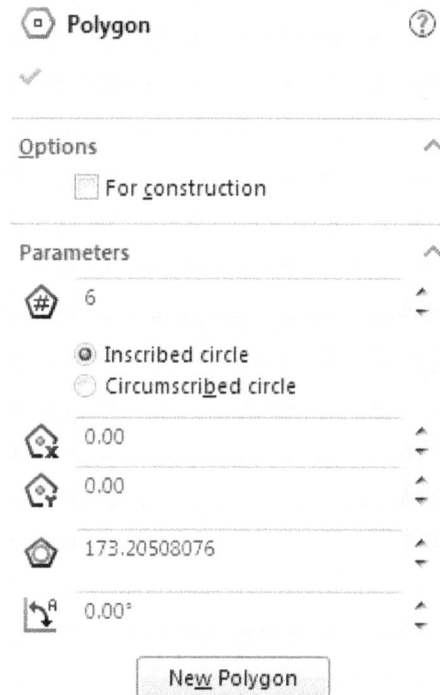

Figure-29. Polygon PropertyManager

- Specify the number of sides of the polygon by using the spinner in the **Parameters** rollout.
- Select the desired radio button from the rollout. If you want to create the polygon inside the circle, then select the **Inscribed circle** radio button. If you select the **Circumscribed circle** radio button, the polygon will be drawn outside the circle. Note that polygon corner points lie on the circumscribed circle if you select the **Circumscribed circle** radio button. On the other hand, sides of polygon are tangent to the circle if you have selected the **Inscribed circle** radio button.
- Click to specify the center point of the circle.
- Click to specify the corner point of the polygon. Figure-30 shows the procedure of creating polygons.

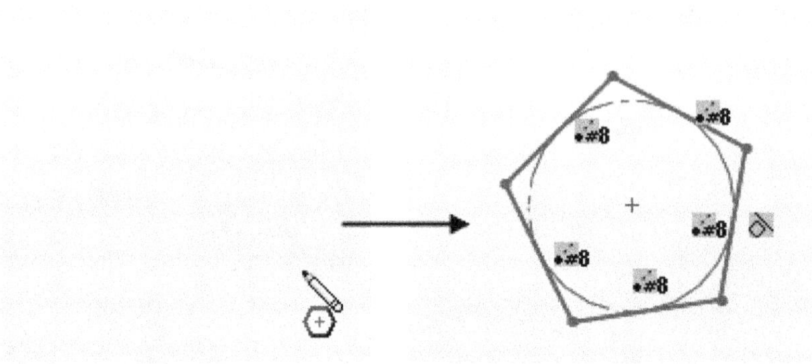

Figure-30. Polygon creation

Spline

The tools in **Spline** drop-down are used to create splines with different methods. Figure-31 show the tools of this drop-down. These tools are discussed next.

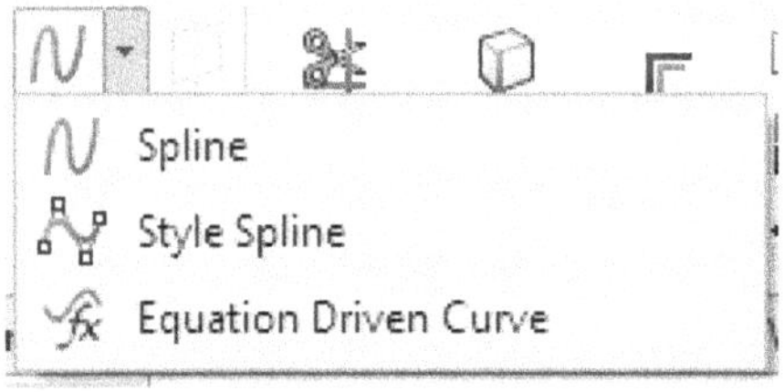

Figure-31. Spline drop-down

Spline

- Click on the **Spline** tool from the drop-down. You are asked to specify the points through which the spline should pass.
- One by one click in the viewport to specify the points of the spline.
- To end specifying points, press the ESC key from the keyboard.

Styled Spline

- Click on the **Styled Spline** tool from the **Spline** drop-down. You are asked to specify the control points for the spline. (Control points control the shape of the spline but they do not lie on the spline).
- One by one click to specify the control points of the spline.
- To end specifying control points, press the ESC key from the keyboard.

Figure-32 shows the splines created.

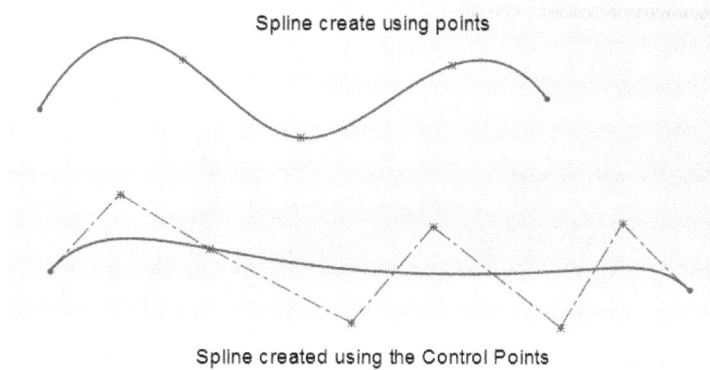

Figure-32. Spline creation

Equation Driven Curve

This tool is very helpful in creating curves using mathematical equations. Follow the procedure given next to create the curve using the equation.

- Click on the **Equation Driven Curve** tool from the **Spline** drop-down. The **Equation Driven Curve PropertyManager** will display as shown in Figure-33.

Figure-33. Equation Driven Curve PropertyManager

- Click on the desired radio button to create the **Explicit** or **Parametric** equation.
- Specify the parameters for y_x. For example, in the Y_x edit box specify **x²/10**.
- Click in the x1 edit box and specify the starting value. For example, specify the value as **0**.
- Click in the x2 edit box and specify the ending value. For example, specify the value as **20**.
- Figure-34 shows the output spline created by specifying the above parameters.

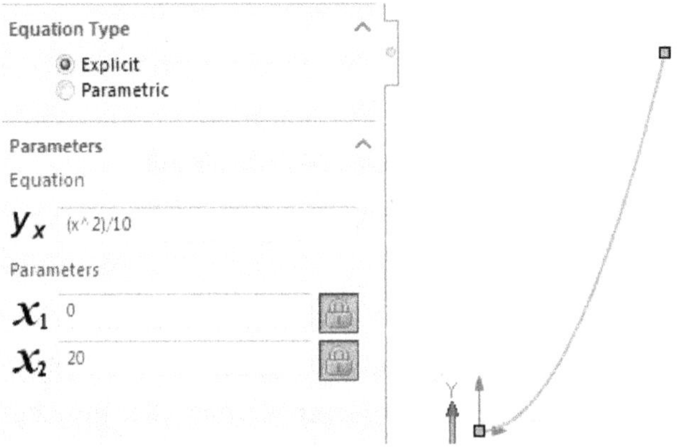

Figure-34. Equation type

Ellipse

The tools in the **Ellipse** drop-down are used to create geometric profiles like; ellipse, partial ellipse, parabola, and conic; refer to Figure-35. The tools in the **Ellipse** drop-down are discussed next.

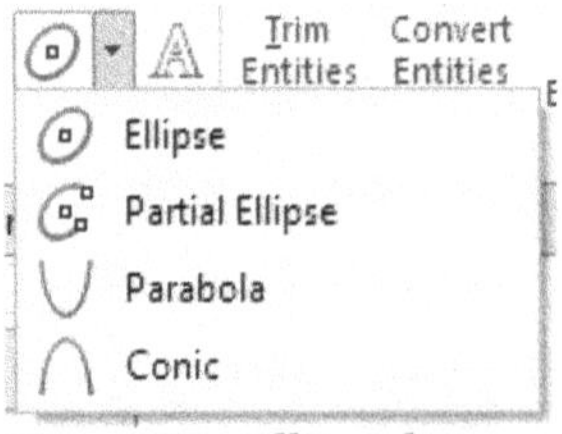

Figure-35. Ellipse drop-down

Ellipse

The **Ellipse** tool is used to create ellipses in the sketch. The procedure to create ellipse is discussed next.

* Click on the **Ellipse** tool. You are asked to specify the center of the ellipse.
* Click to specify the center. You are asked to specify the radius along the major axis.
* Click to specify the radius or specify it by entering value in the **PropertyManager**.
* Click to specify the radius along the minor axis. Figure-36 shows the process of creating ellipse. Once you click to specify the radius of minor axis, the **Ellipse PropertyManager** will be displayed; refer to Figure-37. Note that you can also specify the values in the edit boxes in **PropertyManager**.

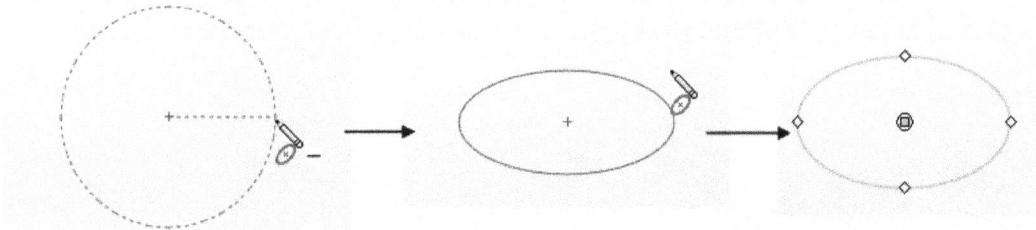

Figure-36. Ellipse creation

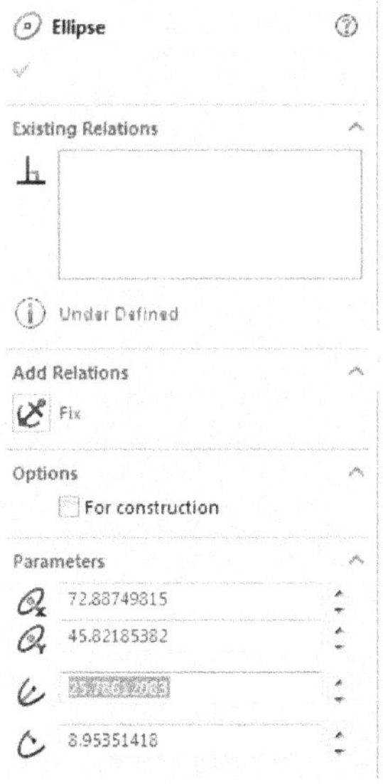

Figure-37. Ellipse PropertyManager

Partial Ellipse

The **Partial Ellipse** tool is used to create partial ellipses in the sketch. The procedure to create partial ellipse is explained next.

- Click on the **Partial Ellipse** tool from **Ellipse** drop-down. You are asked to specify the center point of the ellipse.
- Click to specify the center point. You are asked to specify the radius along major axis.
- Click to specify the radius. You are asked to specify the radius along minor axis.
- Click to specify the radius along minor axis. The point where you specify the radius of minor axis will become the starting point of the partial ellipse.
- Click to specify the end point of the partial ellipse. Figure-38 show the process of creating partial ellipse.
- After specifying the end point of the partial ellipse, the **Ellipse PropertyManager** displays as shown in Figure-39.

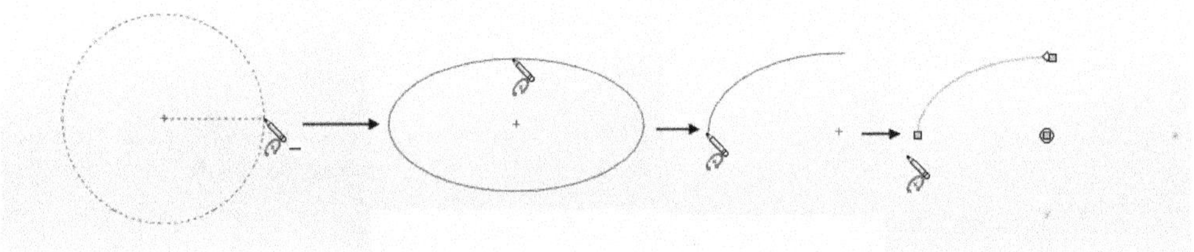

Figure-38. Partial ellipse creation

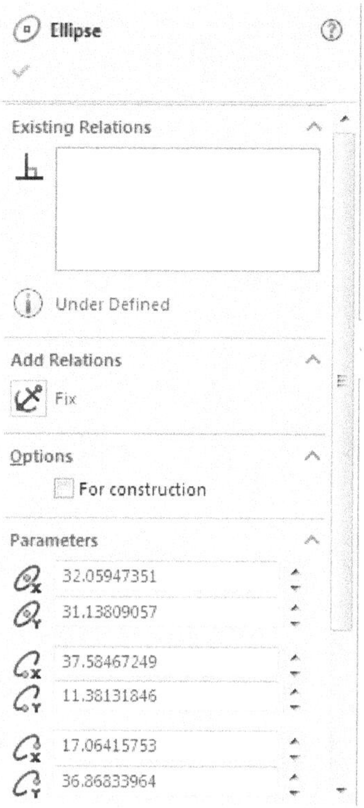

Figure-39. Partial Ellipse PropertyManager

Parabola

The **Parabola** tool is used to create parabola in the sketch. The procedure to create parabola is explained next.

- Click on the **Parabola** tool from the drop-down. You will be prompted to specify the focal center of the ellipse.
- Click to specify the focal center. You are asked to specify the distance of focal center from the directrix.
- Click to specify the distance. You are asked to specify the starting point of the parabola segment.
- Click on the dashed curve to specify the start point and then click to specify the end point of the parabola. Figure-40 shows the process of creating parabola.
- Once the parabola is created, the **Parabola PropertyManager** will be displayed; refer to Figure-41. You can change the parameters as per your requirement by using the options of **PropertyManager**.

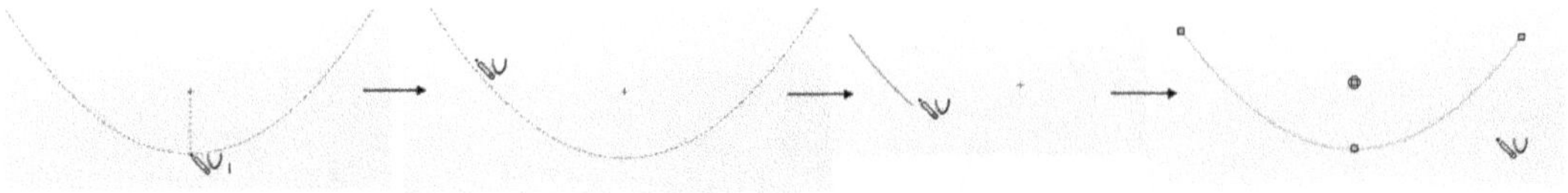

Figure-40. Parabola creation

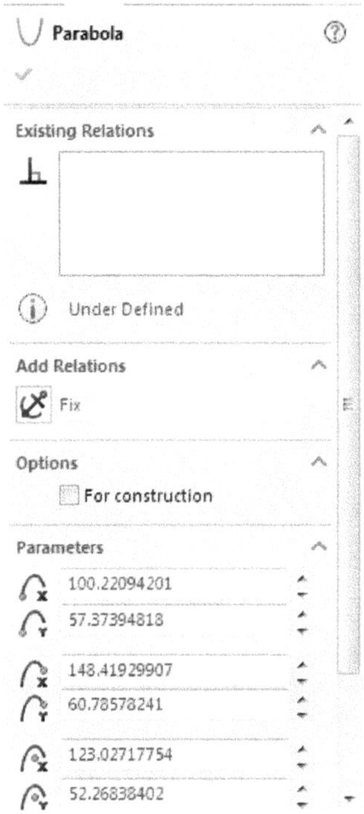

Figure-41. Parabola PropertyManager

Conic

The **Conic** tool is used to create conic curves in the sketch. The procedure to create a conic curve is explained next.

- Click on the **Conic** tool from the drop-down. You will be prompted to specify the start point of the conic.
- Click to specify the start point of the conic curve. You are asked to specify the end point of the conic curve.
- Click to specify the end point. You are asked to specify the top vertex of the curve.
- Click to specify the vertex. You are asked to specify the **Rho** value for the curve. Specify the desired value of **Rho**. Figure-42 shows the process of conic curve creation. Once the curve is created, the **Conic PropertyManager** is displayed as shown in Figure-43.

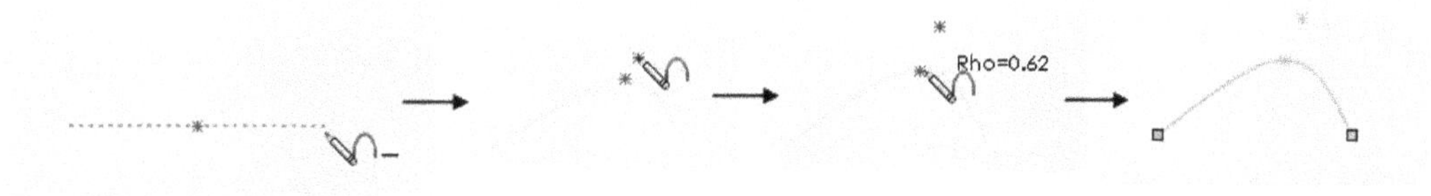

Figure-42. Conic creation

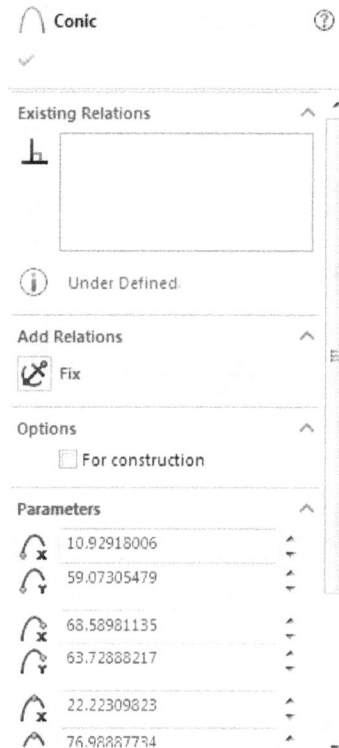

Figure-43. Conic PropertyManager

Sketch Fillet

The **Sketch Fillet** tool is used to create fillet at the corners created by intersection of two entities. Fillet is sometimes also referred to as round. Generally it is not advised to use this tool if there are some major changes going to occur later in the model. The procedure to create fillet is given next.

- Click on the **Sketch Fillet** from the **Sketch Fillet** drop-down. The **Sketch Fillet PropertyManager** displays; refer to Figure-44.

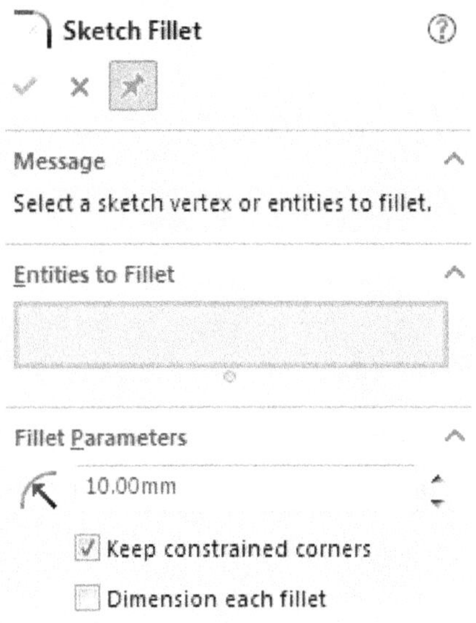

Figure-44. Sketch Fillet PropertyManager

- Enter the radius value in the **Fillet Radius** spinner of **Fillet Parameters** rollout. Select the first entity. You are asked to select the second entity.
- Select the second entity. The preview of fillet is displayed.
- Click on the **OK** button from the **PropertyManager** or select the next two entities between which you want to create the fillet. Figure-45 show the process of fillet creation.

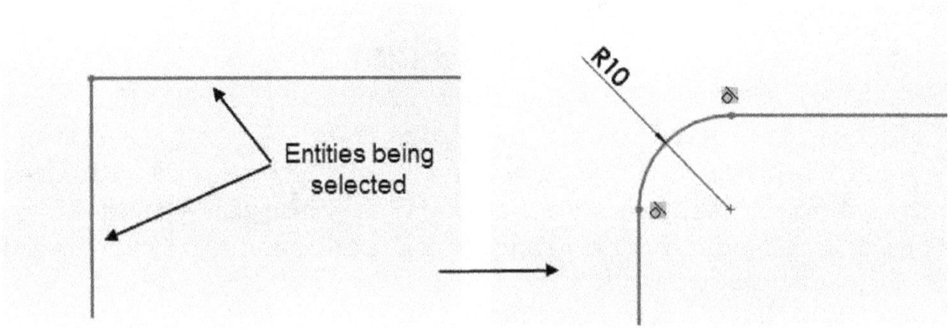

Figure-45. Fillet Creation

Sketch Chamfer

The **Sketch Chamfer** tool is used to create chamfer at the corners created by intersection of two entities. The procedure to create chamfers is explained next.

- Click on the **Sketch Chamfer** from the **Sketch Fillet** drop-down. The **Sketch Chamfer PropertyManager** is displayed; refer to Figure-46.

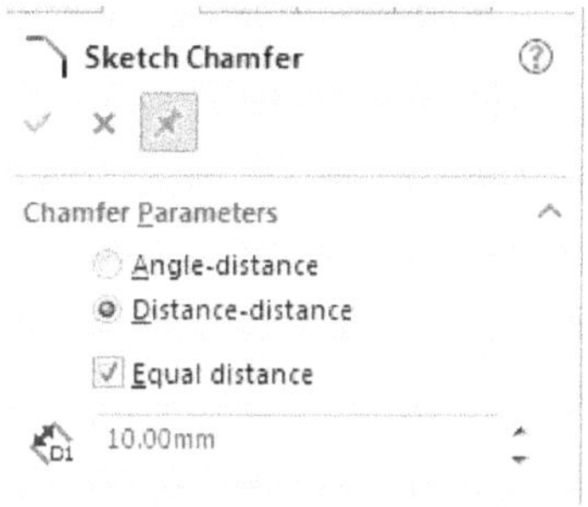

Figure-46. Sketch Chamfer PropertyManager

By default, **Distance-distance** radio button is selected in the **PropertyManager**. Specify the chamfer length for first side. If the **Equal distance** check box is selected then it will be applied for the both sides. If you clear the check box then you can specify the chamfer length from both sides by using the respective edit boxes.

You can also create chamfer by specifying angle and distance. To do so, select the **Angle-distance** radio button. Specify the parameters in the **Chamfer Parameters** rollout.

- Select the first line and then select the second line. The chamfer will be created between both the lines.

Text

The **Text** tool is used to create text which can be used for embossing/engraving on a solid model. The **Text** tool is also used to give notes and other information for the model. The procedure to create text is explained next.

- Click on the **Text** tool. The **Sketch Text PropertyManager** will be displayed as shown in Figure-47.

Figure-47. Sketch Text Property-Manager

- Click in the **Text** box and enter the desired text you want to use in the sketch.
- The text will be created at the default coordinate system.
- Select a curve along which you want to create the text.
- The text will be placed along the curve. Figure-48 show the procedure of creating text.

Figure-48. Text along curve

Note that if the curve length is smaller than the text specified, then the text that can be spaced over the curve will only be displayed.

Point

The **Point** tool is used to create sketch point in the viewport. The point is a very important entity and finds its major usage when you start creating surfaces. The points give the flexibility to parametrically change the surface design. The procedure to create points is explained next.

- Click on the **Point** tool from the **Ribbon.** You are asked to click in the viewport to specify the location of the point.
- Click on the viewport. The **Point PropertyManager** displays as shown in Figure-49.

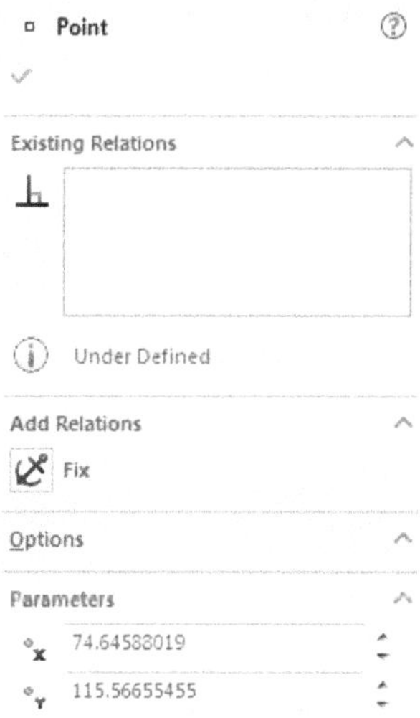

Figure-49. Point PropertyManager

- Click in the spinners of the **PropertyManager** to specify the parameters of the points.

SKETCH EDITING TOOLS

The standard tools to edit sketch entities are categorized in this section. The tools in this section are discussed next.

Trim Entities

The **Trim Entities** tool is available in the **Trim Entities** drop-down of the **Ribbon**. This tool is used to remove unwanted part of a sketch entity. While removing the segments, the tool considers intersection point as the reference for trimming. The procedure to use this tool is explained next.

- Click on the **Trim Entities** tool. The **Trim Entities PropertyManager** displays as shown in Figure-50.

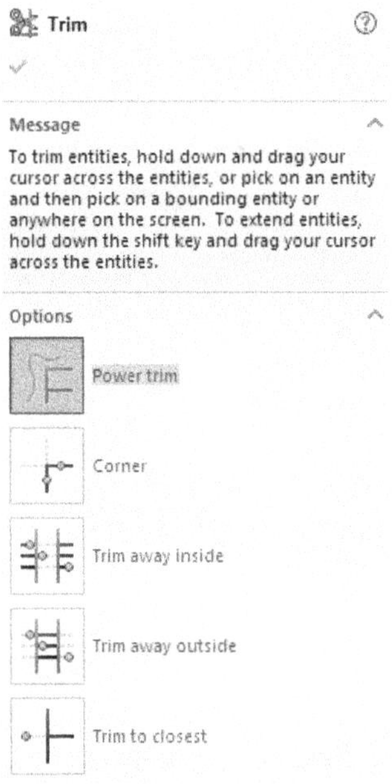

Figure-50. Trim Property-Manager

By default, the **Power trim** button is selected. This button is having the functionality of all the other buttons displayed in the **PropertyManager**. So, we will be explaining the procedure of using this button only.

- Make sure the **Power Trim** button is selected in the **PropertyManager**.
- Click on the entity you want to trim. Note that the selected side of the entity will be removed.
- Click on the entity that you want to use as reference for trimming.
- The entity will be trimmed. Figure-51 shows the procedure of trimming.

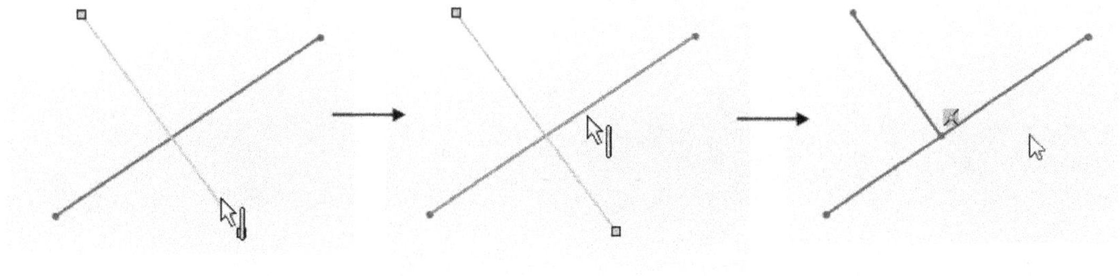

Figure-51. Trimming procedure

Or

- Select the **Trim Entity** tool, make sure the **Power Trim** button is selected.
- Click and hold the mouse button, and drag the cursor over the portion of entities you want to be removed. Refer to Figure-52.

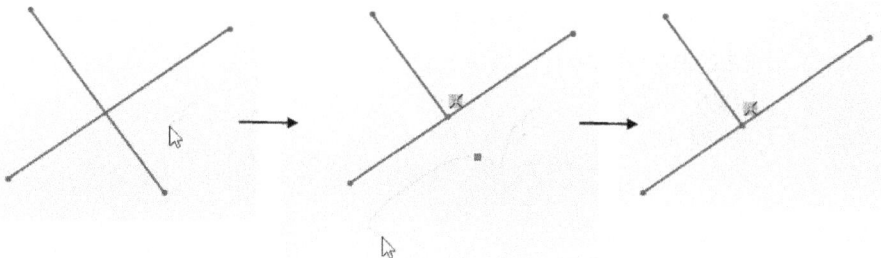

Figure-52. Trimming by dragging

Extend Entities

The **Extend Entities** tool does the reverse of **Trim Entities** tool. This tool is available in the **Trim Entities** drop-down. This tool extends the sketch entities up to the nearest intersecting entity. The procedure to use this tool is discussed next.

- Click on the **Extend Entities** tool from the **Trim Entities** drop-down.
- Hover the cursor on the entity that you want to extend. Preview of the extension will display.
- Click on the entity if the preview is as per your requirement. If you want the entity to be extended in reverse direction then click on the other portion of the entity. Figure-53 shows the process of extending entities.

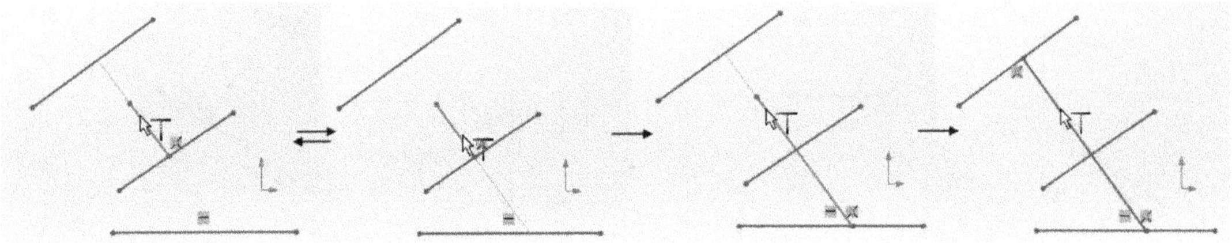

Figure-53. Extending process

The **Convert Entities** and **Intersection Curves** tools will be explained in the Modeling section of the book, later.

Offset Entities

The **Offset Entities** tool is used to create copy of the selected entities at a specified distance from the them. If you are the user of AutoCAD then this tool is the most common tool being used while creating layouts. The procedure to use this tool is given next.

- Click on the **Offset Entities** tool from the **Ribbon**. The **Offset Entities PropertyManager** will be displayed; refer to Figure-54.

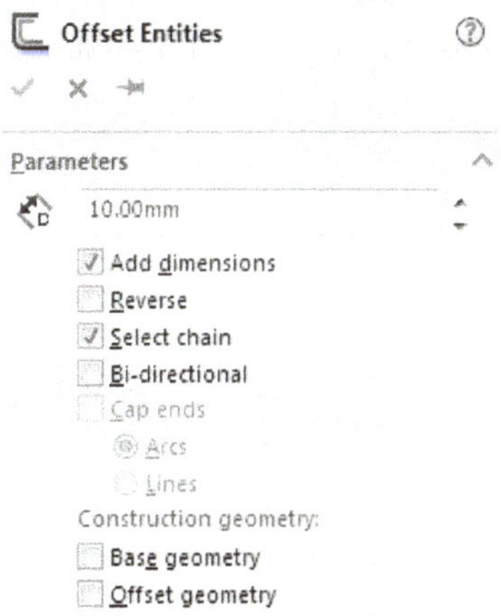

Figure-54. Offset Entities PropertyManager

There are various options in this **PropertyManager** that are linked to the output of the tool. So, we will discuss them one by one.

Offset Distance

This spinner is used to set the distance value for offset. You can also enter the desired value in the edit box.

Add dimensions

Select this check box if you want to create the dimensions while creating offset.

Reverse

Select this check box if you want to reverse the direction of offset being displayed.

Select chain

Select this check box if you want to select the chain of entities connecting to the selected entity.

Bi-directional

Select this check box if you want to create the offset entities on both sides of the selected entity.

Construction Geometry

There are two check boxes in this section, **Base geometry** and **Offset geometry,** to make the base and offset geometry as construction entity. Select the desired check box/boxes from the section.

Cap ends

Select this check box if you want to close the ends of offset entities. This check box is active when the **Bi-directional** check box is selected and entity is selected. After selecting the **Cap ends** check box, you can close the offset entities by using the arcs or lines. For using arcs or lines, select the respective radio button.

- After selecting the desired options, select the entity from the viewport. Preview will be displayed in yellow color.
- Click on OK button to create the offset. Figure-55 shows the process of creating offset entities.

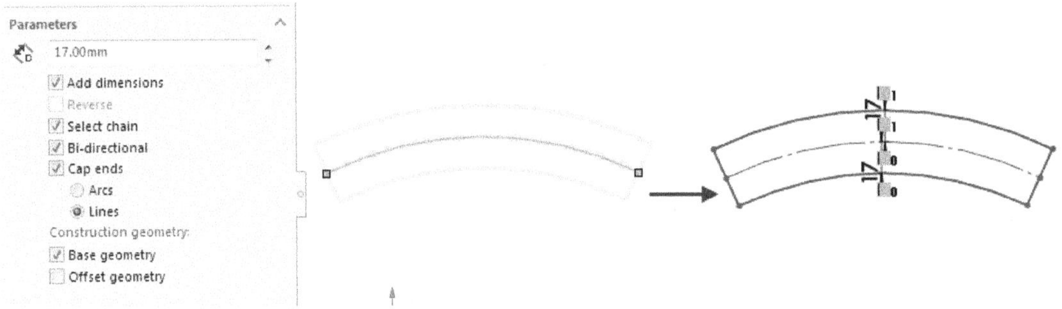

Figure-55. Offset entities creation

Mirror Entities

The **Mirror Entities** tool is used to create mirror copy of the selected entities with respect to a reference called mirror line. The procedure to create mirror entities is given next.

- Click on the **Mirror Entities** tool from the **Ribbon**. The **Mirror PropertyManager** is displayed; refer to Figure-56.

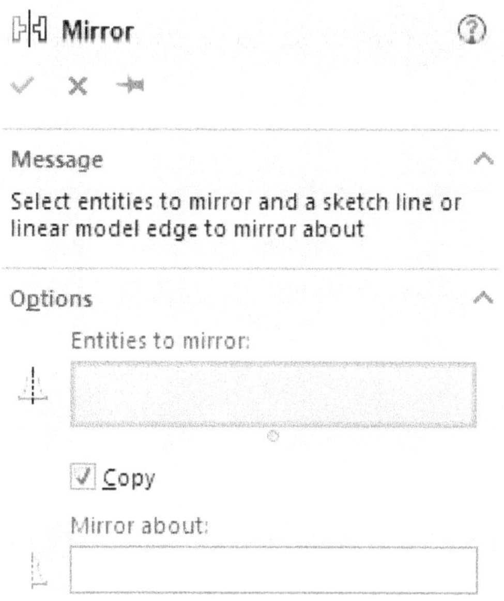

Figure-56. Mirror PropertyManager

- Select the entity/entities you want to create mirror copy of.
- Deselect the **Copy** check box from the **PropertyManager** if you want to delete the original entities after creating mirror copy.
- Click in the **Mirror about** box and select the reference line that you want to use as mirror line. The mirror line can be an edge of a solid, sketch line, or a centerline. Figure-57 shows the process of creating mirror entities.

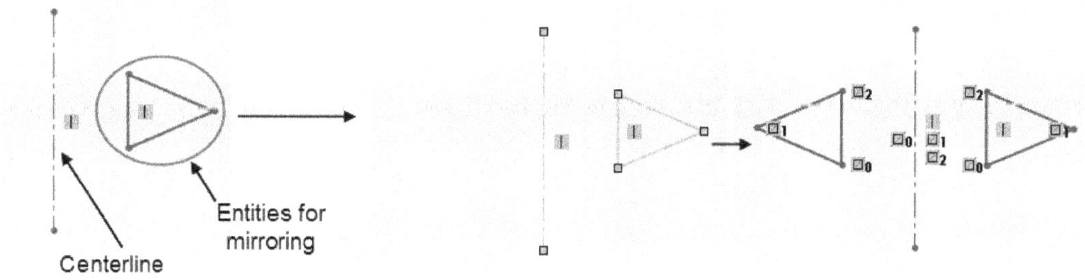

Figure-57. Mirror entities creation

Sometimes, we need to create multiple copies of the sketch entities. Like in sketch of a keyboard or piano. For such cases, SolidWorks provides two tools, **Linear Sketch Pattern** and **Circular Sketch Pattern**. These tools are discussed next.

Linear Sketch Pattern

The **Linear Sketch Pattern** tool is used to create multiple copies of an entity in linear directions. You can create pattern in two linear directions at a time. The procedure to create linear sketch pattern is given next.

*   Click on the **Linear Sketch Pattern** tool from the **Linear Sketch Pattern** drop-down of the **Ribbon**. The **Linear Pattern PropertyManager** is displayed as shown in Figure-58.

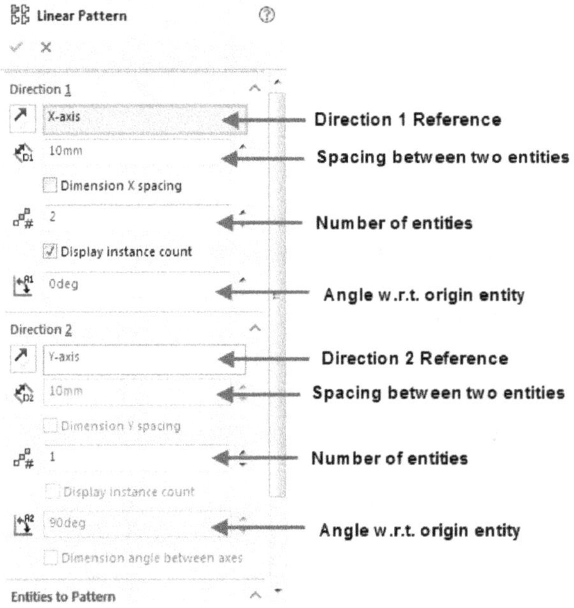

Figure-58. Linear Pattern PropertyManager

*   Select the entity that you want to pattern. Specify the parameters as per your requirement.
*   If you want to create pattern along an axis or line then click in the Direction 1 reference box or Direction 2 reference box as per your need. Now, select the desired axis or line to specify the direction reference.
*   If you want to skip any of the entity created in pattern then expand the **Instances to skip** rollout at the bottom of the **PropertyManager** and click in the **Instances to skip** box. Now, click on the pink dot for the entities from the preview that you do not want to create; refer to Figure-59.

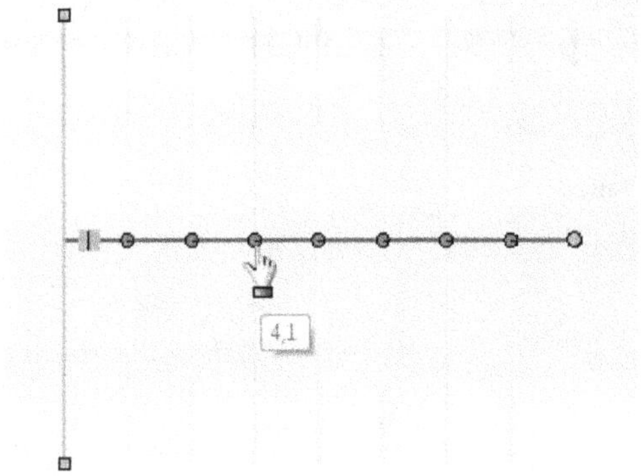

Figure-59. Instances to be removed

- Increase the number of entities in the **Direction 2** rollout to activate the options in the rollout. Figure-60 shows the pattern created for a circle and its respective options in the **PropertyManager**.

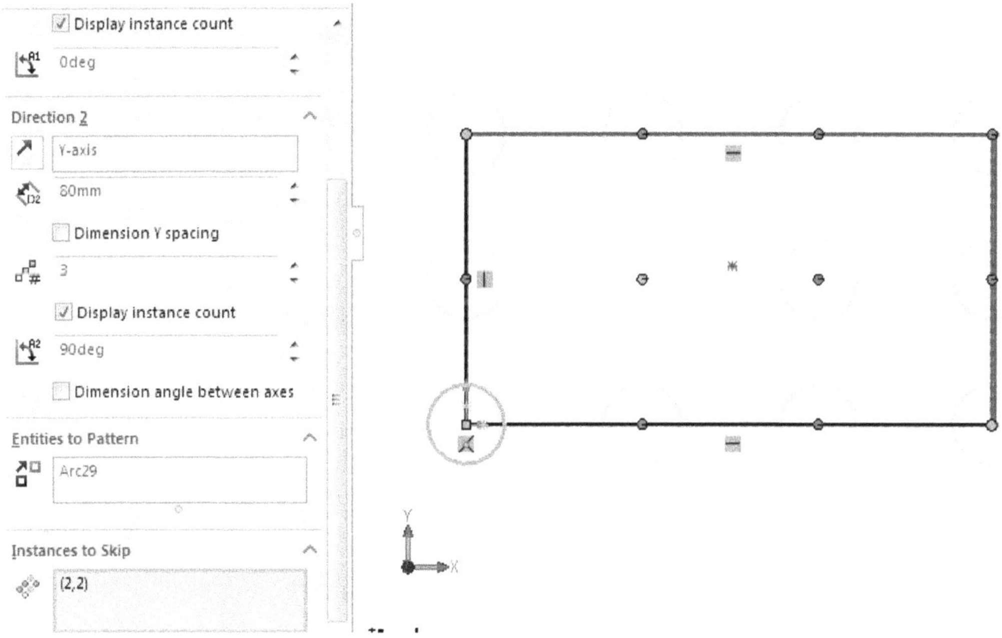

Figure-60. Linear pattern creation

Circular Sketch Pattern

The **Circular Sketch Pattern** tool is used to create multiple copies of an entity in circular fashion. You can create pattern in two linear directions at a time. The procedure to create circular sketch pattern is given next.

- Click on the **Circular Sketch Pattern** tool from the **Linear Sketch Pattern** drop-down. The **Circular Pattern PropertyManager** will be displayed; refer to Figure-61.

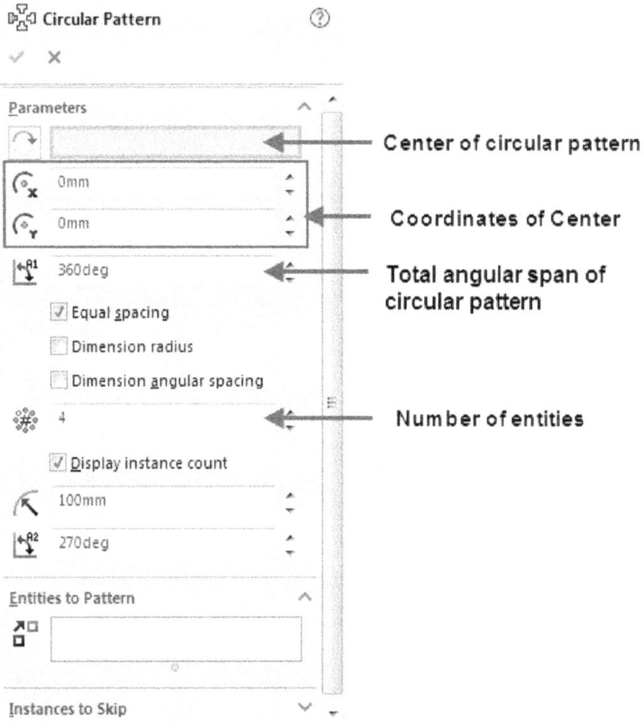

Figure-61. Circular Pattern PropertyManager

- Select the entity that you want to pattern and click in the **Center of circular** pattern box to specify the center of the pattern.
- Select the point to be the origin, specify the number of entities and click on the **OK** button to create the pattern.
- You can skip the entities as you did for Linear pattern. Figure-62 shows a circular pattern created with its parameters specified.

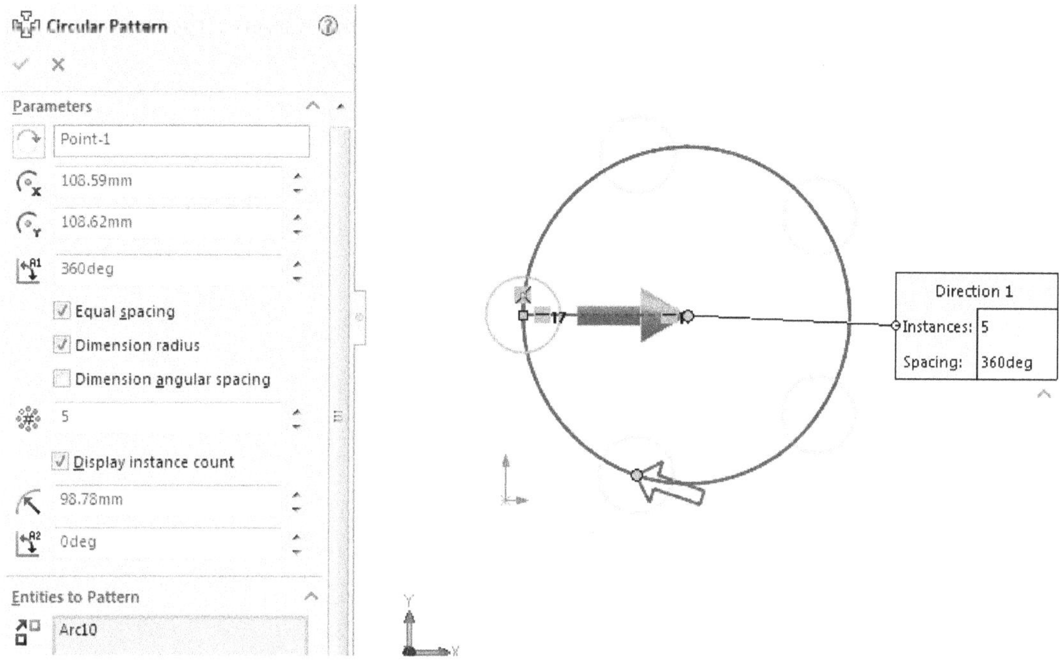

Figure-62. Circular Pattern creation

Move Entities

The **Move Entities** tool is used to move the entities from one position to another position. You can move the entities either by coordinate values or by clicking. The procedure to move entities is discussed next.

- Select the **Move Entities** tool from the **Ribbon**. The **Move Entities PropertyManager** will display as shown in Figure-63.
- Select the entity that you want to move.
- Make sure **From/To** radio button is selected. Now click on the point you want to use as base point.
- Click at the destination point where you want to place the entity. The entity will move to specified place.

*Figure-63. Move Entities Prop-
ertyManager*

Or

- Click on the **X/Y** radio button and enter the distance in X and Y direction to place the sketch entity.

Or

- Select the entity, you want to move and drag it from the key point/curve. Note that circular entities like circle, arc, ellipse are dragged from center for moving them. For the other entities, select the curve from the location which is not a key point. Figure-64 shows a spline and line moved together by dragging.

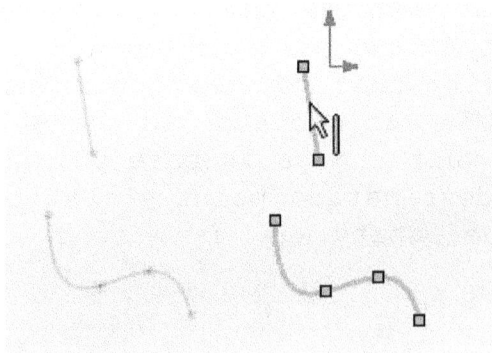

Figure-64. Moving entities by dragging

Copy Entities

The **Copy Entities** tool is available in the **Move Entities** drop-down. This tool is used to copy the entities by specifying position. This tool works in the same way as the **Move Entities** tool does. The only difference is that it does not move the entity but it creates the entities. Figure-65 shows the entities copied by this tool.

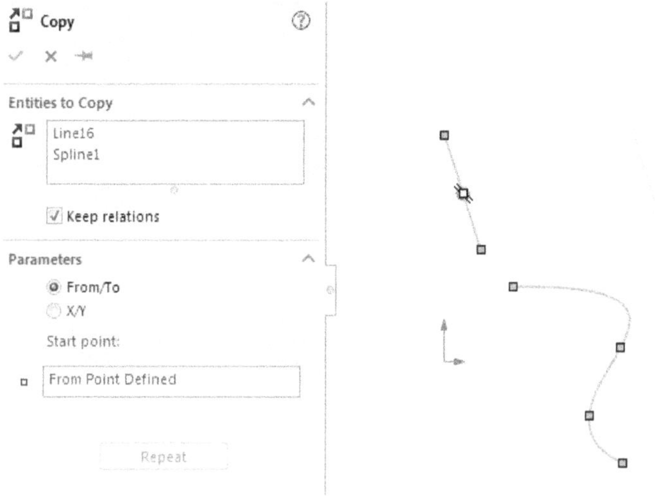

Figure-65. Copying entities using Copy Entities tool

Rotate Entities

The **Rotate Entities** tool is available in the **Move Entities** drop-down. This tool is used to rotate the entities by specifying angle. The procedure to use this tool is given next.

• Click on the **Rotate Entities** tool from the drop-down. The **Rotate PropertyManager** will be displayed as shown in Figure-66.

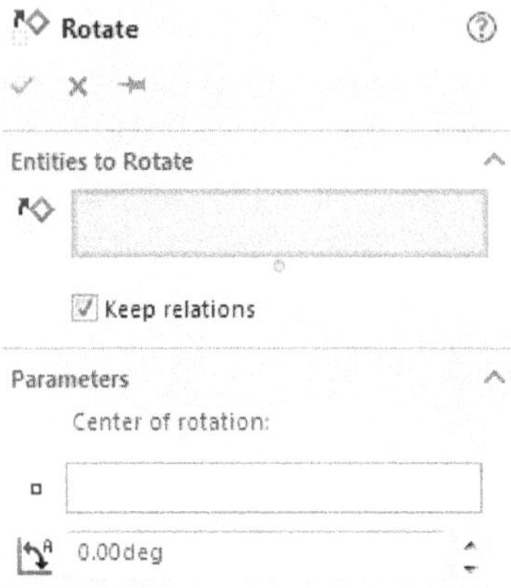

Figure-66. Rotate Property-Manager

- Select the entity that you want to rotate.
- Click in the **Center of rotation** box and click on the point that you want to make center of rotation.
- Specify angle in the **Angle** spinner. The object will rotate by specified value. Refer to Figure-67.

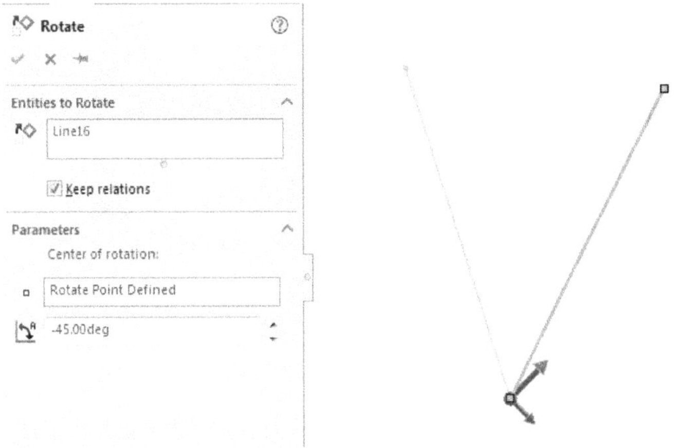

Figure-67. Rotating an entity

Scale Entities

The **Scale Entities** tool is available in the **Move Entities** drop-down. This tool is used to increase or decrease the size of an entity by specified scale value. The procedure to use this tool is given next.

- Click on the tool from the drop-down. The **Scale PropertyManager** will display as shown in Figure-68.

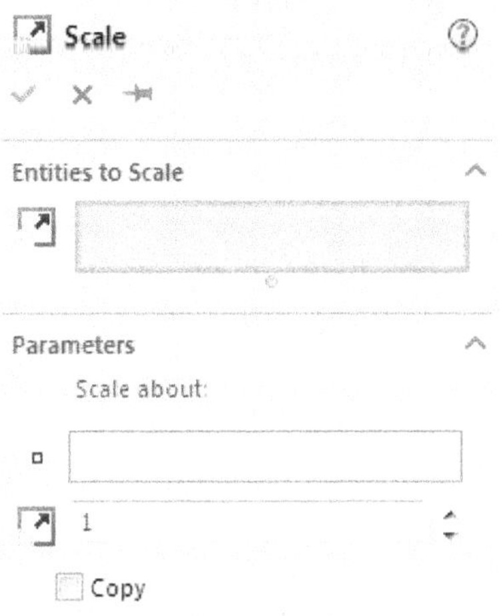

Figure-68. Scale PropertyManager

- Select the entities that you want to scale up or scale down.
- Click in the **Scale about** box and select the base point about which you want to scale the entities.
- Enter the scale value in the **Scale Factor** spinner. You can create a copy of entities scaled to specified value by selecting the **Copy** check box. In that case, the selected entities will remain unchanged. Refer to Figure-69.

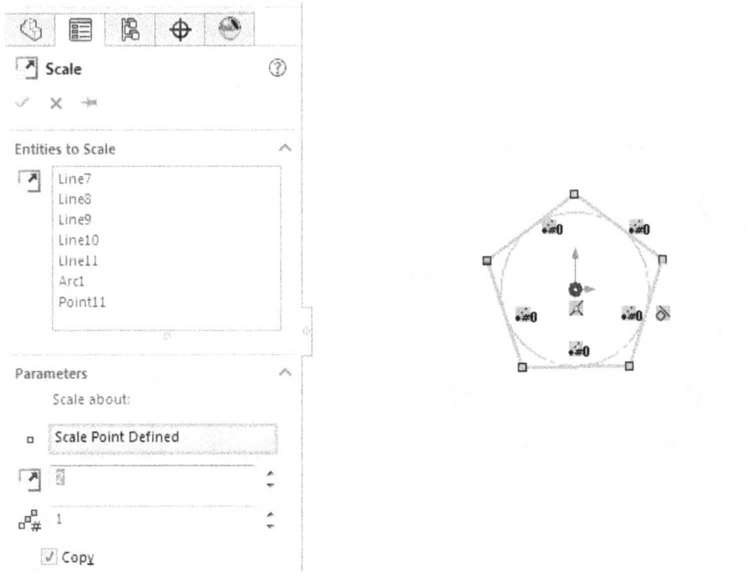

Figure-69. Scaling entities

Stretch Entities

The **Stretch Entities** tool is available in the **Move Entities** drop-down. This tool is used to stretch any sketched entity. The procedure to use this tool is given next.

- Click on the **Stretch Entities** tool from the drop-down. The **Stretch PropertyManager** will be displayed as shown in Figure-70.

Figure-70. Stretch PropertyManager

- By using the cross-rectangle selection, select the portion of entities that you want to stretch.
- Click in the **Stretch about** box and then select the base point on the sketch.
- Move the cursor to the desired position and click to stretch the entities. Figure-71 shows the procedure of stretching the entities.

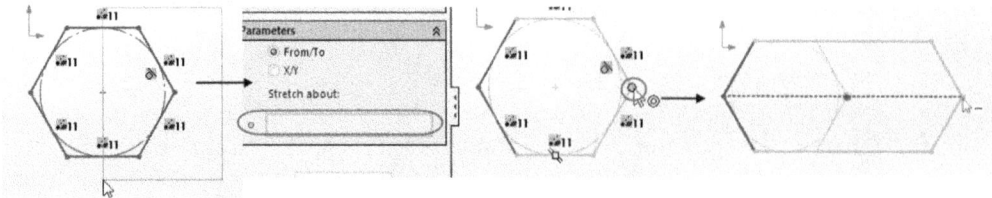

Figure-71. Stretching entities

RELATIONS

Relations are used to constrain the sketch entities dimensionally and/or geometrically. In SolidWorks also you can constrain any sketch entity by using dimensions or geometrical constrains. Both type of constrains are explained next.

Dimensional Constraints (Dimensions)

Dimensions are used to limit size of the sketch entities. For example; specifying length of the line, specifying diameter of the circle, and so on. The tools to dimension sketch entities are given in the **Smart Dimension** drop-down. The tools that are commonly used for dimensioning are discussed next.

Smart Dimension

The **Smart Dimension** tool is used to dimension the entities automatically. This tool can create different type of dimensions like, horizontal, vertical, or inclined dimensions. The procedure to create dimensions are explained next.

- Click on the **Smart Dimension** tool from the **Smart Dimension** drop-down in the **Ribbon**. You are asked to select entities to dimension.
- Select the entity you want to dimension and then click at the desired distance to place dimension.

Selection pattern for dimensioning various entities after selecting the **Smart Dimension** tool is given next.

Dimensioning a line

Click on the line to dimension.

Dimensioning inclined line

Click one by one at the end points of the line. Move the cursor perpendicularly above the line to create inclined dimensions. If you want create horizontal dimension of a line then move the cursor vertically downward or upward beyond the limit of the line. Similarly, move the cursor towards left or right of the line to create vertical dimension.

Dimensioning arcs/circle

Click on the arc/circle to specify its radius/diameter.

Dimensioning elliptical arcs/ellipse

Click on the end points of the elliptical arcs/ellipses to dimension it.

Figure-72 show various entities dimensioned.

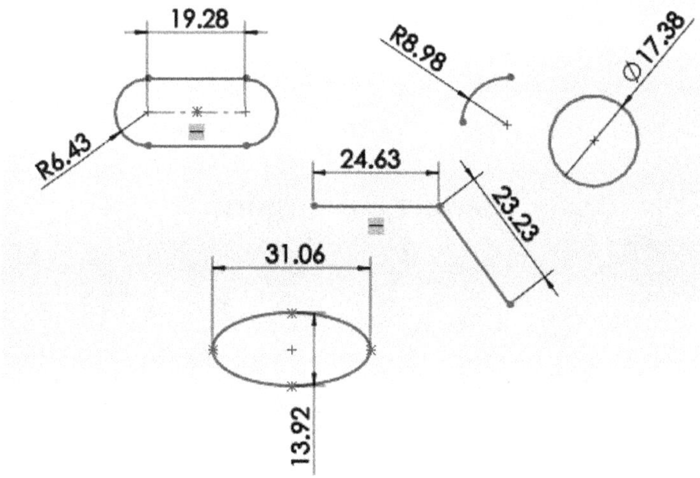

Figure-72. Dimensioning

Ordinate Dimension

You can also use Ordinate dimensioning if you are dimensioning for CNC coordinates. The procedure to create ordinate dimensions is given next.

- Click on the **Ordinate Dimension** tool from the **Smart Dimension** drop-down.
- Select the first reference that you want to make zero reference.
- Click to place the zero reference.
- Select the next line for which you want to specify the dimension.
- After you have specified the dimensions in one direction. Press the **ESC** key and then restart the tool.
- Now, select the next zero reference and repeat the procedure. Figure-73 shows a sketch dimensioned by ordinates.

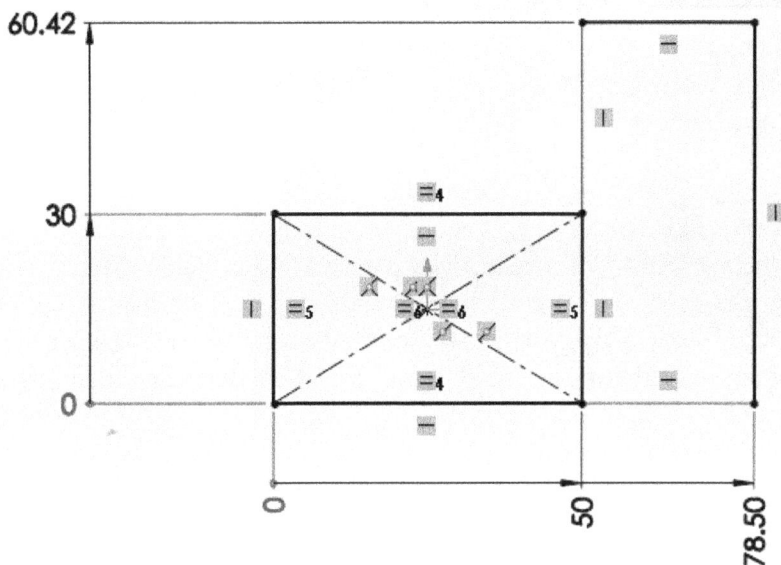

Figure-73. Ordinate dimensioned sketch

Geometric Constraints

These constraints are used to constrain the shape/position of sketch entities with respect to other entities.

- To apply the geometric constraints, click on the **Add Relation** button in the **Display/Delete Relations** drop-down.

The list of constraints that can be applied in SolidWorks is given next.

Horizontal

The **Horizontal** constraint makes one or more selected lines or center lines to become horizontal. You can also select an external entity such as an edge, plane, axis, or sketch curve on an external sketch that will act as a line to apply this constraint. You can also make two or more points to become horizontal using the **Horizontal** constraint. A point can be a sketch point, a center point, an endpoint, a control point of a spline, or an external entity such as origin, vertex, axis, or point in an external sketch. To apply this constraint, invoke the **Add Relations PropertyManager**. Select the entity or entities to apply the **Horizontal** constraint. Choose the **Horizontal** button from the **Add Relations** rollout in the **Add Relations PropertyManager**. You will notice that the name of the horizontal constraint will be displayed in the **Existing Constraints** rollout.

Vertical

The **Vertical** constraint makes one or more selected lines or centerlines to become vertical. You can force two or more points to become vertical using the **Vertical** constraint. To apply this constraint, invoke the **Add Relations PropertyManager** and select the entity or entities to apply the **Vertical** constraint. Choose the **Vertical** button from the **Add Relations** rollout. You will notice that the name of the vertical constraint is displayed in the **Existing Constraints** rollout.

Collinear

The **Collinear** constraint makes the selected lines to lie on the same infinite line; refer to Figure-74. To use this constraint, select the lines to apply the **Collinear** constraint. Choose the **Collinear** button from the **Add Relations** rollout.

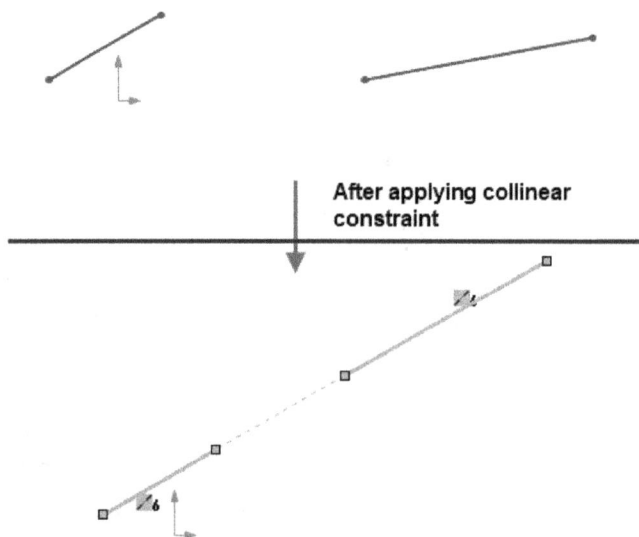

After applying collinear constraint

Figure-74. Applying collinear constraint

Coradial

The **Coradial** constraint makes the selected arcs or circles to share the same radius and the same center points; refer to Figure-75. You can also select an external entity that projects as an arc or a circle in the sketch to apply this constraint. To use this constraint, invoke the **Add Relations PropertyManager**. Select two arcs or circles, or an arc and a circle to apply the **Coradial** constraint. Choose the **Coradial** button from the **Add Relations** rollout.

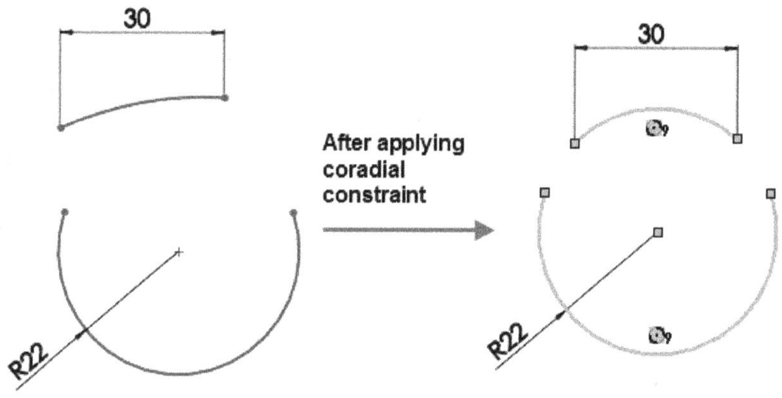

Figure-75. Applying coradial constraint

Perpendicular

The **Perpendicular** constraint makes the selected lines to become perpendicular to each other. To apply this constraint, invoke the **Add Relations PropertyManager**. Select two lines and choose the **Perpendicular** button from the **Add Relations** rollout. Figure-76 shows two lines before and after applying the **Perpendicular** constraint.

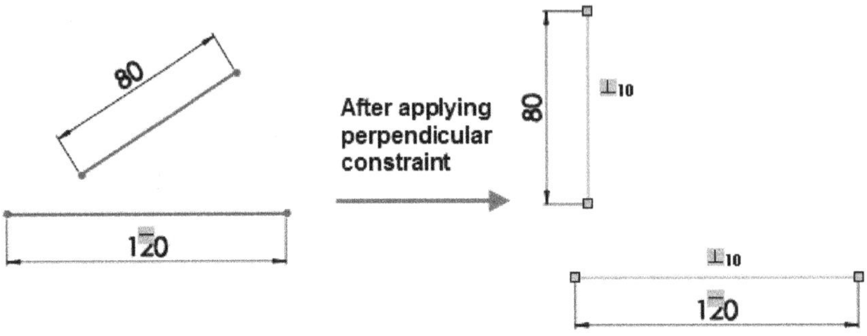

Figure-76. Applying perpendicular constraint

Parallel

The **Parallel** constraint makes the selected lines to become parallel to each other. To apply this constraint, invoke the **Add Relations PropertyManager**. Select two lines and choose the **Parallel** button from the **Add Relations** rollout.

ParallelYZ

The **ParallelYZ** constraint makes a line in the three-dimensional (3D) sketch to become parallel to the YZ plane with respect to the selected plane. To apply this constraint, invoke the **Add Relations PropertyManager**. Select a line in the 3D sketch and then select a plane. Next, choose the **ParallelYZ** button from the **Add Relations** rollout.

ParallelZX

The **ParallelZX** constraint makes a line in the 3D sketch to become parallel to the ZX plane with respect to the selected plane. To apply this constraint, invoke the **Add Relations PropertyManager**. Select a line in the 3D sketch and then select a plane. Next, choose the **ParallelZX** button from the **Add Relations** rollout.

Along X

The **AlongX** constraint makes a line in the 3D sketch to become parallel to the X-axis. To apply this constraint, invoke the **Add Relations PropertyManager**. Select a line in the 3D sketch and then choose the **Along X** button from the **Add Relations** rollout; the selected line will be oriented along the X axis.

Along Y

The **Along Y** constraint makes a line in the 3D sketch to become parallel to the Y-axis. To apply this constraint, invoke the **Add Relations PropertyManager**. Select a line in the 3D sketch and then choose the **Along Y** button from the **Add Relations** rollout; the selected line will be oriented along the Y axis.

Along Z

The **AlongZ** constraint makes a line in the 3D sketch to become parallel to the Z-axis. To apply this constraint, invoke the **Add Relations PropertyManager**. Select a line in the 3D sketch and then choose the **Along Z** button from the **Add Relations** rollout; the selected line will be oriented along the Z axis.

Normal

The **Normal** constraint makes a line in the 3D sketch to become normal to the selected plane; refer to Figure-77. To apply this constraint, invoke the **Add Relations PropertyManager**. Select a line in the 3D sketch and then select a plane. Next, choose the **Normal** button from the **Add Relations** rollout; the selected line will be oriented normal to the plane.

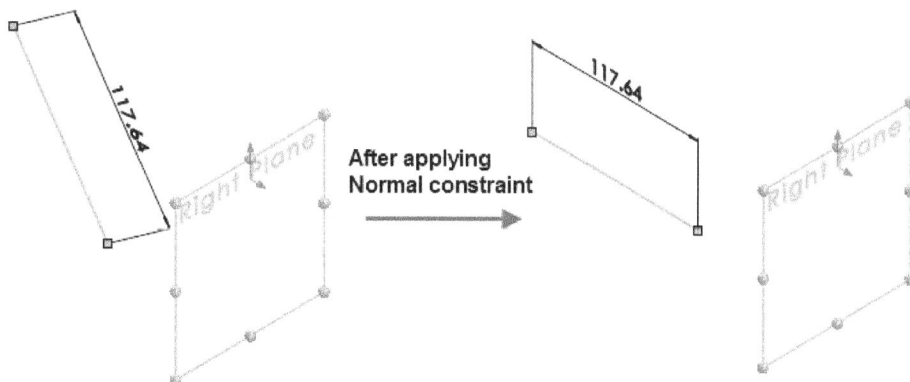

Figure-77. Applying Normal constraint

On Plane

The **On Plane** constraint makes a line in the 3D sketch to become parallel and places on the selected plane.

To apply this constraint, invoke the **Add Relations PropertyManager**. Select a line in the 3D sketch and then select a plane. Next, choose the **On Plane** button from the **Add Relations** rollout; the selected line will be oriented parallel to the selected plane and places on it.

Tangent

The **Tangent** constraint makes a selected arc, circle, spline, or ellipse to become tangent to other arc, circle, spline, ellipse, line, or edge. To apply this constraint, invoke the **Add Relations PropertyManager**. Select two entities and then choose the **Tangent** button from the **Add Relations** rollout.

Concentric

The **Concentric** constraint makes a selected arc or circle to share the same center point with other arc, circle, point, vertex, or circular edge. To apply this constraint, invoke the **Add Relations PropertyManager**. Select the required entity to apply the **Concentric** constraint and then choose the **Concentric** button from the **Add Relations** rollout.

Equal

The **Equal** constraint makes the selected lines to have equal length and the selected arcs, circles, or arc and circle to have equal radii. To apply this constraint, invoke the **Add Relations PropertyManager**. Select the required entity to apply the **Equal** constraint and choose the **Equal** button.

Intersection

The **Intersection** constraint makes a selected point to move at the intersection of two selected lines. To apply this constraint, invoke the **Add Relations PropertyManager**. Select the required entity to apply the **Intersection** constraint. Choose the **Intersection** button from the **Add Relations** rollout.

Coincident

The **Coincident** constraint makes a selected point to be coincident with a selected line, arc, circle, or ellipse. To apply this constraint, invoke the **Add Relations PropertyManager**. Select the required entity to apply the **Coincident** constraint. Choose the **Coincident** button from the **Add Relations** rollout.

Midpoint

The **Midpoint** constraint makes a selected point to move to the midpoint of a selected line. To apply this constraint, invoke the **Add Relations PropertyManager**. Select the point and the line to which the midpoint constraint has to be applied. Choose the **Midpoint** button from the **Add Relations** rollout.

Symmetric

The **Symmetric** constraint makes two selected lines, arcs, points, and ellipses to remain equidistant from a centerline. This constraint also makes the entities to have the same orientation. To apply this constraint, invoke the **Add Relations PropertyManager**. Select the required entity to apply the **Symmetric** constraint and select a center line. Choose the **Symmetric** button from the **Add Relations** rollout.

Fix

The **Fix** constraint makes the selected entity to be fixed at the specified position. If you apply this constraint to a line or an arc, its location will be fixed but you can change its size by dragging the endpoints. To apply this constraint, invoke the **Add Relations PropertyManager**. Select the required entity and choose the **Fix** button.

Merge

The **Merge** constraint makes two sketch points or endpoints to merge in a single point. To apply this constraint, invoke the **Add Relations PropertyManager**. Select the required entities to apply the **Merge** constraint and choose the **Merge** button from the **Add Relations** rollout.

Pierce

The **Pierce** constraint makes a sketch point or an endpoint of an entity to be coincident with an entity of another sketch. To apply this constraint, invoke the **Add Relations PropertyManager**. Select the required entities to apply the **Pierce** constraint and choose the **Pierce** button from the **Add Relations** rollout.

Fully Defined Sketch

Fully defined sketch is the one whose specifications can not be changed unintentionally. In some complex sketches, when you change dimension of one entity, the dimension of other entity gets changed automatically. If you have the sketch fully defined then the dimension of the entities will not change unintentionally. In technical terms, a fully defined sketch is the one in which entities have zero degree of freedom. In SolidWorks, the sketch that is fully defined will be displayed in bold black color; refer to Figure-78.

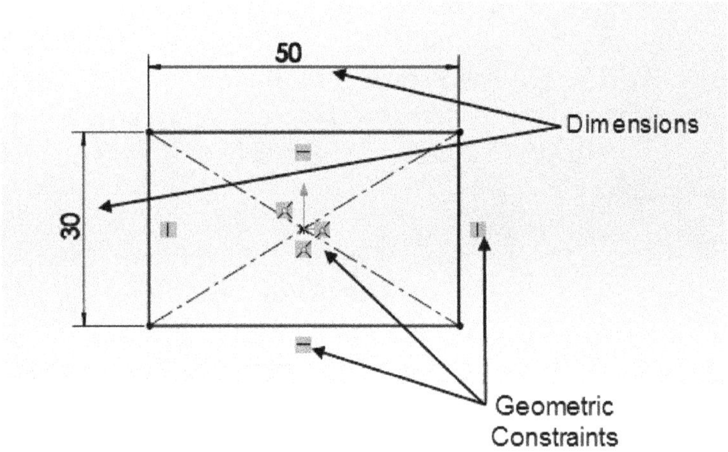

Figure-78. A fully defined sketch

You can fully define a sketch either by manually applying the dimensions and geometric constraints or you can do it by using the **Fully Define Sketch** button. This button is available in the **Display/Delete Relations** drop-down. The procedure to use this tool is explained next.

- Click on the **Fully Defined Sketch** button in the **Display/Delete Relations** drop-down of the **Ribbon**. The **Fully Define Sketch PropertyManager** will display as shown in Figure-79. Note that all the rollouts in the **PropertyManager** are expanded.

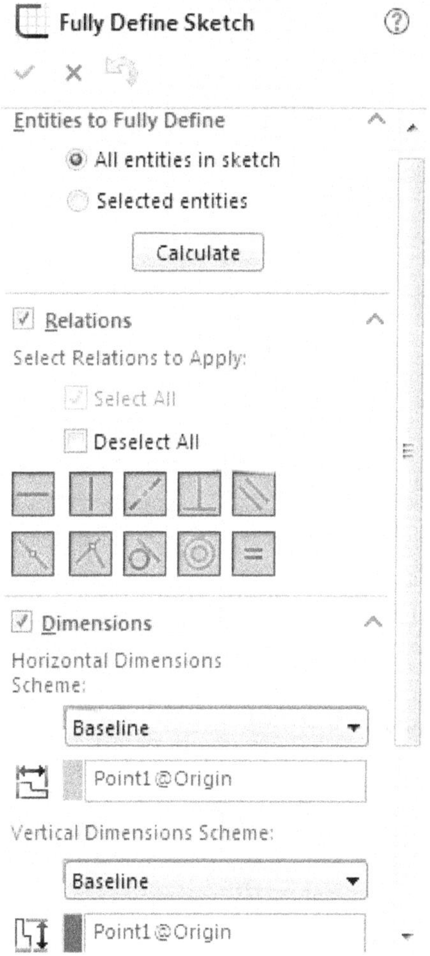

Figure-79. Fully Define Sketch PropertyManager

- Select the type of dimensions and constraints that you want to use while making a sketch fully defined.
- Click on the **Calculate** button and then click on **OK** button to fully define the sketch. The dimensions and constraints will be automatically applied. Now, change the values as per your requirement. Refer to Figure-80.

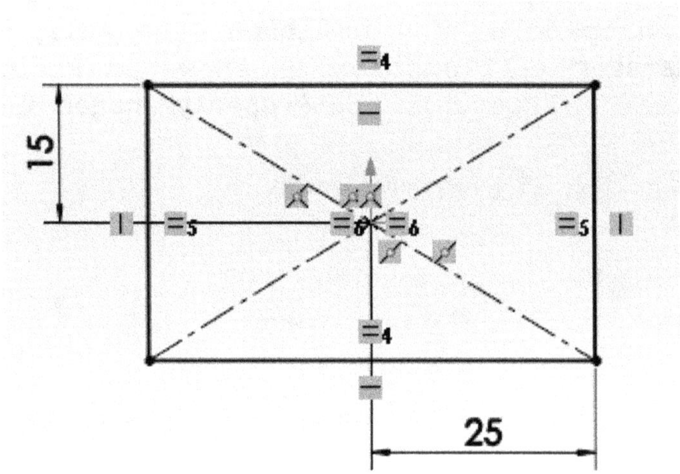

Figure-80. Fully defined sketch

If you add dimensions or constraints that are more than required then such sketches are called over defined sketches. The dimensions, constraints and sketches that are over defined are displayed in yellow color. Refer to Figure-81. In such cases, you need to delete the conflicting dimensions or make them driving. You can also use the **SketchXpert** to do the modifications. The procedure is given as follow:

- Click on the **Over Defined** message in the Information Bar at the bottom of the viewport. Refer to Figure-81.

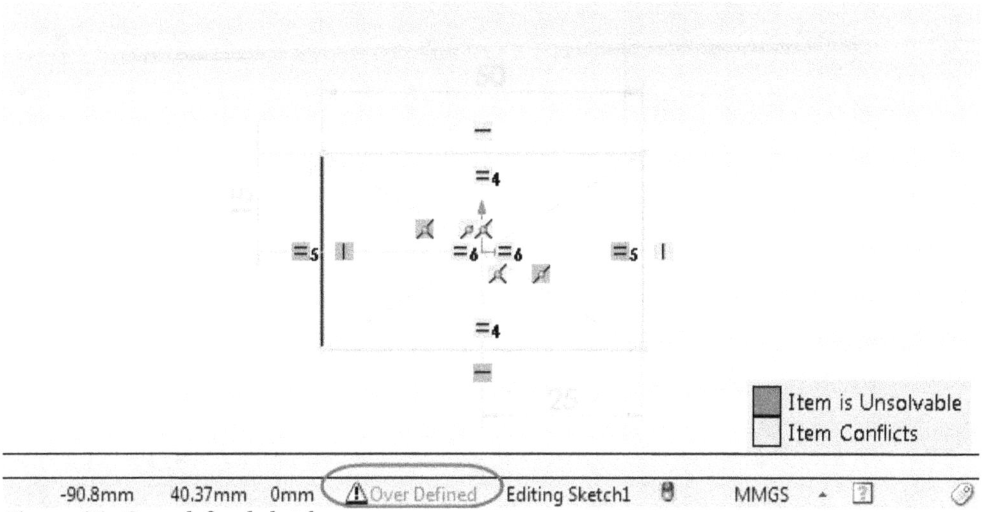

Figure-81. Over defined sketch

- The **SketchXpert PropertyManager** displays as shown in Figure-82.

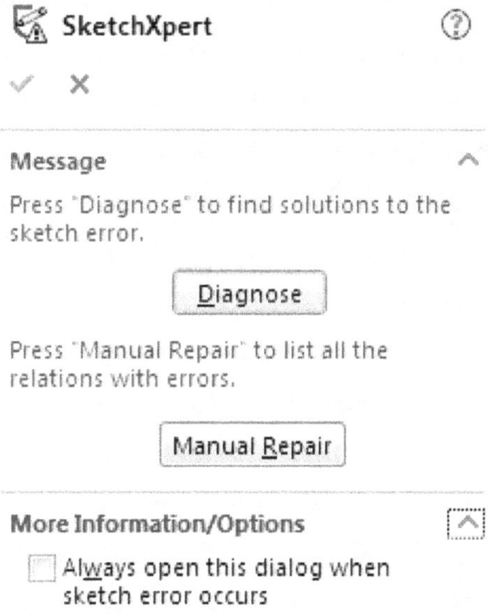

Figure-82. SketchXpert Property-
Manager

- Click on the **Diagnose** button to find automatic solutions. The modified **SketchXpert PropertyManager** will display and interfering dimension will be displayed crossed; refer to Figure-83.

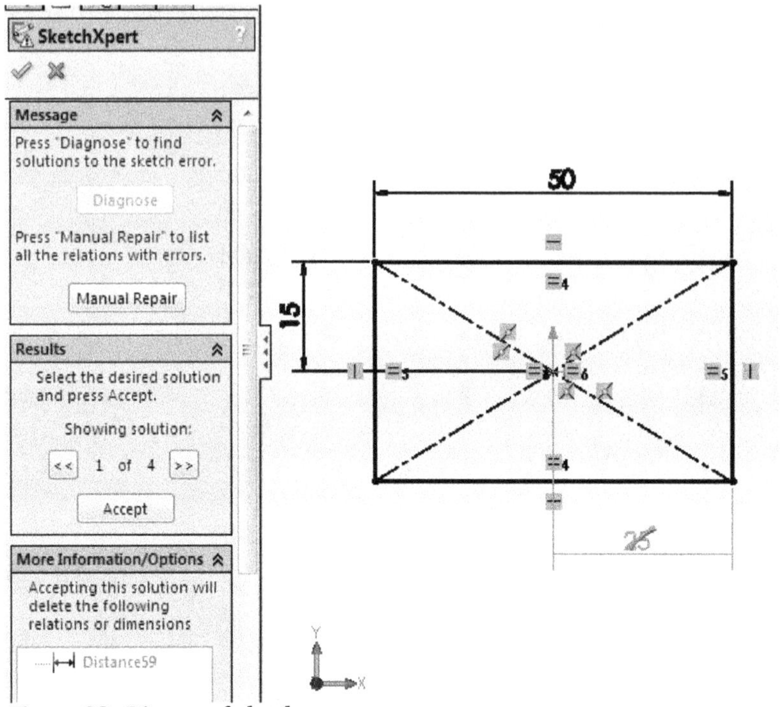

Figure-83. Diagnosed sketch

- Click on the arrow button to check the next solution and once you find the appropriate solution, click on the **Accept** button from the **Results** rollout.
- Click on **OK** button from the **PropertyManager** to accept the changes.

Till this point, we have completed the 2D sketching of SolidWorks. In the next chapter, we will practice on the tools discussed till now using real world examples.

SELF ASSESSMENT

Q1. Discuss about Planes and their placement in 3D space in the classroom. Use the example of a closed room to define different planes.

Q2. Which of the following is not the default place available in SolidWorks?

a. Front Plane
b. Right Plane
c. Top Plane
d. Left Plane

Q3. Which of the following is selected by default in the **Insert Line PropertyManager** while creating a line?

a. As sketched
b. Horizontal
c. Vertical
d. Angle

Q4. How many tools are available in SolidWorks to create a rectangle?

a. 3
b. 6
c. 5
d. 4

Q5. Which of the following tool is not available in SolidWorks to create an arc?

a. Tangent Arc
b. 3 Point Arc
c. Centerpoint Arc
d. 2 Point and Angle Arc

Q6. If you select the **Circumscribed circle** radio button while creating polygon then, the polygon will be drawn outside the circle. (T/F)

Q7. If you want to create the polygon inside the circle, then select the **Circumscribed circle** radio button from the **PropertyManager**. (T/F)

Q8. The **Equation Driven Curve** tool is available in the drop-down.

Q9. Discuss the use of **Trim away inside** button in the **Trim PropertyManager**.

Q10. Discuss the difference between Linear pattern and circular pattern.

Q11. Discuss the use of **Smart Dimension** tool with example.

Q12. What is the difference between dimensional constraining and geometrical constraining?

SPACE FOR STUDENT NOTES

Space for Student Notes

Advanced Dimensioning and Practice

Chapter 3

Topics Covered

The major topics covered in this chapter are:

- *Dimensioning and its relations with drawing.*
- *Dimension Style.*
- *Practical 1.*
- *Practical 2.*
- *Practice Drawings*

DIMENSIONING AND ITS RELATIONS

When we dimension in a sketch it is not confined only to that sketch. You will learn later that it also affects the parameters in Drafting environment. At that time, the style and dimensions that we applied in sketch will be reflected in the draft. So, it is very important to understand dimension styles here. In the previous chapters, we have worked on basic dimensions. In this chapter, we will explain the dimensions and styles in detail.

DIMENSION STYLE

Select a dimension that you have applied in the sketch. The **Dimension PropertyManager** will be displayed; refer to Figure-1.

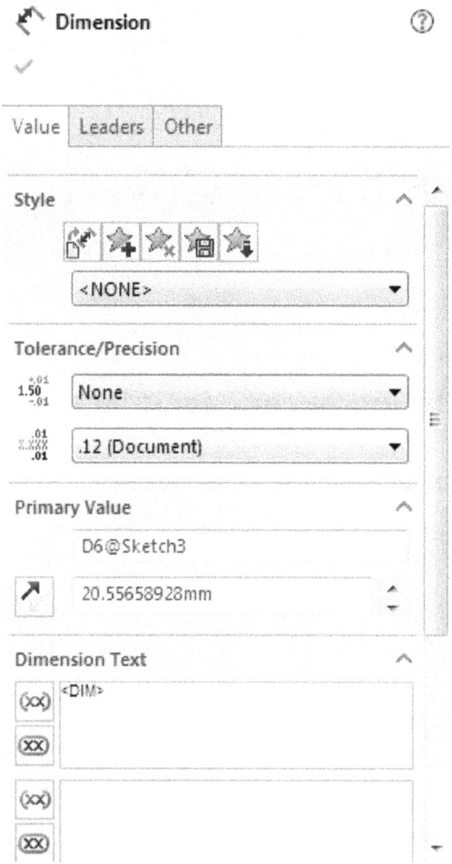

Figure-1. Dimension PropertyManager

The options in the **PropertyManager** are discussed next.

Style Rollout

The **Style** rollout is used to create, save, delete, and retrieve the dimension style in the current document; refer to Figure-2. You can also use the dimension styles saved in other documents using this rollout. The options in this rollout are discussed next.

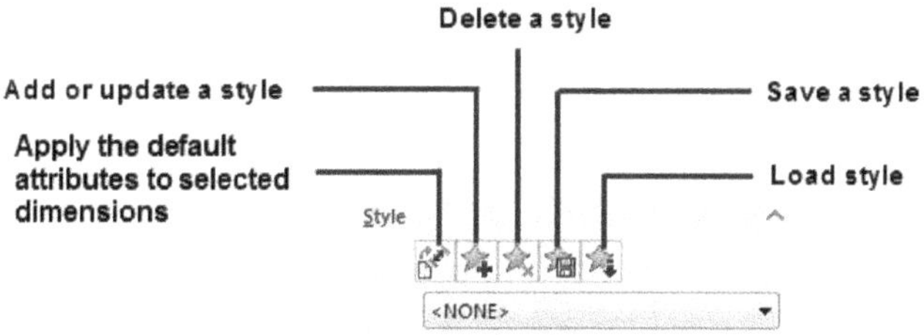

Figure-2. The Style rollout

Apply the default attributes to selected dimensions

The **Apply the default attributes to selected dimensions** button is used to apply the default attributes to the selected dimension or dimensions. The attributes can be dimension text, tolerance, precision, arrow style and so on.

Add or Update a Style

The **Add or Update a Style** button is used to add a dimension style to the current document for a selected dimension. After invoking the **Dimension PropertyManager**, set the attributes using various options provided in this **PropertyManager**. Next, choose the **Add or Update a Style** button. The **Add or Update a Style** dialog box will be displayed, as shown in Figure-3. Enter the name of the dimension style in the edit box and press ENTER; the dimension style will be added to the current document.

You can apply a new dimension style to the selected dimension by selecting a dimension style from the **Set a current Style** drop-down list in the **Style** rollout. You can also update a dimension style. To do so, select the dimension and set the options of the dimension style according to your need. Next, choose the **Add or Update a Style** button to invoke the **Add**

or Update a Style dialog box. Select the dimension style to update from the drop-down list in the dialog box; the two radio buttons in this dialog box will be enabled. Select the **Update all annotations linked to this Style** radio button and choose the **OK** button to update all the dimensions linked to the selected **Style**. If you select the **Break all links to this Style** radio button and choose the **OK** button, then the link between the other dimensions having the same style and the selected **Style** will be broken.

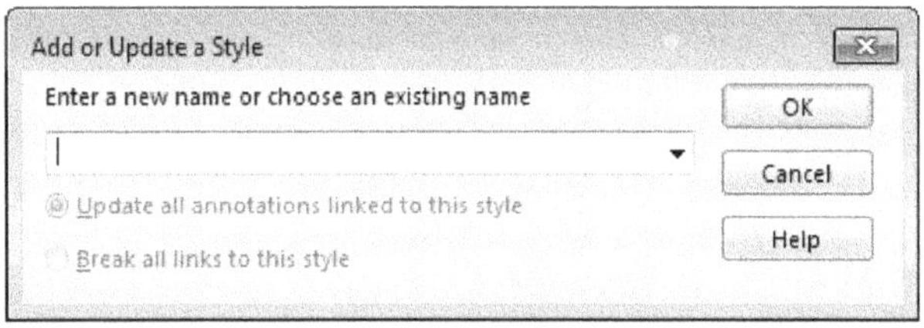

Figure-3. Add or Update a Style dialog box

Delete a Style

The **Delete a Style** button is used to delete a dimension style. Select a dimension style from the **Set a current Style** drop-down list and then choose the **Delete a Style** button.

Save a Style

The **Save a Style** button is used to save a dimension style so that it can be retrieved in some other document. Select a dimension style from the **Set a current Style** drop-down list and choose the **Save a Style** button. The **Save As** dialog box will be displayed. Browse to the folder in which you want to save the style and enter its name in the **File name** edit box. Choose the **Save** button from the **Save As** dialog box. The style file will be saved with the extension *.sldstl*.

Load Style

The **Load Style** button is used to open a saved style in the current document. The properties of that favorite will be applied to the selected dimension. To load a style, choose the **Load Style** button to invoke the **Open** dialog box. Browse to the folder in which the style is saved. Now, select the file with the extension *.sldstl* and choose the **Open** button; the

Add or Update a Style dialog box will be displayed. Choose the **OK** button from this dialog box.

Tolerance/Precision Rollout

The **Tolerance/Precision** rollout shown in Figure-4 is used to specify tolerance and precision in dimensions. The options in this rollout are discussed next.

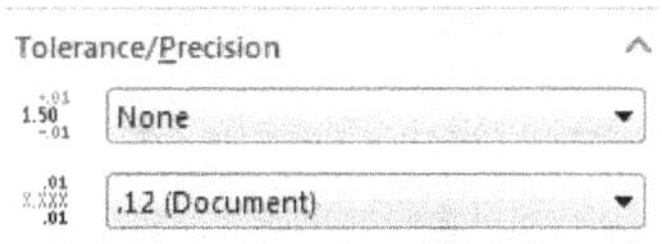

Figure-4. Tolerance/Precision rollout

Tolerance Type

The **Tolerance Type** drop-down is used to apply tolerance to a dimension. By default, the **None** option is selected. Therefore, no tolerance is applied to the dimensions. The other tolerance types available in this drop-down list are discussed next.

Basic

The basic dimension is the dimension taken as reference for other features. To display the basic dimension, select the dimension that you want to display as the basic dimension and then select the **Basic** option from the **Tolerance Type** drop-down list. On doing so, the dimension is enclosed in a rectangle; refer to Figure-5.

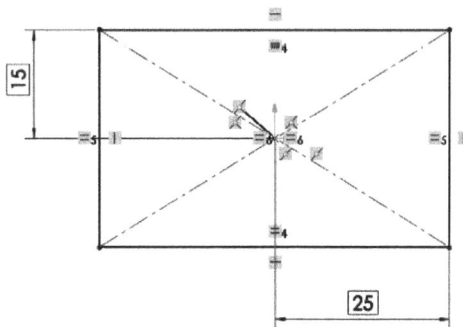

Figure-5. Basic dimensions

Bilateral

The **Bilateral** option is used to display the minimum and maximum limit of tolerance for a dimension. To apply the bilateral tolerance, select the dimension and then select the **Bilateral** option from the **Tolerance Type** drop-down; the **Maximum Variation** and **Minimum Variation** edit boxes will be enabled, where you can enter the maximum and minimum variations for a dimension. Also, the **Show parentheses** check box will be displayed. If you select this check box, the bilateral tolerance will be displayed with parentheses. The dimension with a bilateral tolerance is shown in Figure-6.

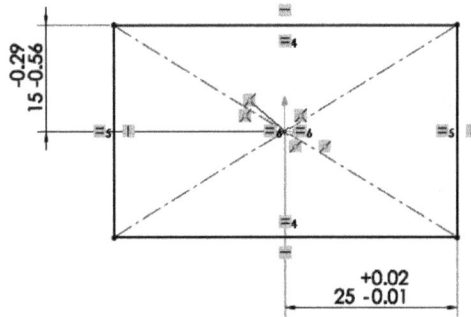

Figure-6. Bilateral tolerance

Limit

The maximum and minimum permissible dimensional values of an entity are displayed on selecting the **Limit** option. To apply this tolerance type, select the dimension to be displayed as the limit dimension and select the **Limit** option; the **Maximum Variation** and **Minimum Variation** edit boxes will be enabled. Enter the values of the maximum and minimum variations. The dimension along with the limit tolerance is shown in Figure-7.

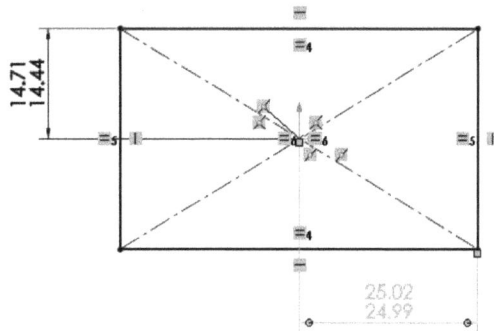

Figure-7. Limit tolerance

Symmetric

The symmetric tolerance when the variation is equal in both positive and negative direction. To use this tolerance, first select the dimension and then select the **Symmetric** option; the **Maximum Variation** edit box will be displayed. Enter the value of the tolerance in this edit box. Also, you can select the **Show parentheses** check box to show the tolerance in parentheses. The dimension along with the symmetric tolerance is shown in Figure-8.

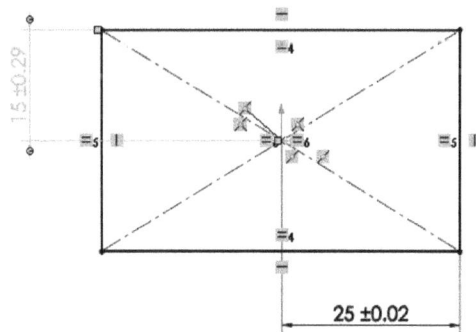

Figure-8. Symmetric tolerance

MIN

In this dimensional tolerance, the **MIN.** symbol is added to dimension as suffix. This implies that the dimensional value is the minimum value that is allowed in the design. To display this dimensional tolerance, select a dimension and then the **MIN** option from the **Tolerance Type** drop-down list. The dimension along with the minimum tolerance is shown in Figure-9.

MAX

In this dimensional tolerance, the **MAX.** symbol is added to dimension as suffix. This implies that the dimensional value is the maximum value that is allowed in the design. To display this dimensional tolerance, select a dimension and then the **MAX** option from the **Tolerance Type** drop-down list. The dimension along with the maximum tolerance is shown in Figure-9.

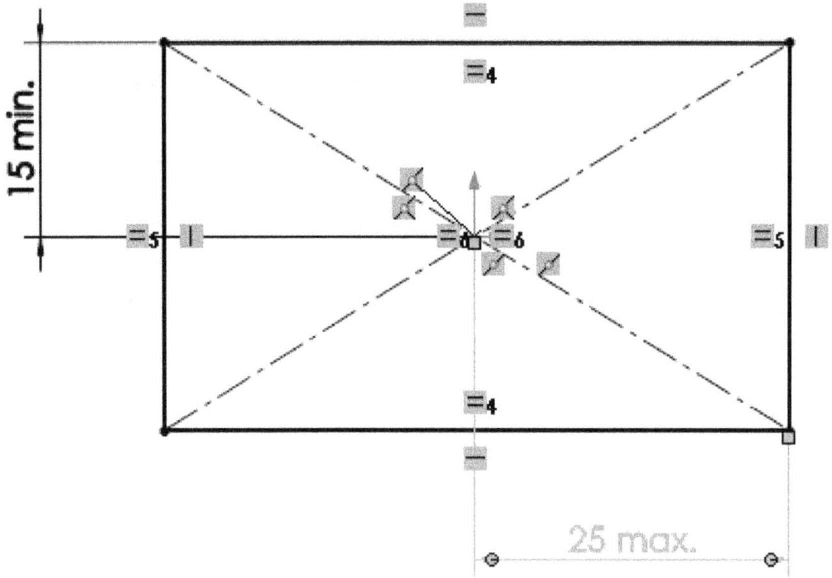

Figure-9. Min max tolerance

Fit

The **Fit** option is used to apply fit according to the Hole Fit and Shaft Fit systems. The **Tolerance/Precision** rollout with the **Fit** option selected in the **Tolerance Type** drop-down list is shown in Figure-10. Select the type of fit from the **Classification** drop-down list. The **Classification** drop-down list is used to define the **User Defined** fit, **Clearance** fit, **Transitional** fit, or **Press** fit. To apply a fit using the Hole Fit system or the Shaft Fit system, select the dimension and then the **Fit** option from the **Tolerance Type** drop-down list. The **Classification, Hole Fit,** and **Shaft Fit** drop-down lists will be displayed below the **Tolerance Type** drop-down list. Select the required fit from the **Classification** drop-down list and select the fit standard from the **Hole Fit** drop-down list or the **Shaft Fit** drop-down list. If you select the **Clearance, Transitional,** or **Press** option from the **Classification** drop-down list and the fit standard from the **Hole Fit** drop-down list, then only the standards matching the selected hole fit will be displayed in the **Shaft Fit** drop-down list and vice versa. However, if you select the **User Defined** option from the **Classification** drop-down list, you can select any standard from the **Hole Fit** and **Shaft Fit** drop-down lists. The **Stacked with line display** button is chosen to display the stacked tolerance with a line. You can also display the tolerance as stacked without a line using the **Stacked without line display** button. If you choose the **Linear display** button, the tolerance will be displayed in the linear form. The dimension along with the hole fit and shaft fit is shown in Figure-11.

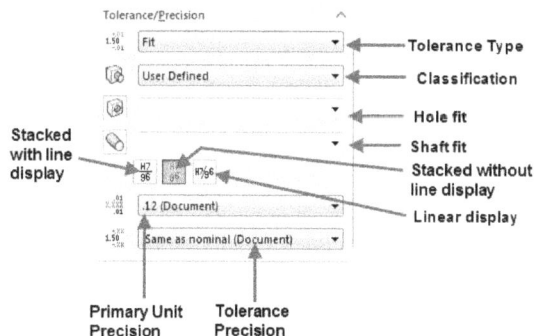

*Figure-10. The **Tolerance/Precision** rollout with the **Fit** option selected in the **Tolerance Type** drop-down list*

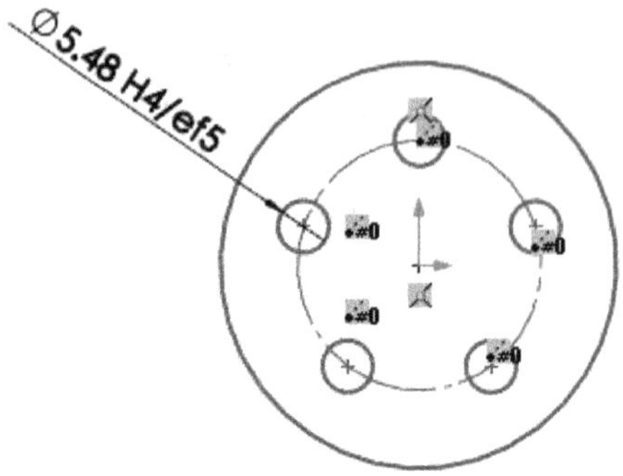

Figure-11. Hole fit and shaft fit

Fit with tolerance
The **Fit with tolerance** option in the **Tolerance Type** drop-down list is used to display tolerance along with the hole fit and shaft fit in a dimension. To apply fit with tolerance, select a dimension, and then select the **Fit with tolerance** option from the **Tolerance Type** drop-down list. Select the type of fit from the **Classification** drop-down list. Next, select the fit standard from the **Hole Fit** or the **Shaft Fit** drop-down list. Tolerance will be displayed with the fit standard only if you select a fit system from the **Hole Fit** or **Shaft Fit** drop-down list. Tolerance will be displayed along with the fit standard in the drawing area. In SolidWorks, tolerance is calculated automatically, depending on the type and standard of the fit selected. The **Show parentheses** check box can be selected to show tolerance in parentheses. The dimension along with the fit and tolerance is shown in Figure-12.

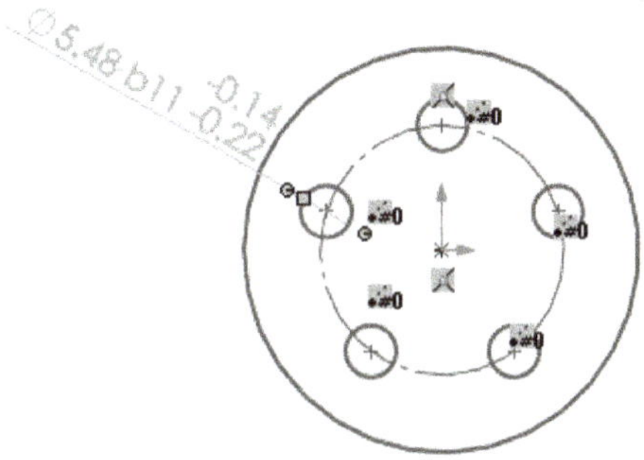

Figure-12. *Dimensioning along with the fit and tolerance*

Fit (tolerance only)

The **Fit (tolerance only)** option in the **Tolerance Type** drop-down list is used to display the tolerance in a dimension based on the hole fit or shaft fit.

None

The **None** option in the **Tolerance Type** drop-down list is used to display the dimensional value without any tolerance.

Unit Precision

The **Unit Precision** drop-down is used to specify the precision of the number of places after the decimal for dimensions. By default, the selected precision is two places after the decimal.

Tolerance Precision

The **Tolerance Precision** drop-down is used to specify the precision of the number of places after the decimal for tolerance. By default, the selected precision is two places after the decimal. This drop-down list will not be available, if the **None** option is selected in the **Tolerance Type** drop-down list.

Dimension Text Rollout

The **Dimension Text** rollout, as shown in Figure-13, is used to add text and symbols to dimension. The text box in this rollout is used to add text to dimension. The **<DIM>** text displayed in the text box symbolizes the dimensional value. You can add text before or after the dimension value. There are two text boxes in SolidWorks 2016 to write text above and below the dimension line. There are four buttons at the left of these text boxes. Choose the **Add Parentheses** button to enclose the dimension text in parentheses. Choose the **Inspection Dimension** button to enclose the dimension text in an obround shape and this dimension will be checked during inspection. Choose the **Center Dimension** button ⟷ to place the dimension at the center of the dimension line. If you need to place the text at a distance from the dimension line, choose the **Offset Text** button ⟿ and drag the dimension to the required location. Note that you can also choose all four buttons for a specific dimension.

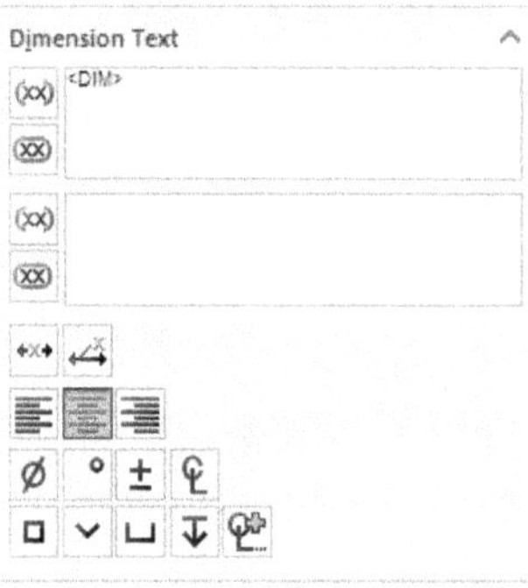

Figure-13. Dimension Text rollout

This rollout also provides buttons to modify text justification and add symbols such as diameter, degree, plus/minus, centerline, and so on to the dimension text. You can add more symbols by choosing the **More Symbols** button from the **Dimension Text** rollout and click on the **More Symbols** button from the flyout displayed. On choosing this button, the **Symbol Library** dialog box will be displayed, as shown in Figure-14.

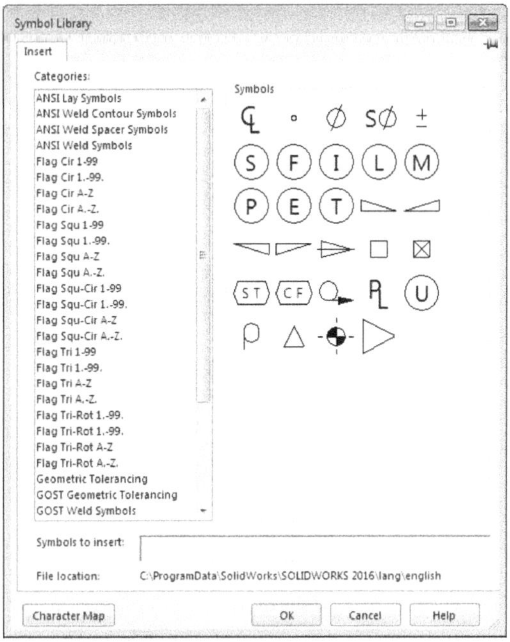

Figure-14. Symbol Library *dialog box*

Dual Dimension Rollout

You need to select the check box in the **Dual Dimension** rollout to enable the options in this rollout, refer to Figure-15. The options in this rollout are used to display the alternative dimension value. Note that the alternative dimension value is displayed in square brackets, as shown in Figure-16. The options in this rollout are similar to those discussed in the earlier sections. Note that the alternative unit is set in the **Dual Dimension Length** cell in the **Document Property - Units** dialog box. To invoke this dialog box, choose **Tools > Options** from the SolidWorks menus; the **System Options** dialog box will be displayed. Choose the **Document Properties** tab; the name of this dialog box will be changed to the **Document Property - Drafting Standard** dialog box. In this dialog box, select the **Units** option from the area on the left to display the options for setting units. Note that on selecting the **Unit** option, the name of this dialog box will be changed to the **Document Property - Units** dialog box.

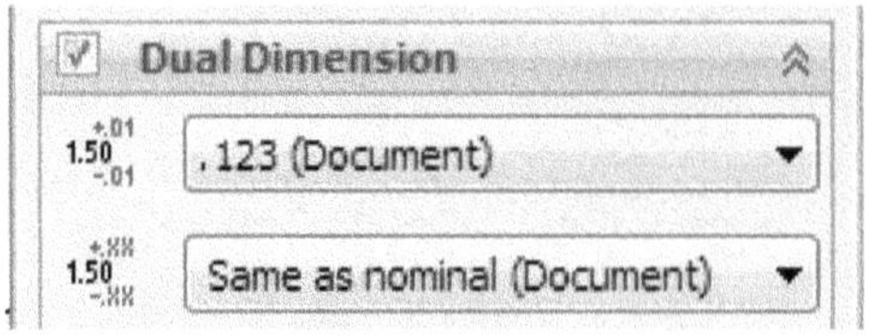

Figure-15. Dual Dimension rollout

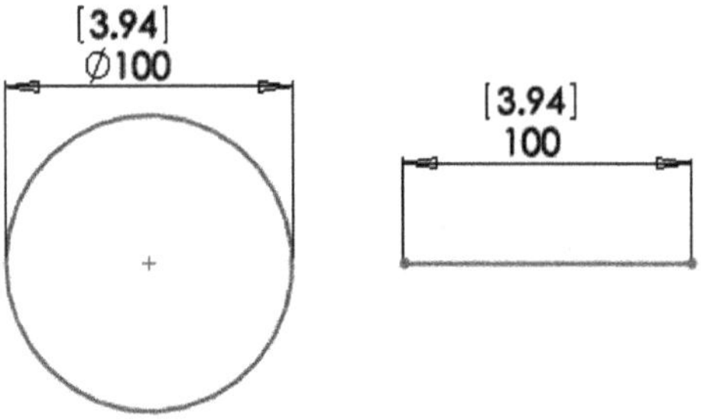

Figure-16. Entities with dual dimension

Sometimes, you may need to change the type of arrowheads or place the dimension at a distance from the entity because of space constraint. In SolidWorks, these actions can be performed by choosing the **Leaders** tab in the **Dimension PropertyManager**. The rollouts in this tab are discussed next.

Witness/Leader Display Rollout

The **Witness/Leader Display** rollout is used to specify the arrowhead style in dimensions, refer to Figure-17. The options in this rollout are discussed next.

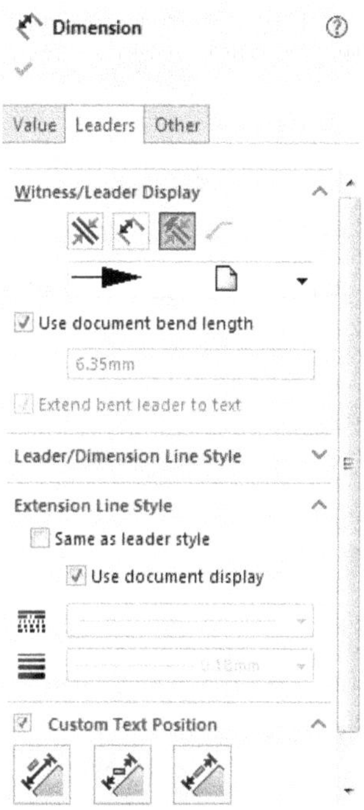

Figure-17. The rollouts in the **Leaders** *tab*

Outside

The **Outside** button is used to display the arrows outside the extension line. To do so, select a dimension from the drawing area and choose the **Outside** button from the **Witness/ Leader Display** rollout.

Inside

The **Inside** button is used to display the arrows inside the extension line. To do so, select a dimension from the drawing area and choose the **Inside** button. You can also click on the control point displayed on the arrowhead to reverse its direction.

Note that you can also click on the control point displayed on the arrowhead to reverse the direction of the arrowhead.

Smart

The **Smart** button is chosen by default and the arrows are displayed inside or outside the extension line, depending on the space available between the extension lines.

Directed Leader

This button is chosen to change the leader style of a dimension created on a surface by using the **DimXpert**.

Style

The **Style** drop-down list is used to select the style of the arrowhead. The unfilled triangular arrow is selected by default. You can select any arrowhead style for a particular dimension or dimension style. To change the arrowhead style, select a dimension from the drawing area and then the arrowhead style from the **Style** drop-down list.

Use document bend length

To change the length of a leader line after the bend, clear this check box and specify the length in the edit box below the check box. By default, the value specified in the **Document Properties - Detailing - Dimensions** dialog box will be displayed in the edit box given below this check box.

Leader/Dimension Line Style Rollout

To enable the options in the **Leader/Dimension Line Style** rollout, you need to clear the **Use document display** check box in it. After clearing this check box, the **Leader Style** and **Leader Thickness** drop-down lists will be enabled. The **Leader Style** drop-down list is used to specify the leader style and the **Leader Thickness** drop-down list is used to specify the thickness of the leader. You can also select the **Custom size** option from the **Leader Thickness** drop-down list and specify the thickness of the leader in the spinner available below this drop-down list as per your requirement.

Extension Line Style Rollout

To enable the options in the **Extension Line Style** rollout, you need to clear the **Use document display** check box and **Same as leader style** check box in it. The options in this rollout are same as discussed for **Leader/Dimension Line Style** rollout. These options modify the display of extension lines in dimensions.

Custom Text Position Rollout

The options in this rollout are used to specify the position of the text on a dimension line. Select the check box in the **Custom Text Position** rollout to enable the options in this rollout. The options in this rollout are discussed next.

Solid Leader, Aligned Text

If you choose this button, the leader line will be placed parallel to the dimension line along with the text.

Broken Leader, Horizontal Text

On choosing this button, leader line will be placed parallel to the horizontal axis along with the text.

Broken Leader, Aligned Text

Choose this button to place the dimension line to the center of the text. In this case, the text and the leader line will be placed parallel to the dimension line.

In SolidWorks, you can change the units and font of the dimension text by choosing the **Other** tab. The rollouts in this tab are discussed next.

Override Units Rollout

If you need to change the existing units of the dimension, select the check box in this rollout to expand it and select the units from the **Length Units** drop-down list.

Text Fonts Rollout

The font style set in the **Document Property - Detailing - Annotations Font** dialog box will be the default font style. To change the font style, clear the **Use document font** check box and change the font style by choosing the **Font** button in the **Text Fonts** rollout.

Options Rollout

If you select the **Read only** check box in this rollout, the dimensional value cannot be changed. If you select the **Driven** check box, the value will be the driven value.

Horizontal/Vertical Dimensioning between Points

As mentioned earlier, you can add a horizontal or vertical dimension between two points. To add any of these dimensions, choose the required button from the **Dimensions/Relations** toolbar or the **Smart Dimension** flyout in the **Sketch CommandManager**. Select the first point and then the second point. Next, specify a point to place the dimension; the **Modify** dialog box will be displayed. Enter a new dimension value in this dialog box and press ENTER.

Now, we will deal with real-world problems related to sketches.

Practical 1

Create the sketch as shown in Figure-18. Also dimension the sketch as per the figure.

Steps to be performed:

Below is the step by step procedure of creating the sketch shown in the Figure-18.

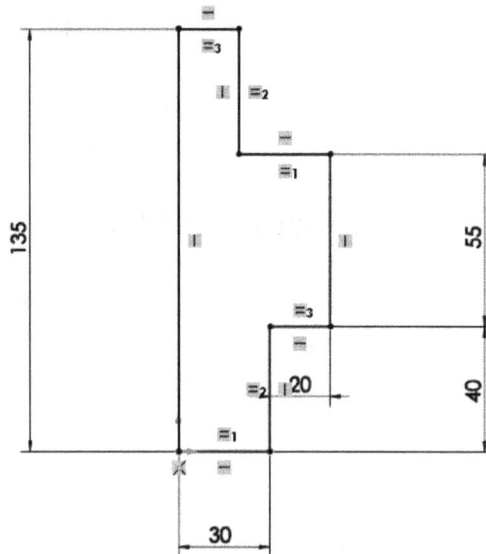

Figure-18. Practical 1

Starting Sketching Environment

- Start SolidWorks if not started already.
- Click on the **New** button from the Menu Bar. The **New SolidWorks Document** dialog box will display; refer to Figure-19.

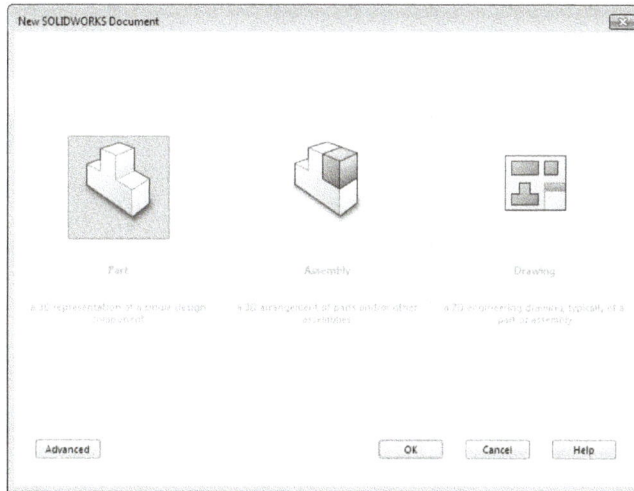

Figure-19. New SolidWorks Document dialog box

- Double click on the Part button. The Part environment of SolidWorks will display as shown in Figure-20.

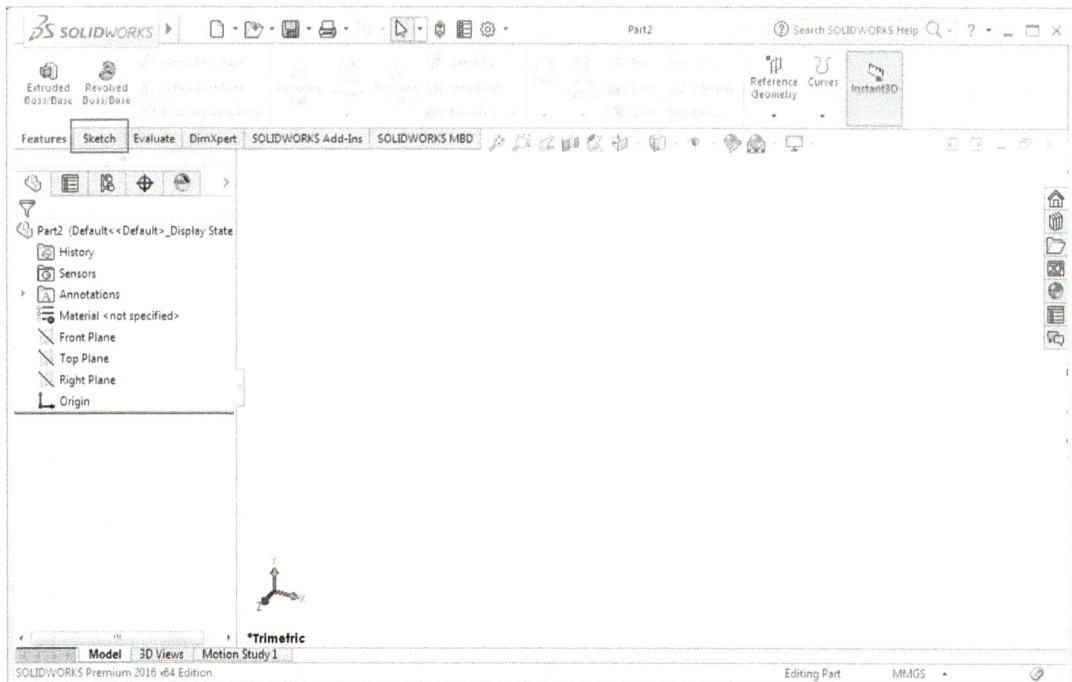

Figure-20. SolidWorks application window

- Click on the **Sketch** tab of the **Ribbon**; refer to Figure-20. The tools related to sketch will display in the **Ribbon**.

Starting Sketching

- Click on the **Sketch** button. Three default planes will be displayed.
- Select the **Front** plane from the viewport, refer to Figure-21. The viewport will become parallel to the view screen.

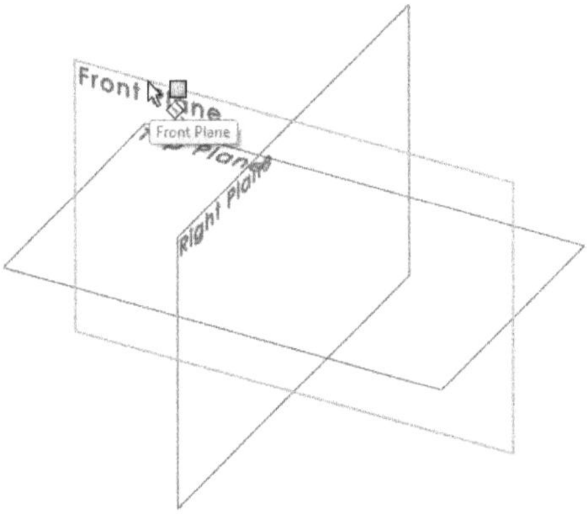

Figure-21. Selecting front plane

Creating Lines

- Click on the **Line** button from the Ribbon. The Line tool will become active and you are asked to select the start point.
- Click on the coordinate system and move the cursor towards right; refer to Figure-22.

Figure-22. Starting creation of line

- Enter **30** in the dimension box displayed below the line.
- Move the cursor vertically upwards and enter the value **40** in the dimension box.
- Move the cursor towards right and enter **20** in the dimension box. Refer to Figure-23.

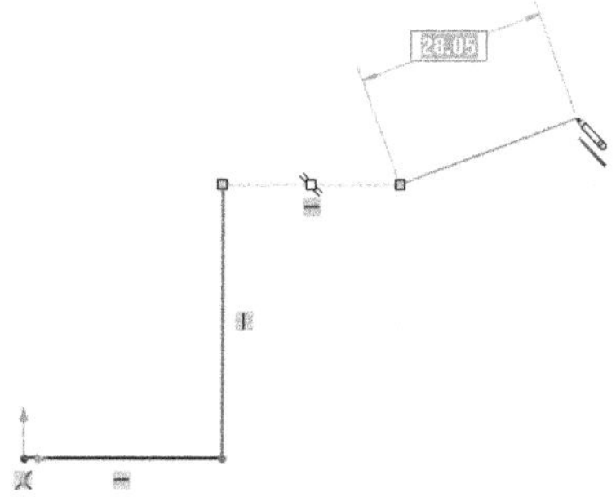

Figure-23. Sketch after specifying 20
value

- Move the cursor upward and specify the value as **55** in the dimension box.
- Move the cursor towards left and specify the value as **30** in the dimension box.
- Move the cursor upward and specify the value as **40** in the dimension box.
- Move the cursor towards left and specify **20** in the dimension box.
- Move the cursor downward and click on the coordinate system to close the sketch.

The sketch after performing the above steps is displayed as shown in Figure-24.

Dimensioning the Sketch

- Click on the **Smart Dimension** button from the Ribbon. You are asked to select entities.

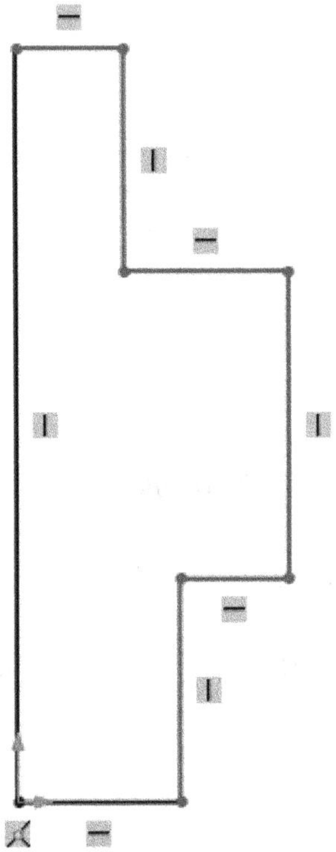

Figure-24. Completed Sketch

- Click on the bottom line joining with the coordinate system and move the cursor downwards; refer to Figure-25.
- Click at the appropriate distance to place the dimension. The Modify input box will display.
- Enter the dimension if you want to change. In this case, we will press ENTER to apply the default value.
- Click on the vertical line joining the end point of the bottom line selected earlier; refer to Figure-26.

Figure-25. Dimensioning bottom line

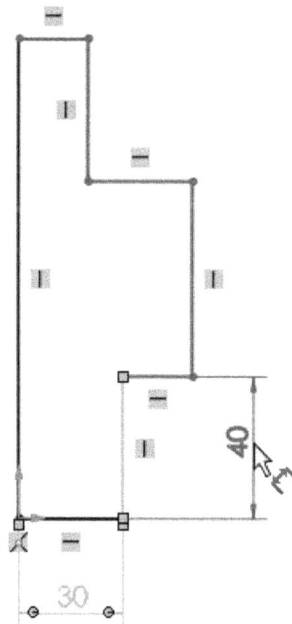

Figure-26. Vertical dimension to selected

• Place the dimension at proper distance from the line and press ENTER at the **Modify** input box.

• Similarly, dimension other entities of the sketch.

The sketch after applying all the dimensions will be displayed as shown in Figure-27.

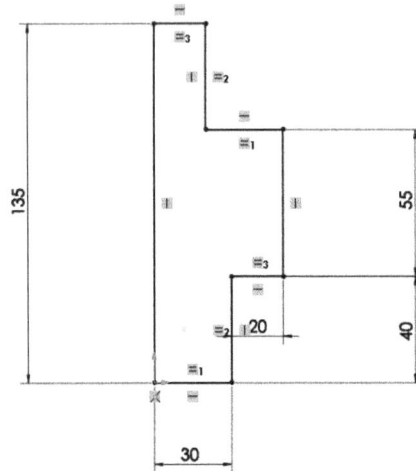

Figure-27. Final sketch after applying dimensions

Practical 2

In this practical, we will create the sketch as shown in Figure-28.

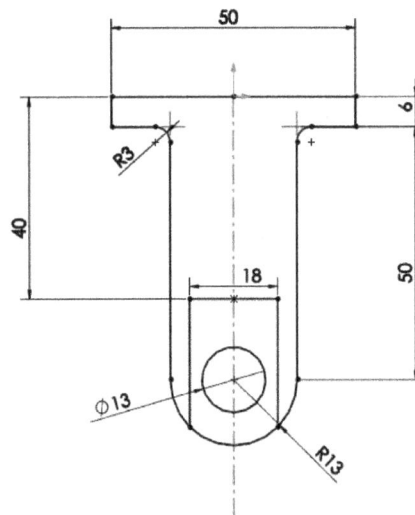

Figure-28. Practical 2

Steps to be performed:

Below is the step by step procedure of creating the sketch shown in the Figure-28.

Starting Sketching Environment

- Start SolidWorks if not started already.
- Click on the **New** button from the Menu Bar. The **New SolidWorks Document** dialog box will display.
- Double click on the **Part** button. The Part environment of SolidWorks will display.
- Click on the **Sketch** tab of the **Ribbon**. The tools related to sketch will display in the **Ribbon**.

Starting Sketching

- Click on the **Sketch** button from the **Ribbon**. Three default planes will be displayed.
- Select the **Front** plane from the viewport. The viewport will become parallel to the view screen.

Creating Lines

- Click on the **Line** button from the **Ribbon**. You are asked to specify the start point of the line.
- Click on the coordinate system and enter **25** in the dimension box.
- Move the cursor down perpendicular to the previous line and enter **6** in the dimension box.
- Move the cursor to left and enter **12** in the dimension box.
- Move the cursor downwards and enter **50** in the dimension box. Till this point, our sketch should display like Figure-29

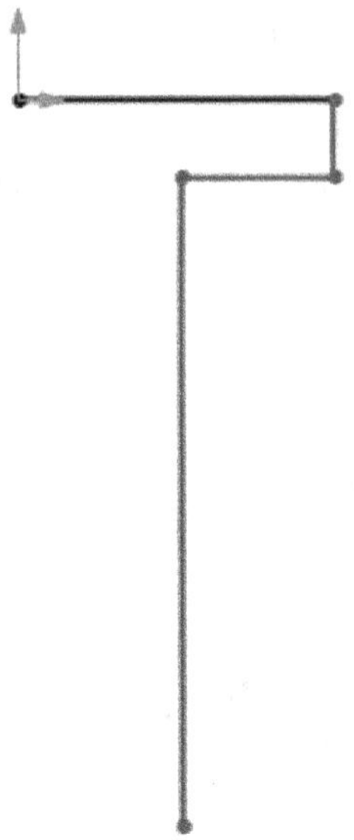

Figure-29. Sketch after creating lines

Creating Arcs

- Click on the down arrow of **Arcs** drop-down and select **Tangent Arc** button from the list. You asked to click on an end point.
- Click on the end point of vertical line recently created and move the cursor downwards and towards right until you get the preview as shown in Figure-30.
- When you get the preview like the Figure-30, click to create the arc.

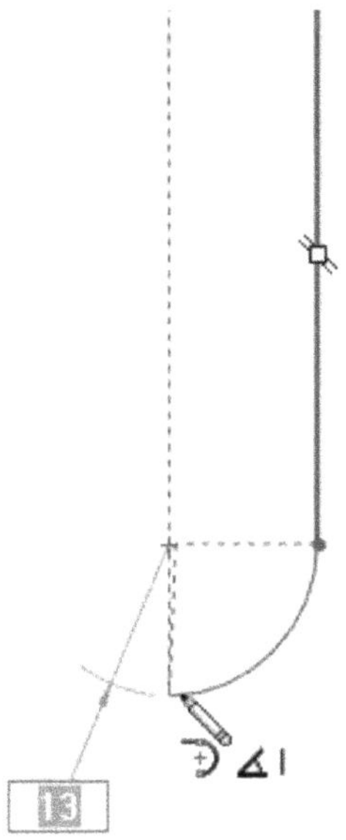

Figure-30. Arc creation

Creating Fillet

- Click on the **Sketch Fillet** tool from the **Ribbon** . The **Sketch Fillet PropertyManager** will display as shown in Figure-31.

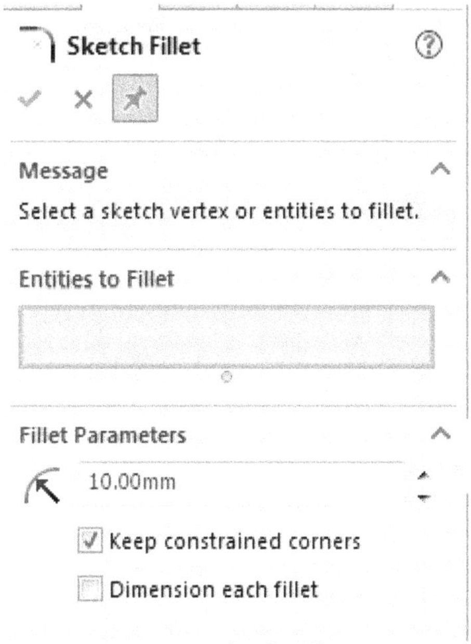

Figure-31. Sketch Fillet PropertyManager

- Click in the **Radius** spinner edit box in the **Parameter**s rollout of the **PropertyManager** and enter the value as **3**.
- Select the lines as shown in Figure-32 for applying fillet. The fillet will be created between the two lines.
- Click on the **OK** button from the **PropertyManager**.

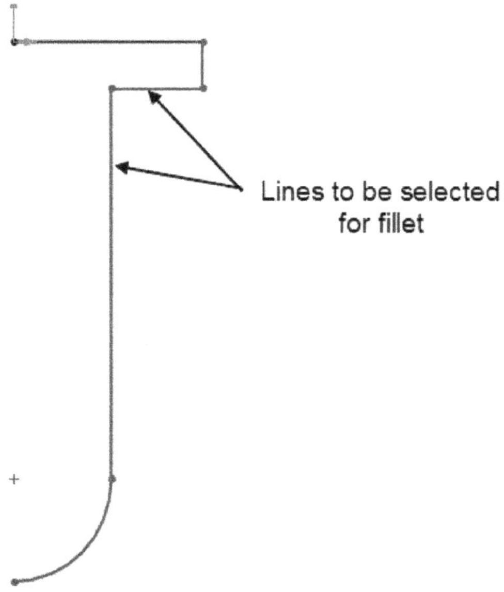

Lines to be selected for fillet

Figure-32. Lines selection for fillet

Creating Mirror Copy

- Click on the **Centerline** tool from the **Line** drop-down. The **Line PropertyManager** will display as shown in Figure-33.
- Click on the coordinate system (start point of the sketch line) and then the end point of the arc. A center line will be created.
- Press **ESC** from **Keyboard** and exit the tool.

Figure-33. Line PropertyManager

- Select the **Mirror Entities** tool from the **Ribbon** and select all the entities we have sketched except center line; refer to Figure-34.

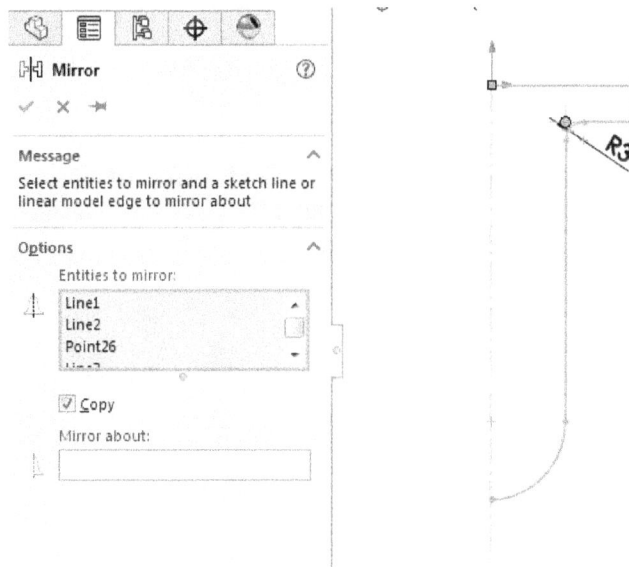

Figure-34. Entities selected for creating mirror

- Click in the **Mirror about** box and select the center line from the sketch. Preview of mirror will be displayed; refer to Figure-35.

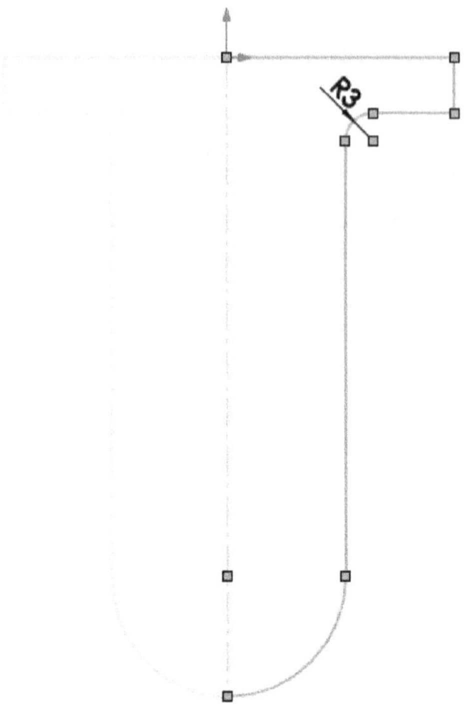

Figure-35. Mirror preview

- Click on the **OK** button from the **PropertyManager** to create the mirror.

Creating Circle and lines to complete

- Click on the **Circle** tool from the **Circle** drop-down in the **Ribbon**. The **Line PropertyManager** will be displayed; refer to Figure-36.

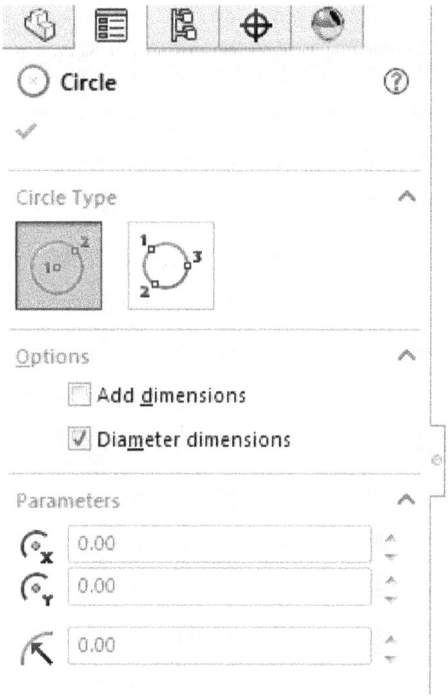

Figure-36. Circle PropertyManager

- Click on the **Diameter dimensions** check box from the **PropertyManager**.
- Click at the center point of the bottom arc and drag the cursor; refer to Figure-37.

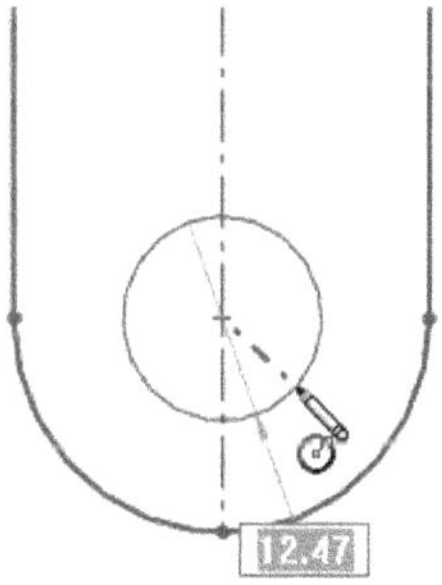

Figure-37. Circle creation

- Enter the dimension as **13** in the input box.
- Click on the Tick mark from the **PropertyManager** to close it.
- Click on the **Line** tool from the **Ribbon** and click at any point on the center line to specify start point of the line; refer to Figure-38.

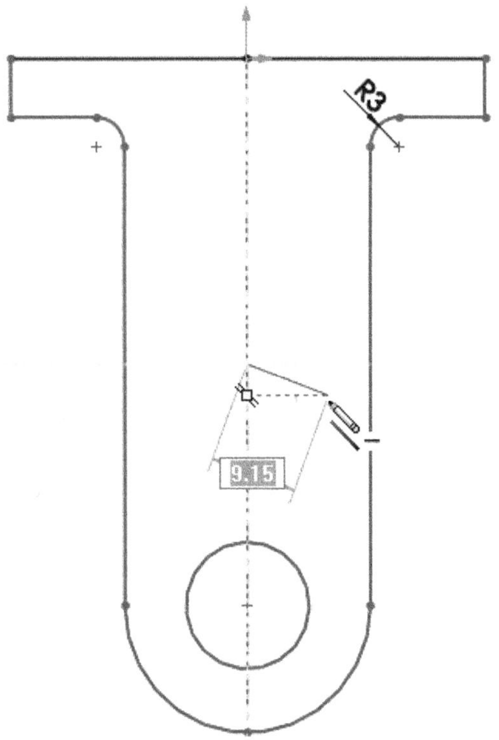

Figure-38. Point selected on centerline

- Move the cursor horizontally towards right and specify the value as 9.
- Move the cursor vertically downwards and click when the cursor is on arc.
- Press ESC to exit the tool.
- Mirror both the lines as we did earlier. The sketch should display as shown in Figure-39.

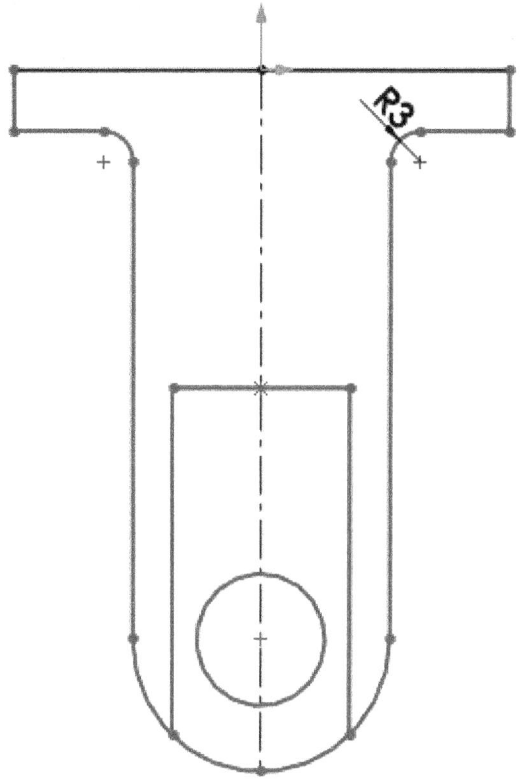

Figure-39. Sketch after all sketching operations

Dimensioning Sketch

- Click on the **Smart Dimension** tool from the Ribbon and select the arc. Dimension will get attached to cursor.
- Place the dimension at proper spacing. Press **ENTER** at the **Modify** input box.
- Click on the circle and place the dimension at proper place. Press **ENTER** at the **Modify** input box.
- Click on the two lines as shown in Figure-40.

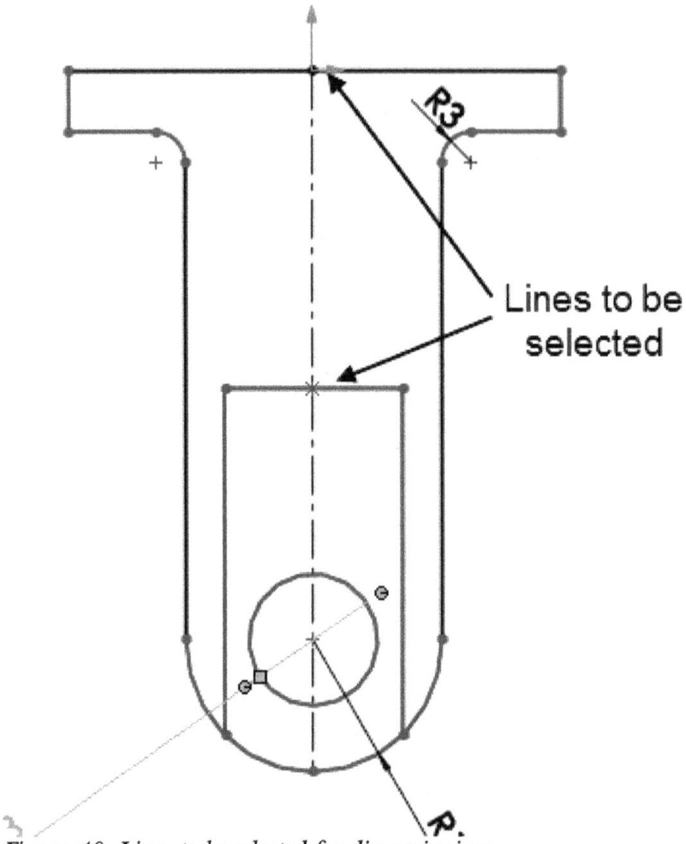

Figure-40. Lines to be selected for dimensioning

- Click to place the dimension at its proper place. In the **Modify** input box, enter the value as **40.**

In the same way, dimension all the entities in the sketch until it is fully defined. The final sketch after dimensioning will be displayed as shown in Figure-41.

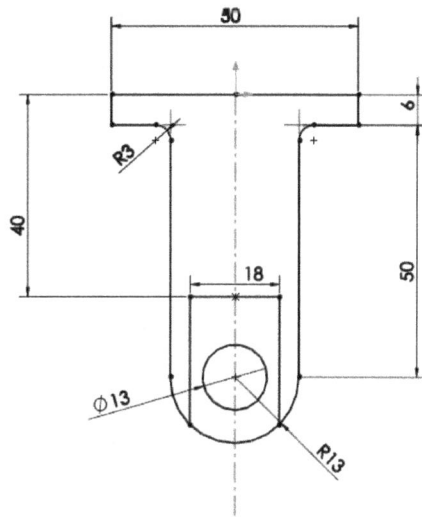

Figure-41. Final sketch

Practical 3

Create the sketch as shown in Figure-42. Also dimension the sketch as per the figure.

Figure-42. Practical 3

Steps to be performed:

Below is the step by step procedure of creating the sketch shown in the Figure-28.

Starting Sketching Environment

- Start SolidWorks if not started already.
- Click on the **New** button from the Menu Bar. The **New SolidWorks Document** dialog box will display.

- Double-click on the **Part** button. The **Part** environment of SolidWorks will display.

- Click on the **Sketch** tab of the **Ribbon**. The tools related to sketch will display in the **Ribbon**.

Starting Sketching

- Click on the **Sketch** button from the **Ribbon**. Three default planes will be displayed.
- Select the **Top** plane from the viewport. The viewport will become parallel to the view screen.

Creating Circles

- Click on the **Circle** button from the **Circle** drop-down in the **Ribbon**. You are asked to specify the center point of the circle.
- Select the **Diameter dimensions** check box and **Add dimensions** from the **PropertyManager**.
- Click at the center of coordinate system to specify center of the circle.
- Enter the diameter as **1.125** in the edit box displayed.
- Again, click on the center of the coordinate system and specify the diameter value as **1.75** in the edit box.
- Click at the top left of the circles created to specify the center point of other circle; refer to Figure-43.
- Specify the diameter value as **0.75**.

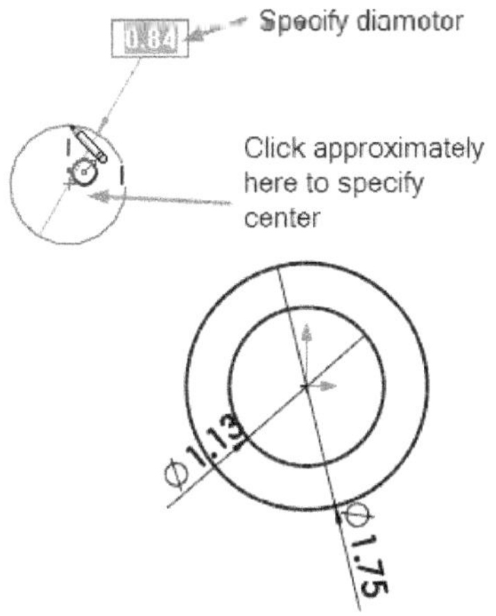

Figure-43. Center position for circle

- Click at the center of newly created circle and specify the diameter value as **1.625**.
- Clear the **Diameter dimensions** check box from the PropertyManager and click on the center of coordinate system to specify center of a circle.
- Specify the radius of circle as **1.375** in the edit box.

Creating Slots

- Click on the **Centerpoint Arc Slot** button from the **Slot** drop-down. You are asked to specify the center point for construction circle of arc slot.
- Select the **Add dimensions** check box from the **Slot PropertyManager**.
- Click at the center of the coordinate system. Specify the radius of construction circle as **2.312**. You are asked to specify the starting point of the slot arc.
- One by one click at the two positions displayed in Figure-44. Move the cursor and you are asked to specify the width of the slot.

- Click in the screen when the reading of width is approximately **0.9**; refer to Figure-45.

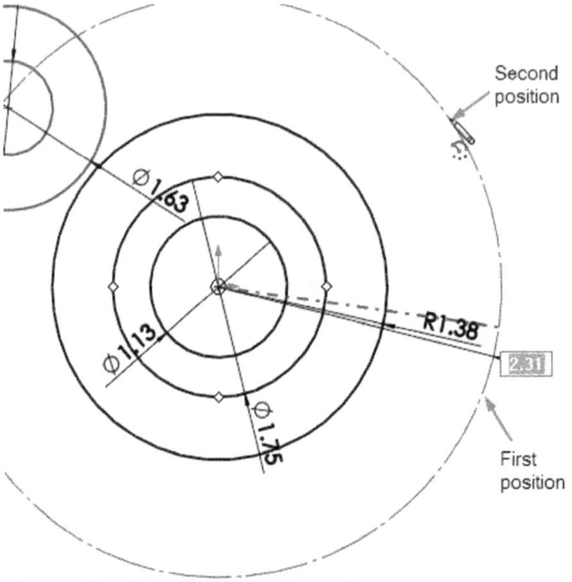

Figure-44. Positions selected sor slot arc

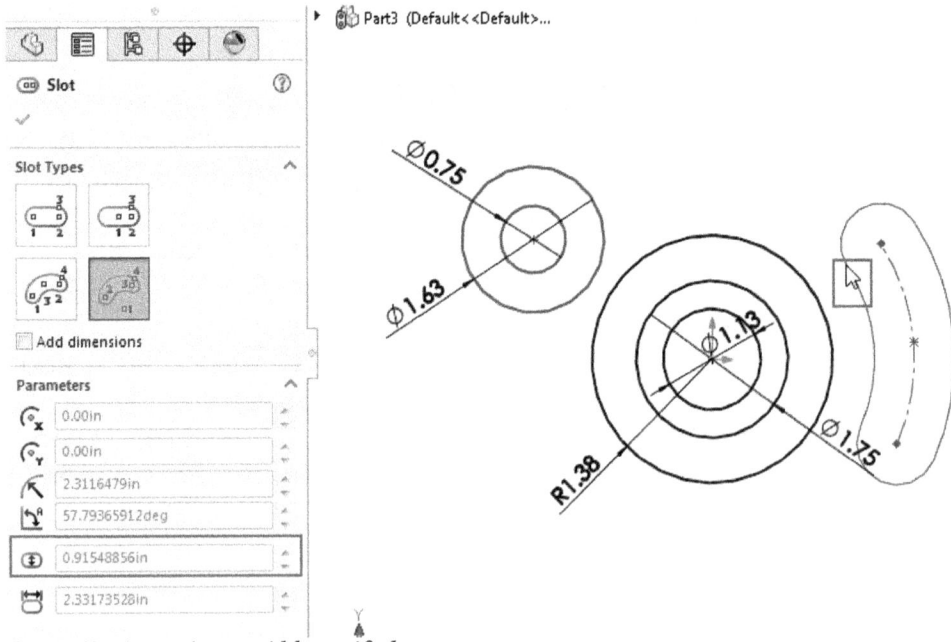

Figure-45. Approximate width specified

- Press the **Esc** key from the Keyboard and change the dimensions as shown in Figure-46.

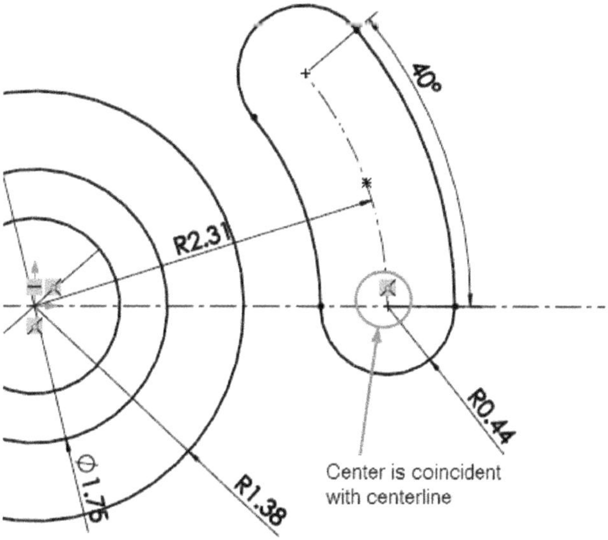

Figure-46. Dimensioned arc slot

- Similarly, create other arc slot at the same location but with the corner radius of **0.875**; refer to Figure-47.

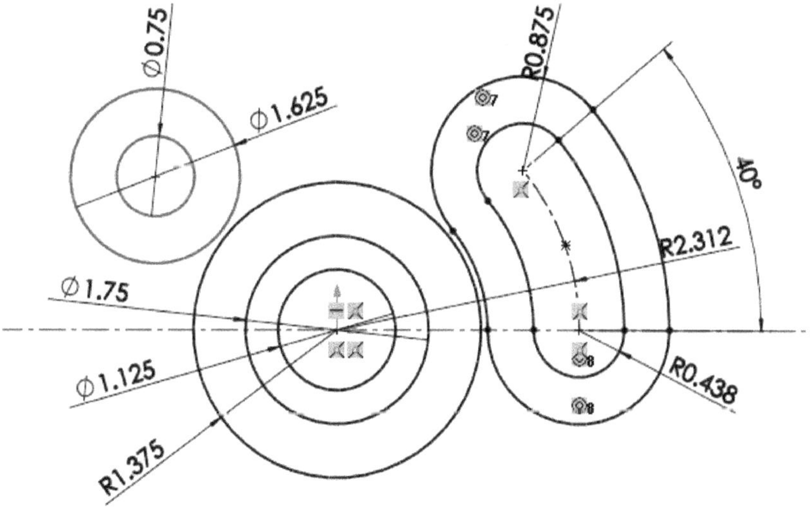

Figure-47. Sketch after creating second slot

- Click on the **Straight Slot** button from the **Slot** drop-down and create the slots as shown in Figure-48.

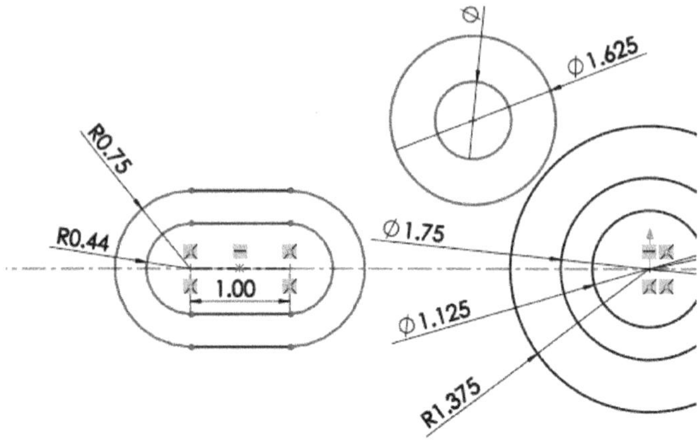

Figure-48. Straight slots created

Dimensioning not-dimensioned entities

- Click on the **Smart Dimension** tool and click on the center of circle with diameter **1.625.**
- Next, click on the center of coordinate system. Move the cursor vertically upward and place the dimension; refer to Figure-49.

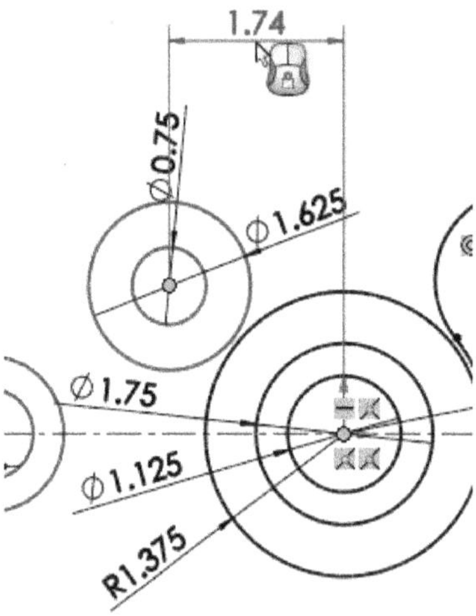

Figure-49. Placing dimension

- Enter the distance value as **0.75.**

- Similarly, apply other dimensions; refer to Figure-50.

Figure-50. Sketch after applying dimensions

Creating Arcs

- Click on the **3 Point Arc** tool from the **Arc** drop-down in the **Ribbon**; refer to Figure-51. The **Arc PropertyManager** will be displayed; refer to Figure-52.

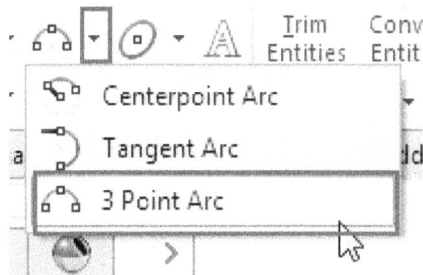

Figure-51. Arc drop-down

Figure-52. Arc PropertyManager

- Click one by one at the locations displayed in the Figure-53. Click on the **OK** button from the **PropertyManager** and change the value of radius to **1.75**.

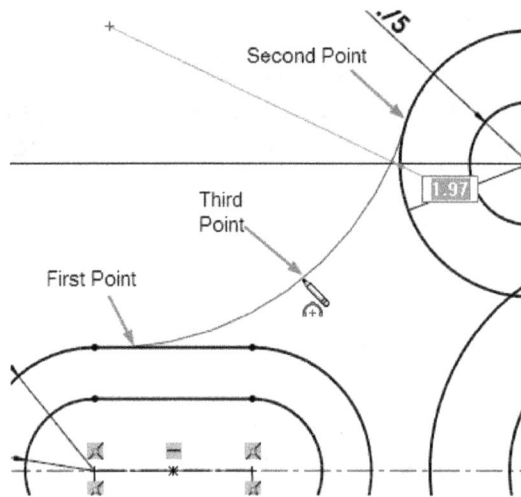

Figure-53. Points to click for creating arc

- Select the arc and connected circle by holding the CTRL key, and select the **Tangent** button from the **Properties PropertyManager**; refer to Figure-54.

Figure-54. Tangent button to be selected

- Similarly, make the arc and slot tangent at the connecting point; refer to Figure-55.

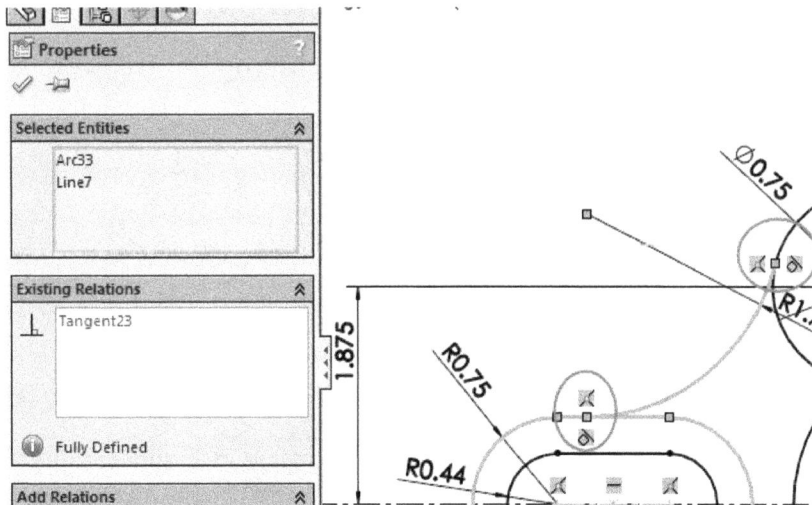

Figure-55. Tangent arc created

- Click on the **Three Point Arc** tool again and similarly create the other arcs; refer to Figure-56.

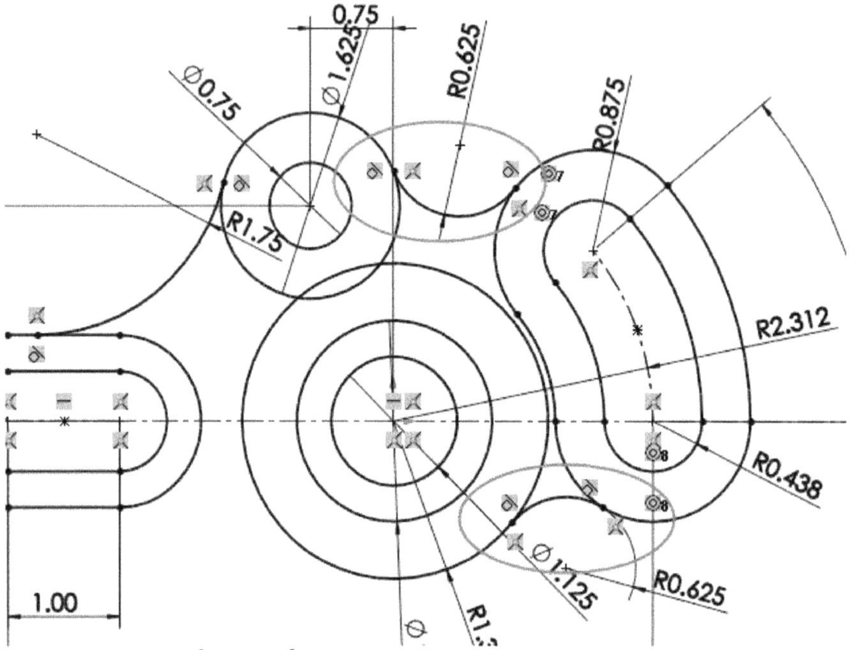

Figure-56. Arcs to be created

Creating Line and Trimming

- Click on the **Line** tool from the **Line** drop-down. The **Insert Line PropertyManager** will be displayed.
- One by one click on the points shown in Figure-57.

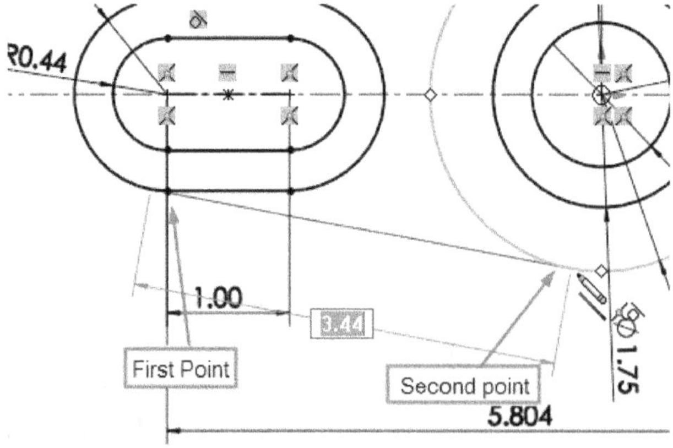

Figure-57. Points selected for line

- Make the end points of the line tangent to slot and circle.
- Click on the **Trim Entities** tool from the **Trim Entities** drop-down and remove the extra sketched entities; refer to Figure-58.

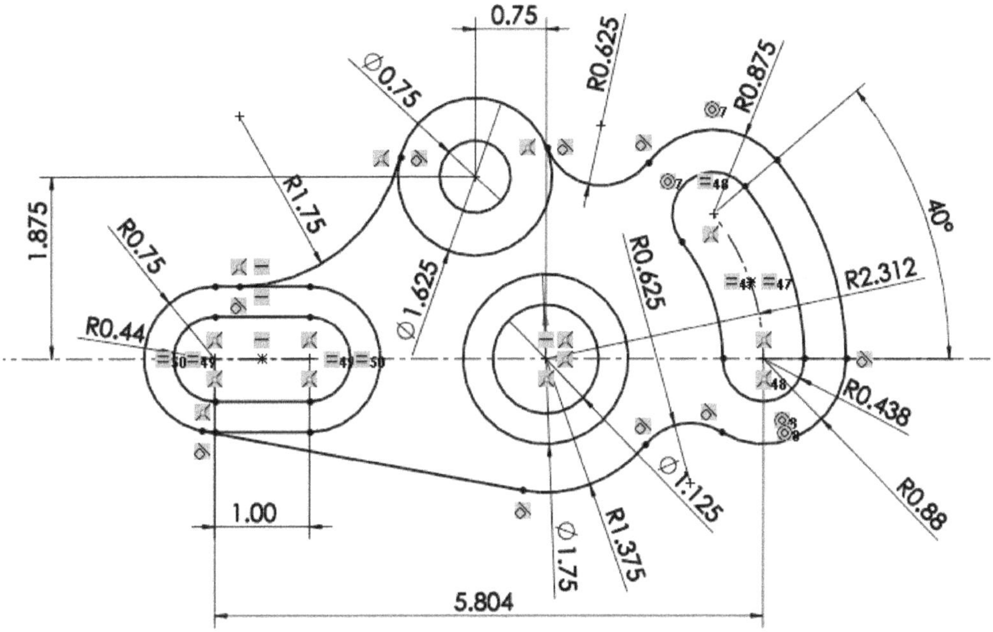

Figure-58. Sketch after trimming extra entities

Following are some sketches for practicing.

Practice 1

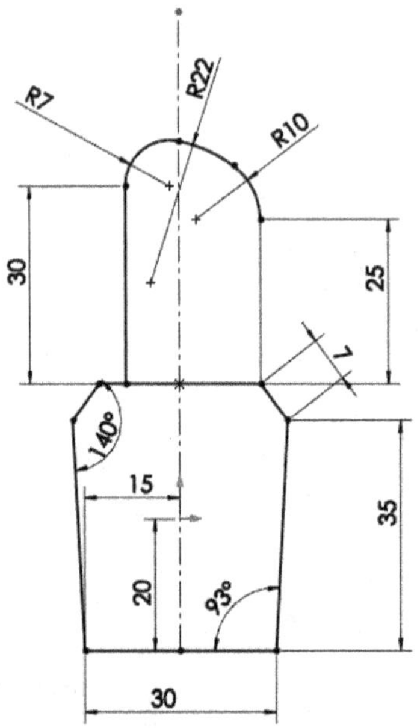

Figure-59. Practice1

Practice 2

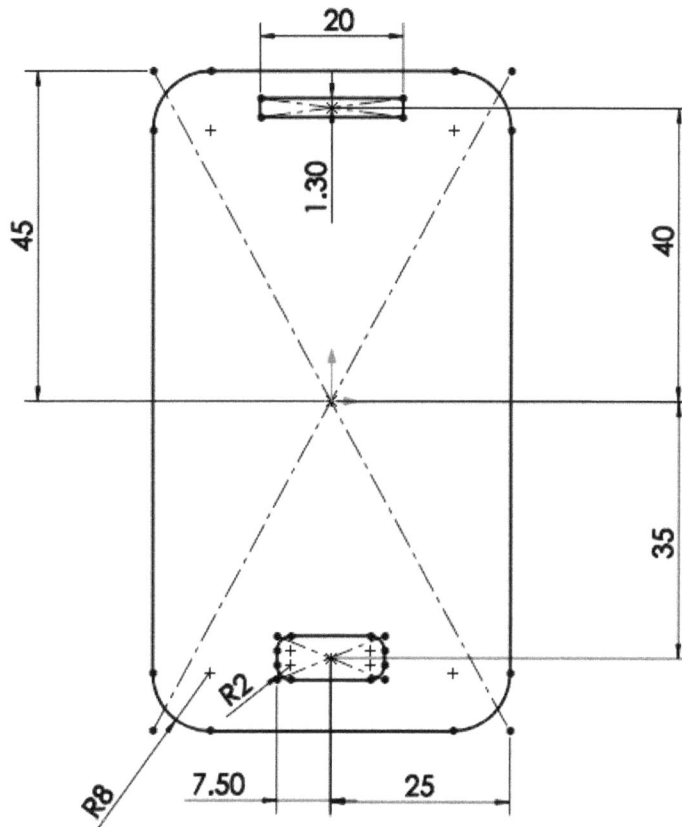

Figure-60. Practice2

In Next problems, we will create sketches from engineering drawings. Note that now onwards we will use engineering drawings in place of SolidWorks sketches.

Practice 3

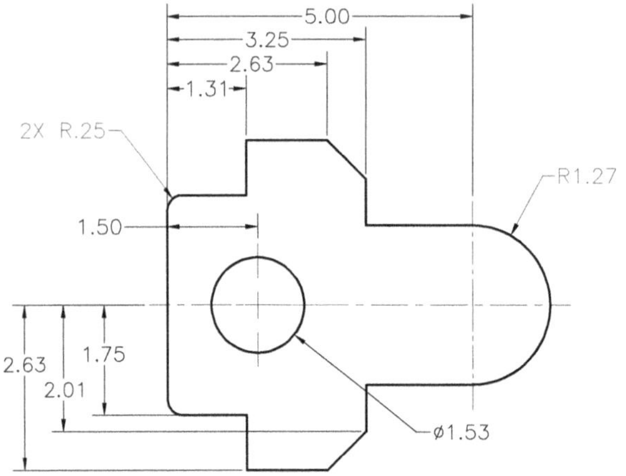

Figure-61. Practice 3

Practice 4

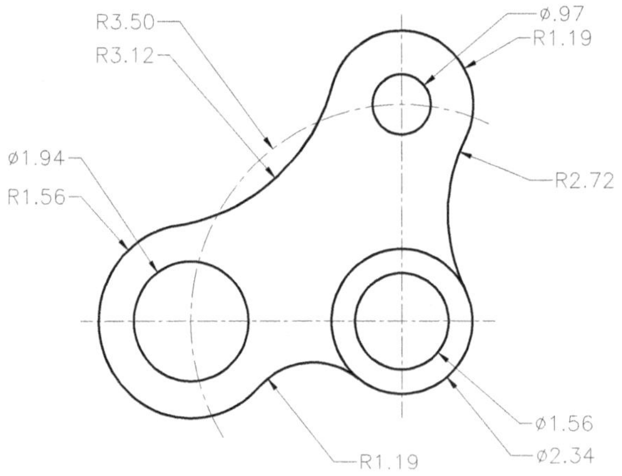

Figure-62. Practice 4

Practice 5

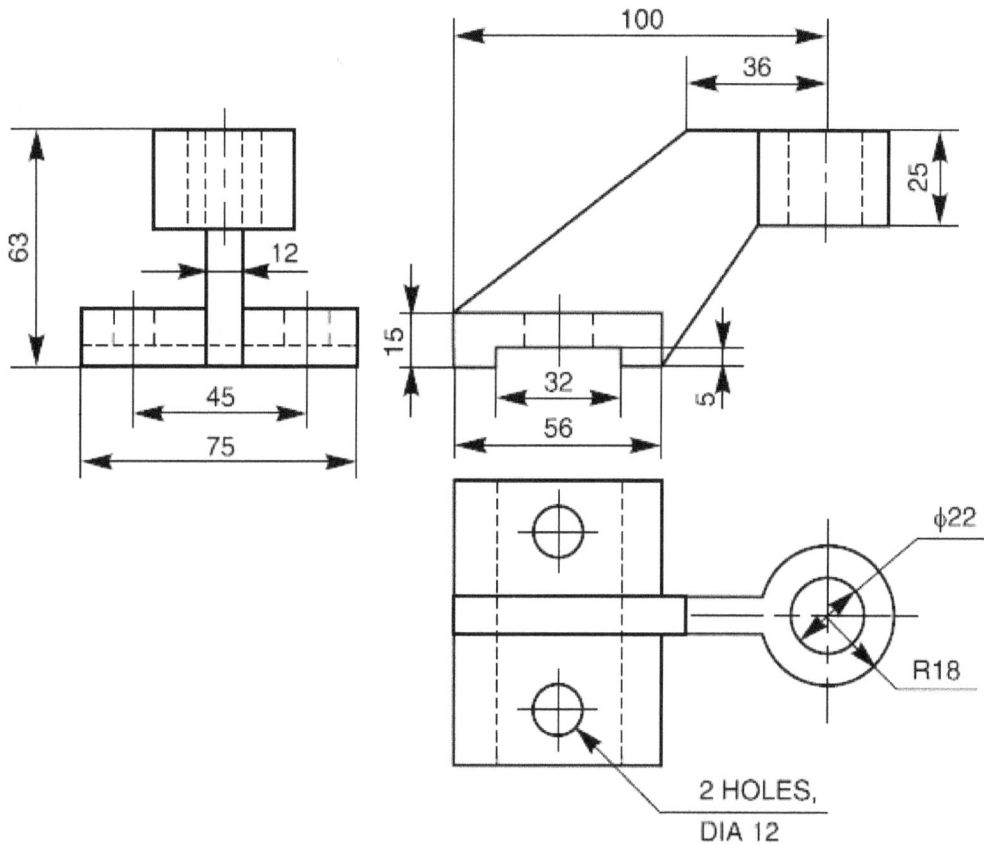

Figure-63. Practice 5

To get more exercises, mail us at cadcamcaeworks@gmail.com

SELF ASSESSMENT

Q1. The _____ **PropertyManager** is used to create, save, delete, and retrieve the dimension style in the Part environment.

Q2. On selecting a dimension, the _____ **PropertyManager** is displayed.

Q3. The _____ button is used to delete a dimension style.

Q4. The style file will be saved with the extension _____ .

a. .sldstl
b. .sldprt
c. .slddrw
d. .prt

Q5. Which of the following is a type of tolerance?

a. Limit
b. MIN
c. None
d. Both a and b

Q6. The inspection dimensions are enclosed in _____ shaped box.

Q7. You can display dual dimensions in sketches of SolidWorks. (T/F)

Q8. The arrows are displayed inside or outside the extension line, depending on the space available between the extension lines by using the **Smart** button in **Leader Display** rollout. (T/F)

FOR STUDENT NOTES

FOR STUDENT NOTES

3D Sketch and SolidModeling

Chapter 4

Topics Covered

The major topics covered in this chapter are:

- *3D Sketching and Plane Selection.*
- *Extruded Boss/Base tool.*
- *Revolved Boss/Base tool.*
- *Swept Boss/Base tool.*
- *Creating Extra references for modeling.*
- *Lofted Boss/Base tool.*
- *Boundary Boss/Base tool.*
- *Hole Wizard*
- *Thread Tool*
- *Removing material using the above tools*

3D SKETCHING

Till this point, we have worked on 2D drawings and have created them on Front plane. Now, we will come out of the 2D window and will explore the 3D world.

A 3D sketch is the sketch which is not confined to one plane only. The 3D sketch can be in all the planes available in the viewport. To start with the 3D sketching, we are required to again open the **Sketch** tab in **Ribbon**. The steps to create a 3D sketch are given next.

- Click on the down arrow below the **Sketch** button. A list of tools will display.
- Click on the **3D Sketch** button from the list; refer to Figure-1. The coordinate system will display inclined; refer to Figure-2.

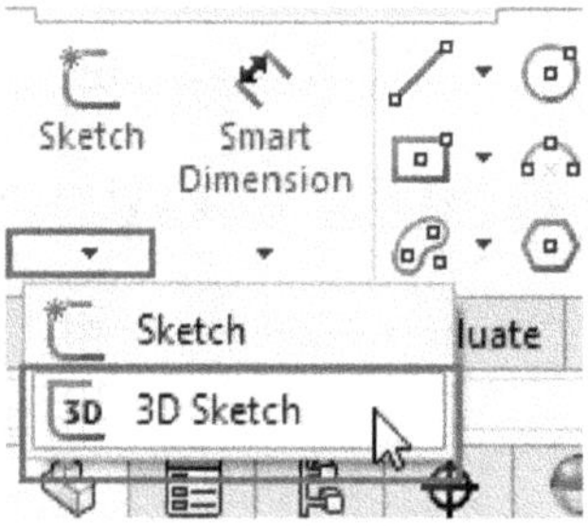

Figure-1. 3D Sketch button

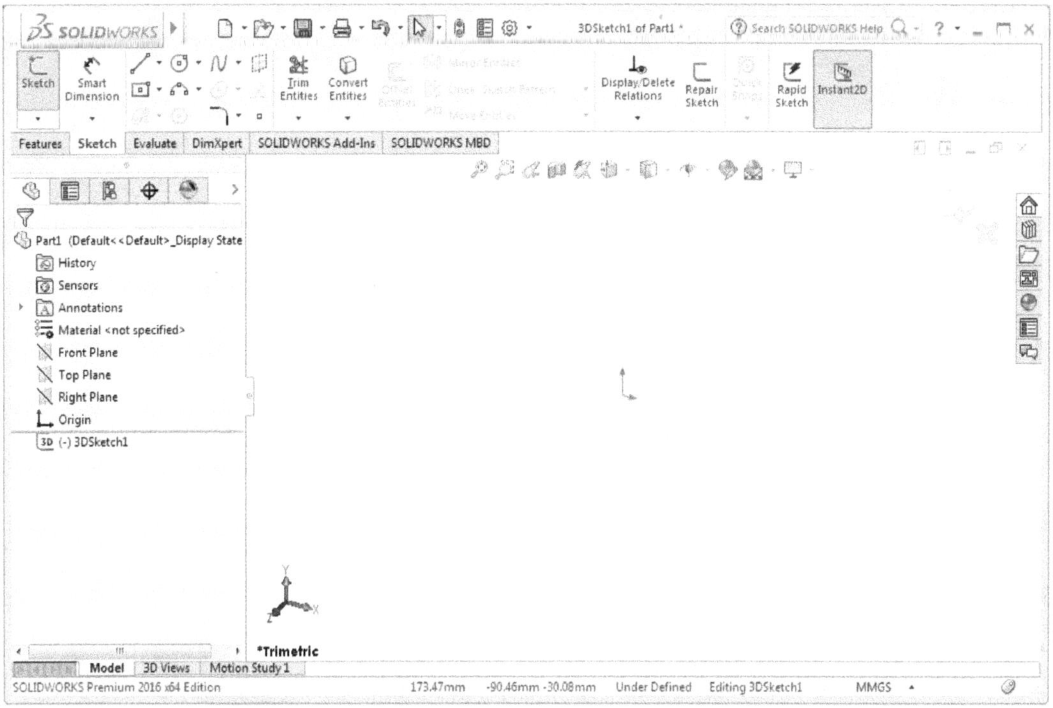

Figure-2. 3D Sketching environment

- Click on any of the sketching tool to create entities like, line, circle, rectangle and so on. (In our case, the **Line** tool is selected.)
- The cursor will start annotating the current sketching plane. Figure-3 shows the cursor which denotes that the current sketching plane is XY.

Figure-3. Current Sketching
plane

- Click to specify the start point of the line in the XY plane.
- The point in 3D space where you click will become the current sketching plane.

- Enter the parameters for the tool; refer to Figure-4.

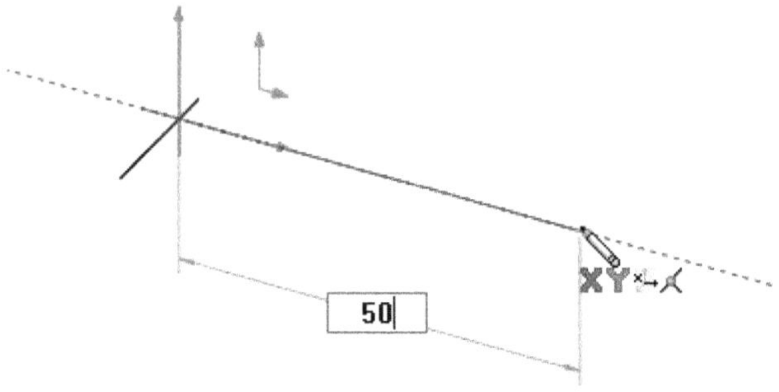

Figure-4. Entering parameters of line

- Press **TAB** from the **Keyboard** to toggle between three standard planes; refer to Figure-5.

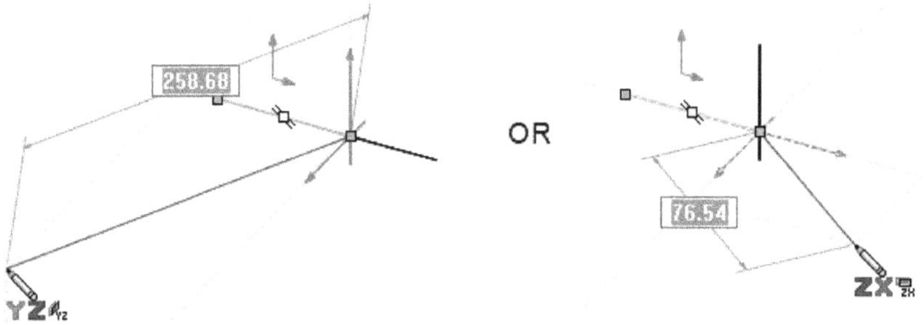

Figure-5. Toggling between planes

- Click or enter the parameters in the desired planes to create the sketch.
- You can also join two points in different planes by using the sketching tools; refer to Figure-6.

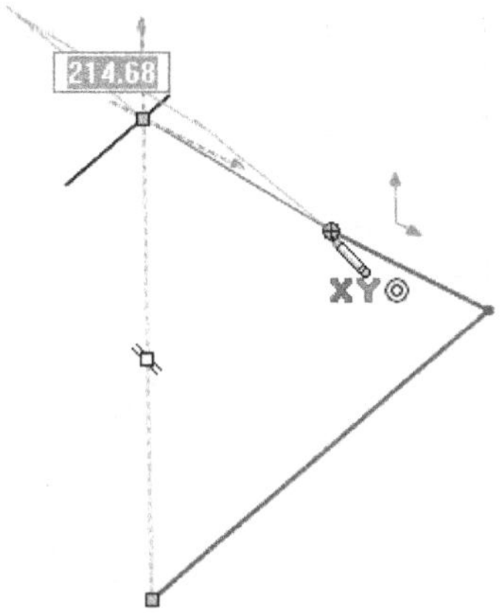

Figure-6. Connecting entities in 3D space

When we will be creating surfaces, we will use the 3D sketch again.

Convert Entities

The **Convert Entities** tool is used when you need the projection of any face, edge or sketch entity in another feature. In this way, you can create the sketch entities from the projection of other features. The tool is available in the **Convert Entities** drop-down in the **Sketch** tab of **Ribbon**. The procedure to use this tool is given next.

• Click on the **Convert Entities** tool from the **Convert Entities** drop-down in the **Ribbon**. The **Convert Entities PropertyManager** will be displayed; refer to Figure-7.

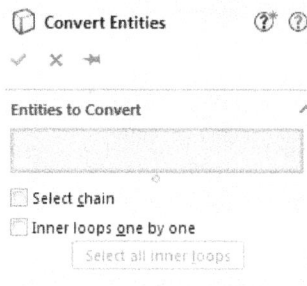

Figure-7. Convert Entities PropertyManager

- Select the face, edge or curve that you want to use in your current sketch; refer to Figure-8.

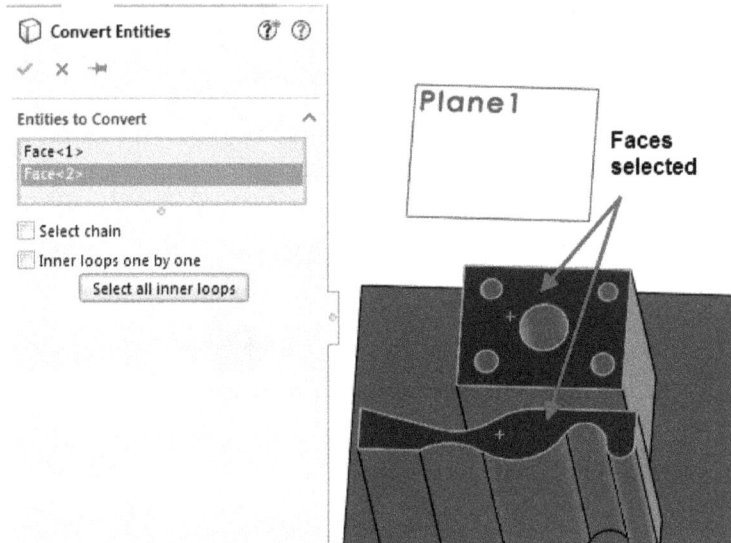

Figure-8. Selecting faces for converting entities

- Click on the **OK** button from the **PropertyManager**. The sketch entities will be generated; refer to .

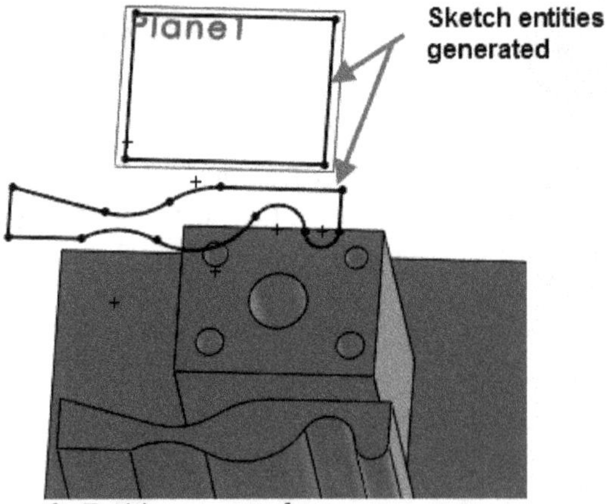

Figure-9. Entities converted

For any kind of analysis, simulation, assembly, or CAM; we need solid models. In SolidWorks, we are provided various tools to convert sketch into solid. Name of some of such tools are **Extruded Boss/Base** tool, **Revolved Boss/Base** tool, **Lofted Boss/Base** tool, and so on. These tools are explained one by one as follows.

EXTRUDED BOSS/BASE TOOL

Extruded Boss/Base tool is used to create a solid volume by adding height to the selected sketch. In other words, this tool adds material (by using the boundaries of sketch) in the direction perpendicular to the plane of sketch. In the term Boss/Base; the Base denotes the first feature and Boss denotes the feature created on any other feature. The steps to create extruded feature is given next.

- Click on the **Features** tab of the **Ribbon**. The tools related to solid modeling will display; refer to Figure-10.

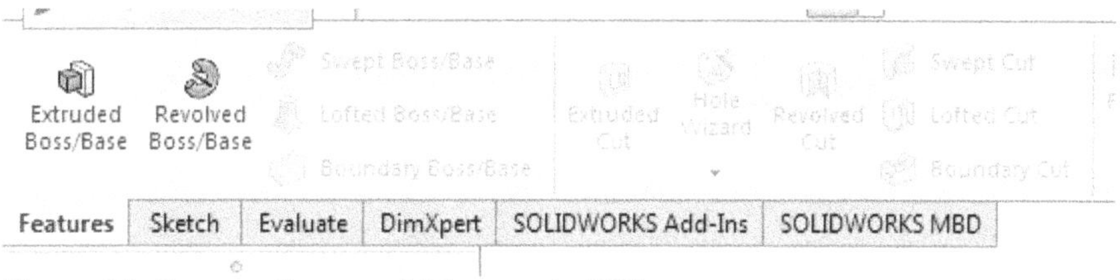

Figure-10. Features Command Manager in Ribbon

- Click on the **Extruded Boss/Base** tool from the **Ribbon**. The **Extrude PropertyManager** will display; refer to Figure-11.

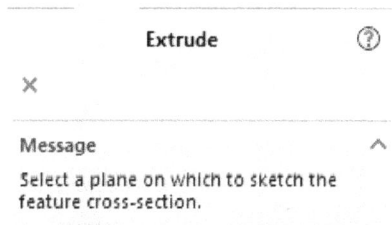

Figure-11. Extrude PropertyManager

- Select a plane from the planes displayed in the viewport. The sketching environment will display with the sketch tools activated.
- Create a closed sketch and then click on the **Exit Sketch** button from the viewport as shown in Figure-12. You can also select the **Exit Sketch** button from the **Ribbon**.

The **Boss-Extrude PropertyManager** will display as shown in Figure-13.

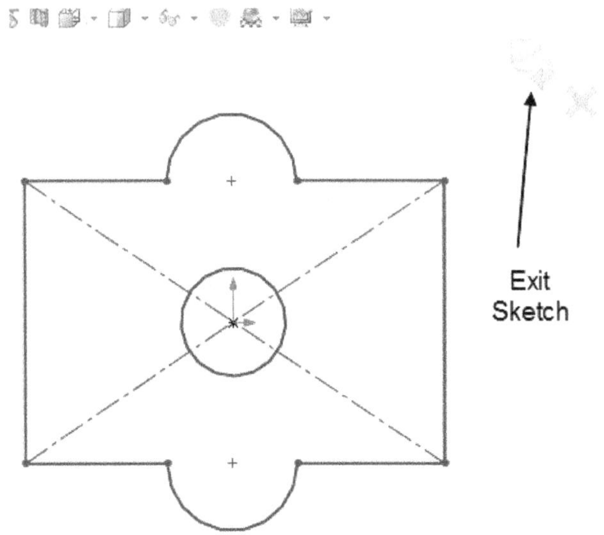

Figure-12. Sketch environment of extrude

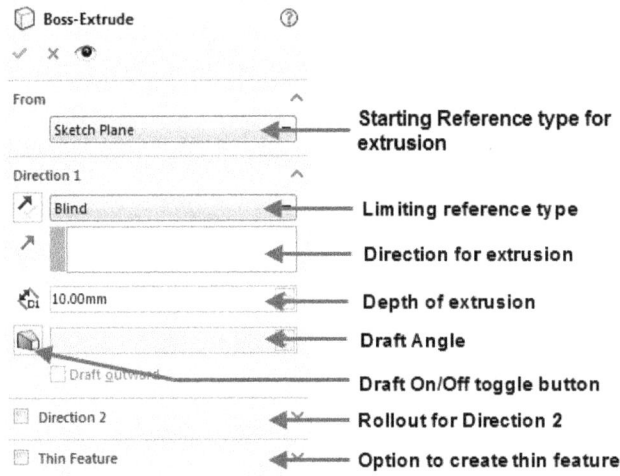

Figure-13. Boss-Extrude PropertyManager

• Click on the **Starting reference** drop-down and select the desired option.

There are four options in this drop-down; **Sketch Plane, Surface/Face/Plane, Vertex,** and **Offset**. The **Sketch Plane** is selected by default. Select this option if you want the extrusion to start from sketching plane. Select the **Surface/ Face/Plane** option to start the extrusion from the selected surface/face/plane. Select the **Vertex** option to start extrusion from selected vertex. Select the **Offset** option if you want to start at specified distance from the sketching plane; refer to Figure-14.

Figure-14. Preview of offset option

• Click in the **Limiting reference type** drop-down and select the reference for end of extrusion.

There are six options in the drop-down; **Blind, Up To Vertex, Up To Surface, Offset From Surface, Up To Body,** and **Mid Plane**. If you have selected **Blind** or **Mid Plane** option, then you need to specify the distance value in the **Height of extrusion** spinner. If you have selected any of the other option then select the respective reference from the viewport. Figure-15 shows preview of extrusion by using the **Mid Plane** option. Note that if **Mid Plane** option is selected then the Direction 2 rollout will not display.

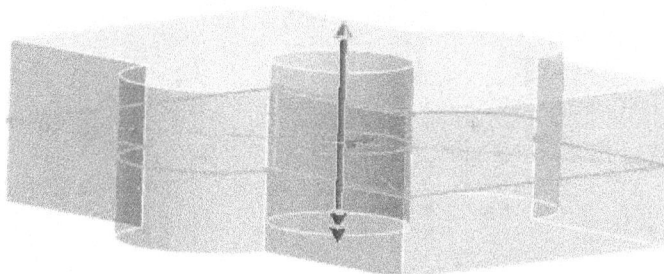

Figure-15. Mid-Plane extrusion

- Click in the **Direction of extrusion** selection box and select the reference if you do not want to extrude perpendicular to the sketching plane and want to extrude along selected axis/plane
- Click in the edit box for extrusion height and enter the desired extrusion height or you can set the value by using spinner.
- Click on the **Draft On/Off** button to apply draft angle on the vertical faces of the model. On selecting this button, **1°** draft will be applied by default taking the sketching plane as reference. Select the **Draft outward** check box to apply draft angle outwards on the vertical faces of extrusion. Specify the draft angle in the **Draft Angle** spinner.

The parameters you specified above can also be applied to the opposite direction. To apply these parameters, select the **Direction 2** check box. The parameters for the opposite direction will display.

- Select the **Thin Feature** check box to create the thin walled extrusion. Enter the thickness in the **Thickness** edit box of the **Thin Feature** rollout. Figure-16 shows a thin featured extrusion. **Note that if open sketch is selected for extrude then this option gets selected automatically.**

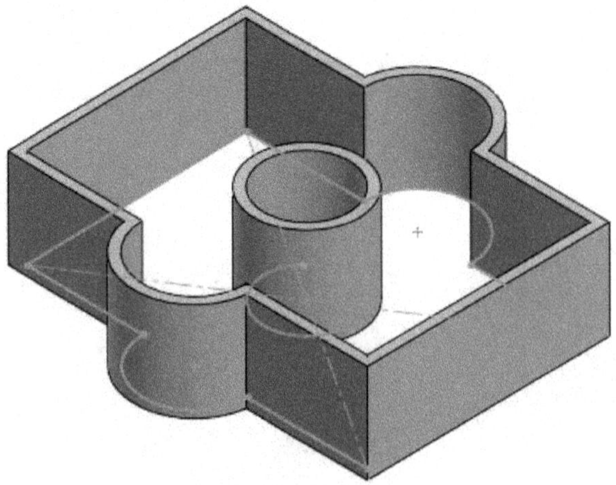

Figure-16. Thin Feature extrusion

- If you want to close the start and end face of the extrusion then select the **Cap Ends** check box; refer to Figure-17.

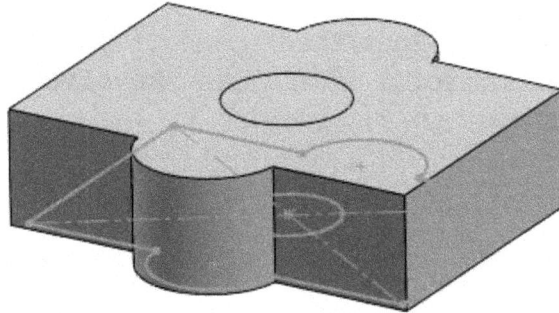

Figure-17. Extrusion with cap ends

- Once you have finished creating the feature, click on the **Detailed Preview** button from the **PropertyManager** to verify the feature; refer to Figure-18.

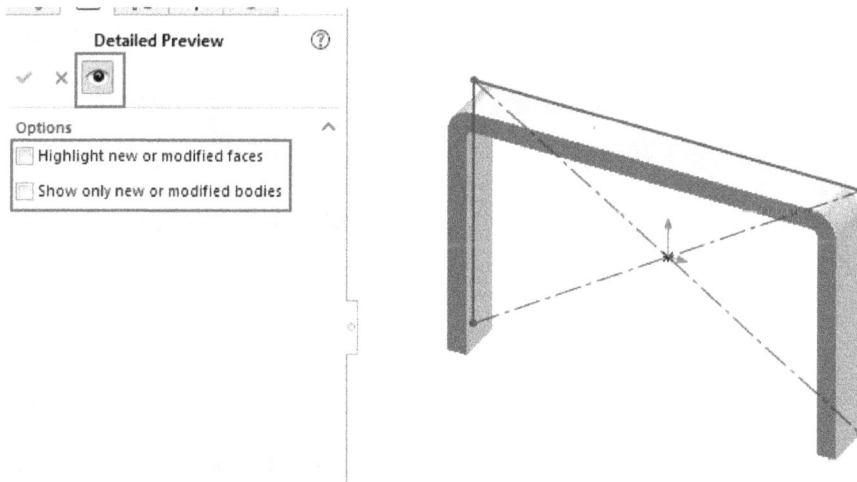

Figure-18. Detailed Preview button

- Select the **Highlight new or modified faces** check box if you want to highlight the new/modified faces only. Similarly, you can select the **Show only new or modified bodies** check box if you want to display only new or modified objects.

REVOLVED BOSS/BASE TOOL

Revolved Boss/Base tool is used to create a solid volume by revolving a sketch about selected axis. In other words, if you revolve a sketch about an axis then the volume that is covered by revolved sketch boundary is called revolved boss/base feature. The steps to create revolved boss/base feature are given next.

- Click on the **Revolved Boss/Base** tool. If you have not selected any existing sketch, then the **Revolve PropertyManager** displays as shown in Figure-19.

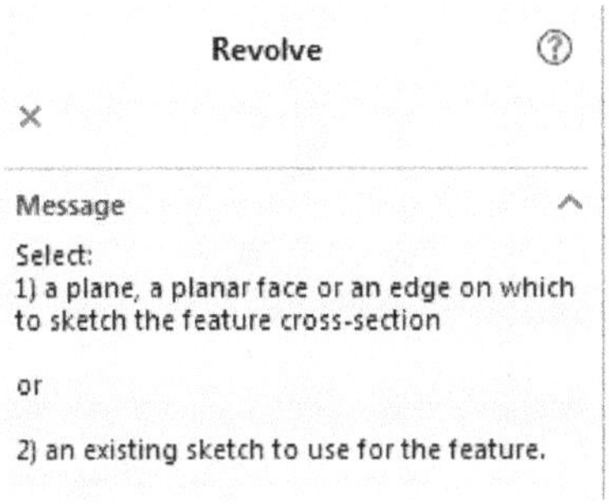

Figure-19. Revolve PropertyManager

- Select a plane if you want to create a new sketch or select an already created sketch. In our case, we are selecting an already created sketch.
- Select the inside region of the sketch that you want to revolve; refer to Figure-20. The updated **Revolve ProperyManager** will display as shown in Figure-21.

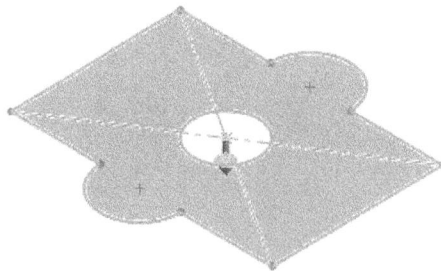

Figure-20. Region selected for revolve

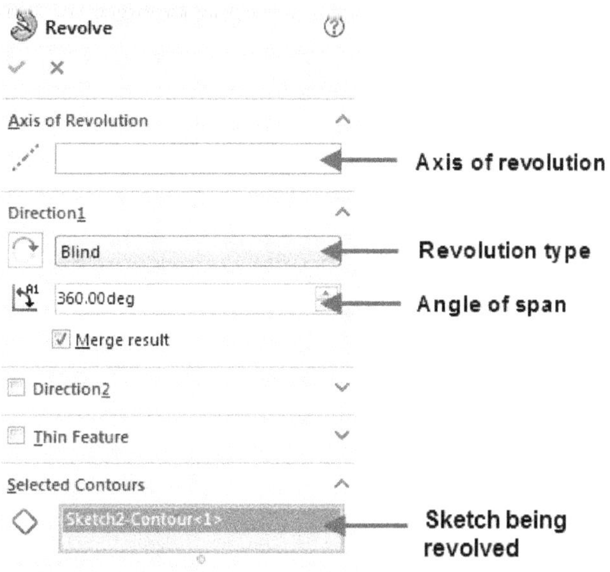

Figure-21. Updated Revolve Property Manager

- Click in the **Axis of Revolution** selection box to select the axis. Select the edge, line, or center line about which you want to revolve the sketch. Preview of the revolve feature will display; refer to Figure-22.

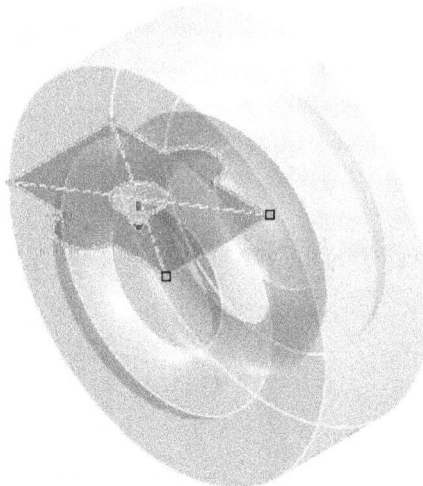

Figure-22. Preview of revolve feature

- Click on the **Revolve Type** drop-down and specify the revolution limiting reference. The options in this drop-down are same as discussed for **Extruded Boss/Base** tool.
- If you have selected Blind option in the Revolve Type drop-down then specify the degrees of revolution by using the Angle spinner.
- Click on the **Direction 2** check box to revolve in the direction opposite the earlier on. The options in the **Direction 2** rollout are same as discussed earlier.
- You can also create thin feature by selecting the **Thin Feature** check box. **Note that if you select an open sketch then this option is automatically selected**.
- Click in the **Selected Contours** box to add more sketches for revolution and select the sketches you want to revolve.

Figure-23 shows a sketch, axis of revolution, and resulting revolve feature preview.

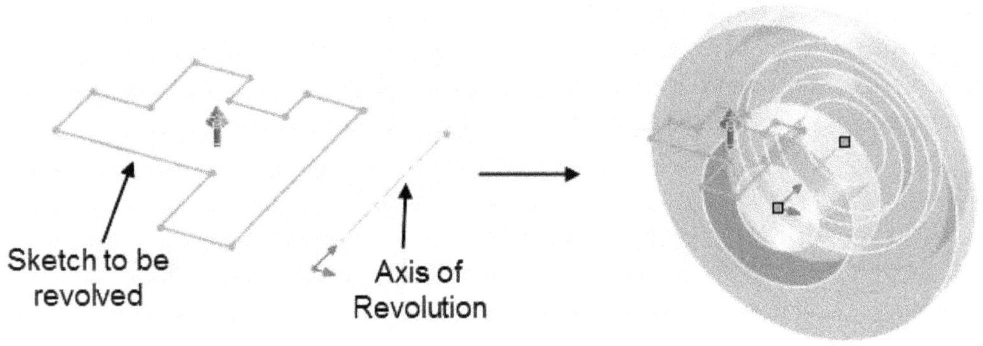

Figure-23. Revolved feature

SWEPT BOSS/BASE TOOL

The **Swept Boss/Base** tool is used to create a solid volume by moving a sketch along the selected path. In other words, if you move a sketch along a path then the volume that is covered by moving sketch boundary is called swept boss/base feature. Note that to use this tool, you must have a sketch section and a path, then only the tool will be active. The steps to create swept boss/base feature are given next.

- Click on the **Swept Boss/Base** tool. The **Sweep PropertyManager** will display as shown in Figure-24.

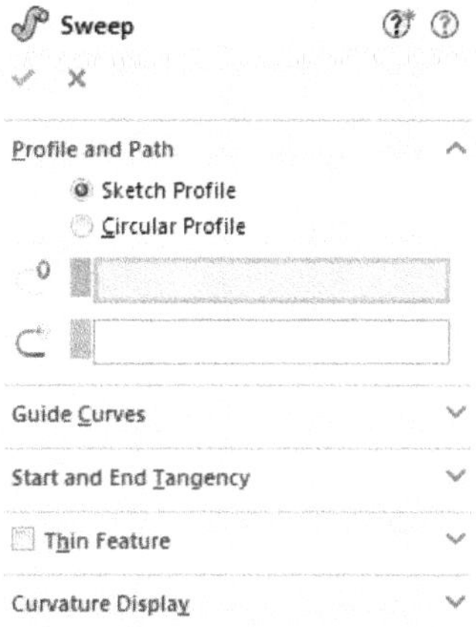

Figure-24. Sweep Property-Manager

- By default, the **Sketch Profile** radio button is selected and you are asked to select a sketch section (profile).
- Select the close section from viewport that you want to sweep.
- On selecting the section, the **Path** selection box becomes selected automatically and you are asked to select a path.
- Select the curve that you want to use as path. Preview of the sweep feature will be displayed.
- If your path is extended on both sides of section then three buttons will be displayed below the **Path** selection box; refer to Figure-25. Select the **Direction 1, Bidirectional**, and **Direction 2** button as per the requirement. Preview of sweep with these buttons is shown in Figure-26.

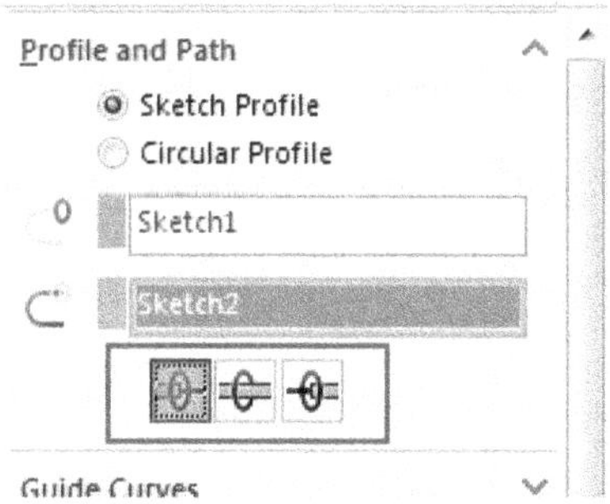

Figure-25. Buttons for bidirectional sweep

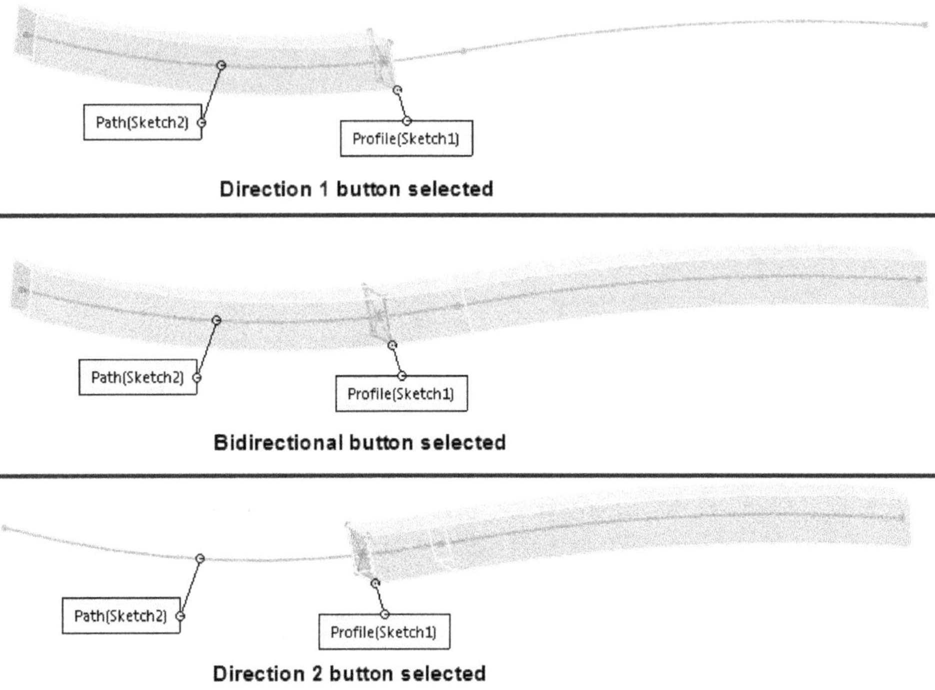

Direction 1 button selected

Bidirectional button selected

Direction 2 button selected

Figure-26. Sweep with directional options

- If your path is not perpendicular to the section plane then click on the **Options** rollout. The options in the rollout will display as shown in Figure-27.

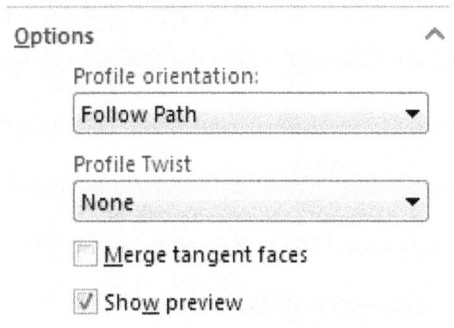

Figure-27. Options rollout

- Click on the **Orientation/twist type** drop-down and select **Keep normal constant** option to create the sweep feature. Figure-28 shows a sweep feature created by this method.

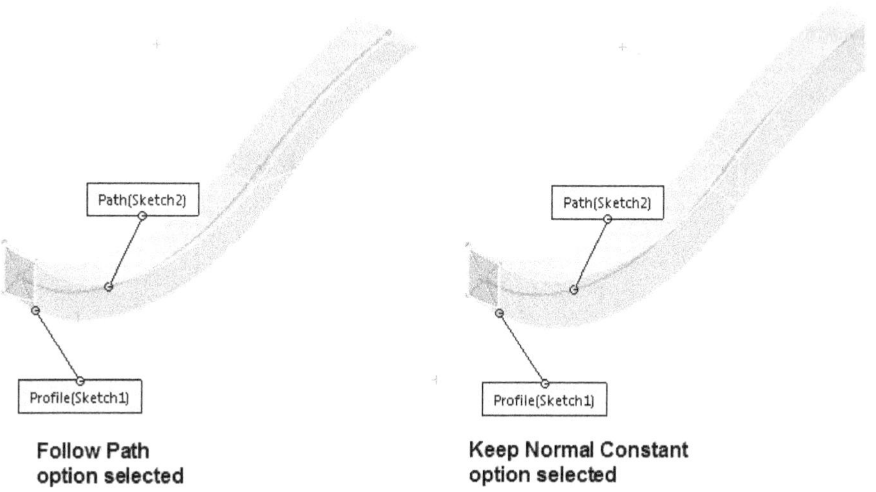

Figure-28. Sweep with Follow Path and Keep Normal Constant options

Using Guide Curves

The **Follow Path** option is selected by default in the **Profile Orientation** drop-down and as a result the swept feature is created by making the sketch section exactly follow the curve's curvature. If you want to influence the shape of swept feature with the help of a guide curve then expand the **Guide Curves** rollout and select the guide curve/curves; refer to Figure-29. Select the **Follow path and First guide curve** option from the **Profile Twist** drop-down if you want to sweep feature to follow the path and first guide curve for its shape; refer to Figure-30. Similarly, you can select the **Follow First and Second Guide Curves** option from the drop-down to make the sweep feature follow both the guide curves for its shape.

Note that you need to have individual sketches for section, path, and guide curves although they can be on same plane.

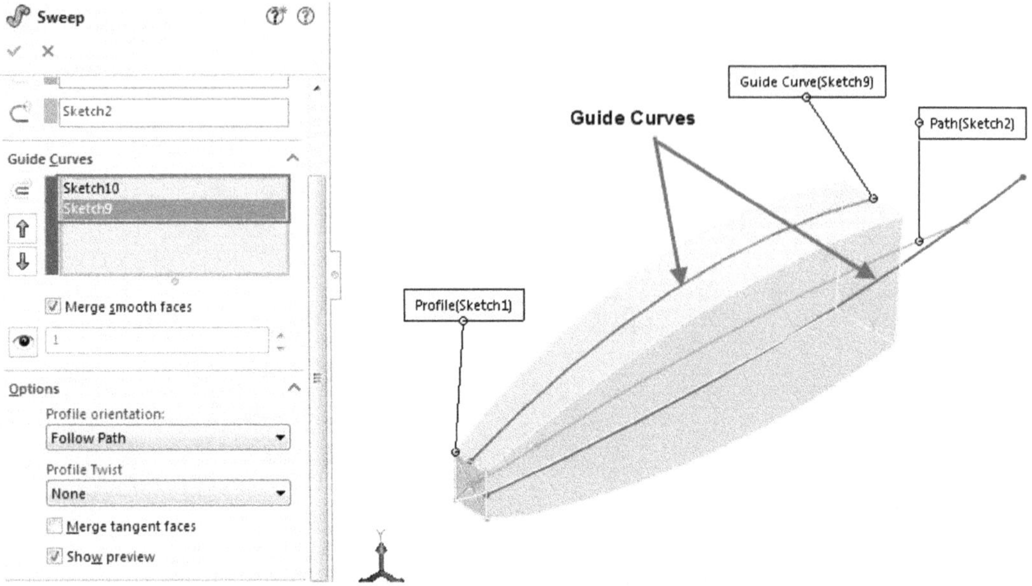

Figure-29. Sweep with two guide curves

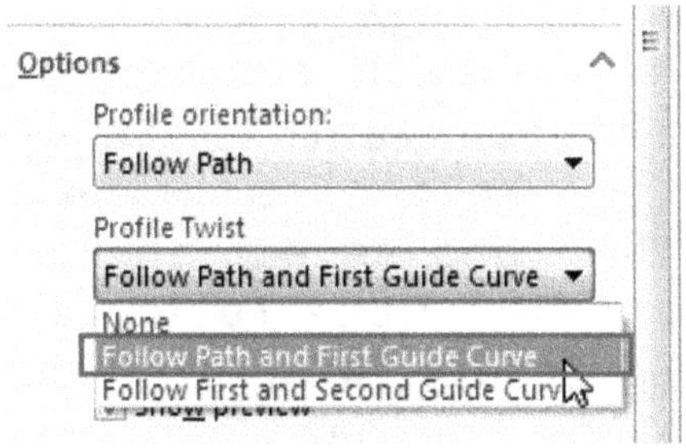

Figure-30. Profile Twist drop-down

Applying Twist in Sweep Feature

You can twist the section while sweeping along the path to create drill bit type of shape of conduits. To do so, select the **Specify Twist Value** option from the **Profile Twist** drop-down in the **Options** rollout after selecting the section and path. The **Options** rollout will display as shown in Figure-31.

Figure-31. Options rollout with twist along path

Click on the **Twist Control** drop-down and select the desired unit to twist the section. In this case, we have selected **Revolutions** option. Specify the desired number of revolutions in the **Direction 1** spinner. Click on the flip button to reverse the twisting. Figure-32 shows the preview of twisting along the path while sweeping.

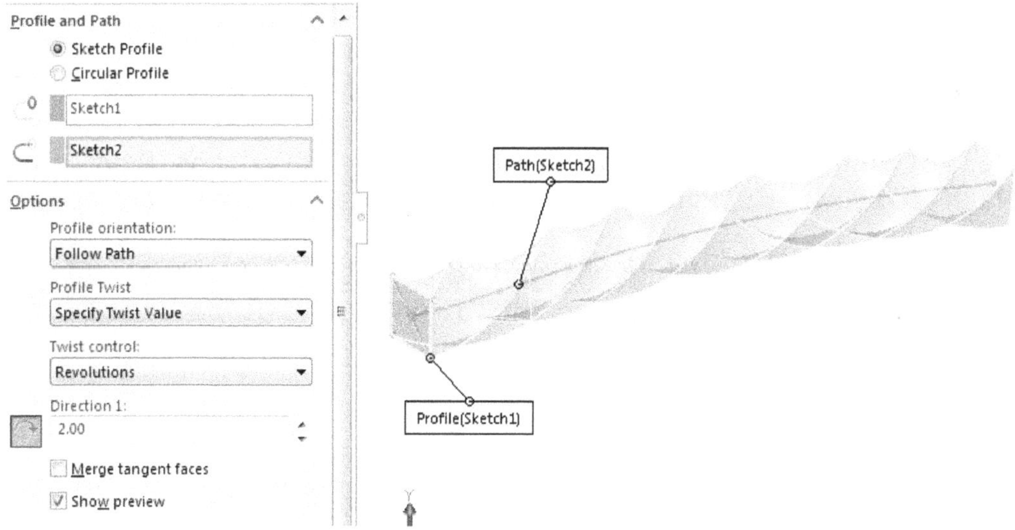

Figure-32. Preview of twist along path

You can also make a spring by using this option; refer to Figure-33.

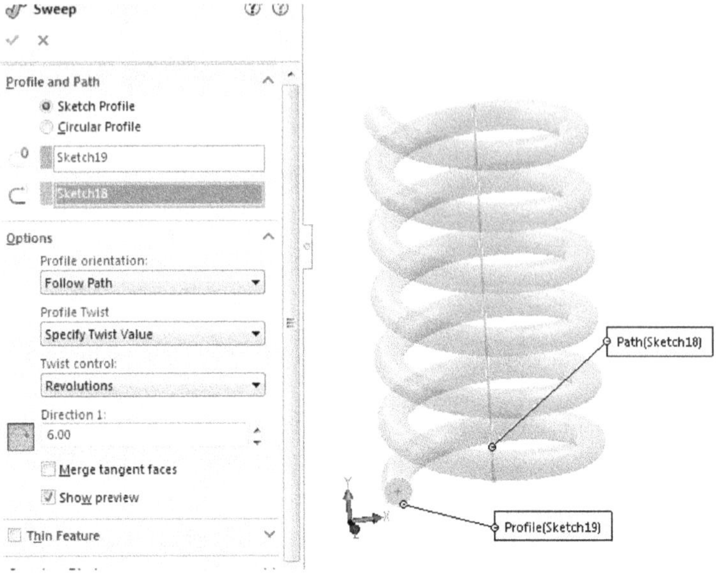

Figure-33. Spring created by twist slong path

Circular Profile Sweep

This is a new option in SolidWorks 2016. Using this option, you can create round bar/rod. The procedure to use this option is given next.

- Select the **Circular Profile** radio button from the **Profile and Path** rollout of **PropertyManager**. You are asked to select a path.
- Select the path along which the circular sweep should be created. Preview of the circular sweep will be displayed; refer to Figure-34.

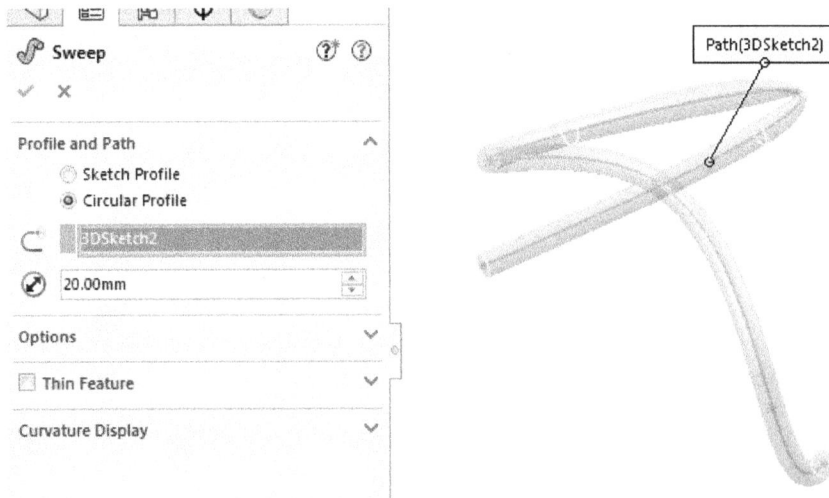

Figure-34. Preview of circular profile sweep

- Specify the desired diameter in the **Diameter** spinner and click on the **OK** button from the **PropertyManager**.

Before we move to **Lofted Boss/Base** tool. We need to understand the procedure to create reference planes, axis, coordinate system, and other references as they are important for advanced modeling.

REFERENCE GEOMETRY

There are various types of reference geometries that can be created in SolidWorks. All the tools to create these reference geometries are available in the **Reference Geometry** drop-down; refer to Figure-35. The tools available in the drop-down are:

- Plane
- Axis
- Coordinate System
- Point
- Center of Mass
- Mate Reference

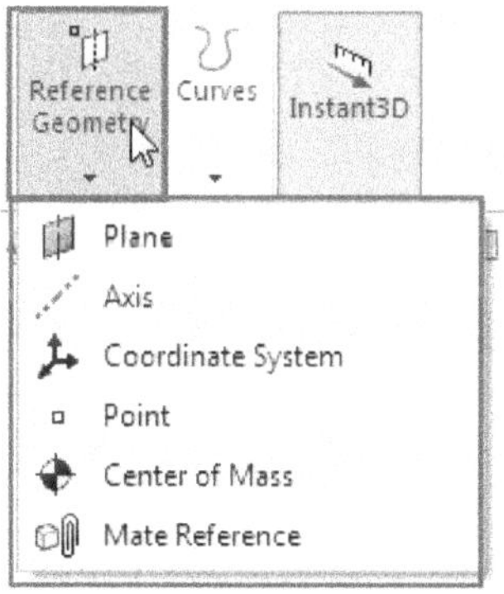

Figure-35. Reference drop-down

These tools are discussed next.

Plane

The **Plane** tool is used to create reference planes. By default there are three planes available in SolidWorks: **Front**, **Top**, and **Right**. To create more planes follow the steps given next.

- Click on the **Plane** tool from the **Reference Geometry** drop-down. The **Plane PropertyManager** will display as shown in Figure-36.

Figure-36. Plane PropertyManager

- You can select maximum three references to create a plane. You can select plane/face, edge/axis/curve, or vertex/point. The ways in which you can create planes by using these references are discussed next.

Creating plane at a distance from plane/face

- To create a plane at a distance from a plane/face, select the plane/face. The updated **Plane PropertyManager** will display as shown in Figure-37.
- Specify the desired distance in the spinner.
- Click on the **OK** button to create the plane.

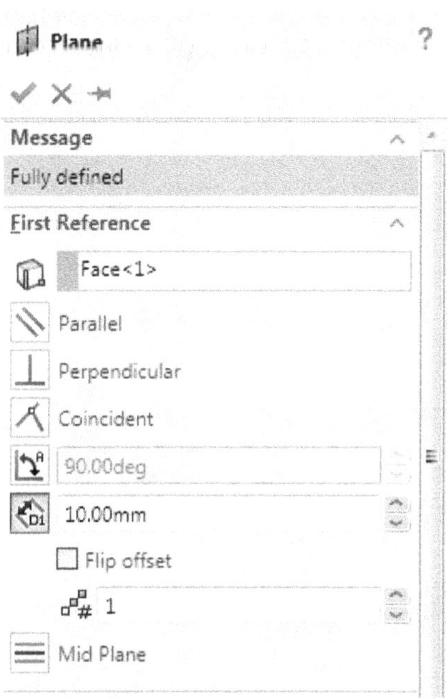

Figure-37. Updated Plane PropertyManager

Creating plane at an angle to plane/face

- Activate the **Plane PropertyManager** and select a plane/face to which you want to specify the angle.
- Click on the **At Angle** button to specify the angle.
- Click in the **Second Reference** box and select the edge or axis to which you want to make the plane coincident or select the two planar point through which you want the plane to pass. Figure-38 shows the plane create by both the methods discussed.

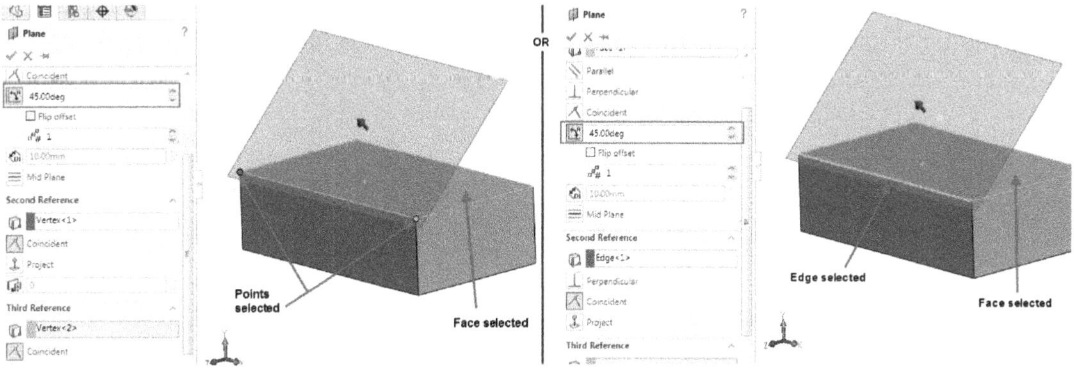

Figure-38. Plane creation at angle

Creating plane passing through points

- Activate the **Plane PropertyManager** and one by one click three points through which you want the plane to pass through. Figure-39 shows the plane passing through three points.

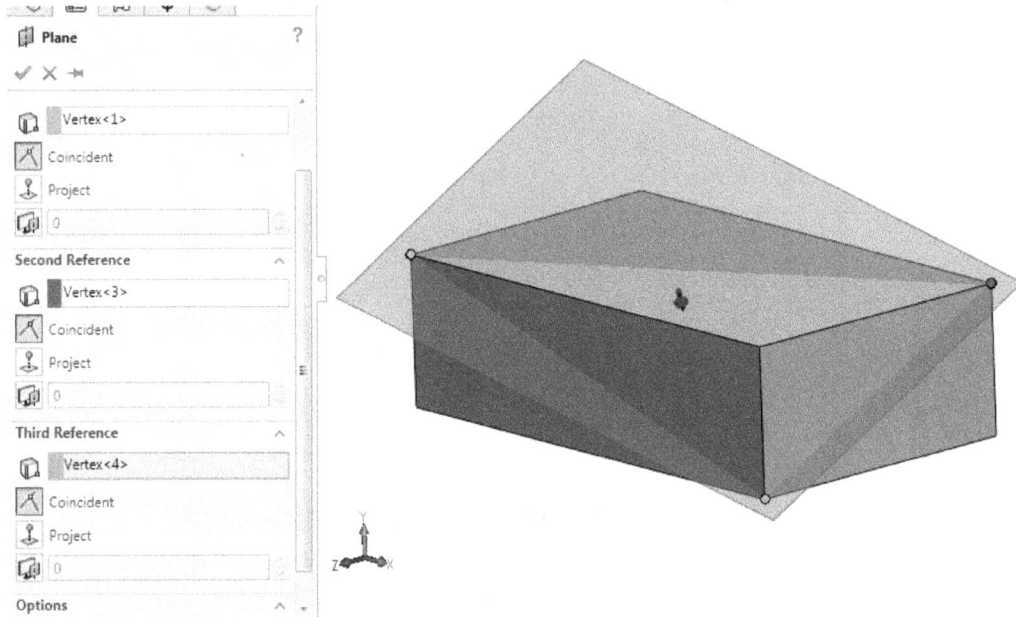

Figure-39. Plane passing through three points

Plane Parallel to Screen

This option is a new feature of SolidWorks 2016. The procedure to create plane parallel to screen is given next.

- Right-click on any face, edge, or vertex of the model. A shortcut menu will be displayed; refer to Figure-40.

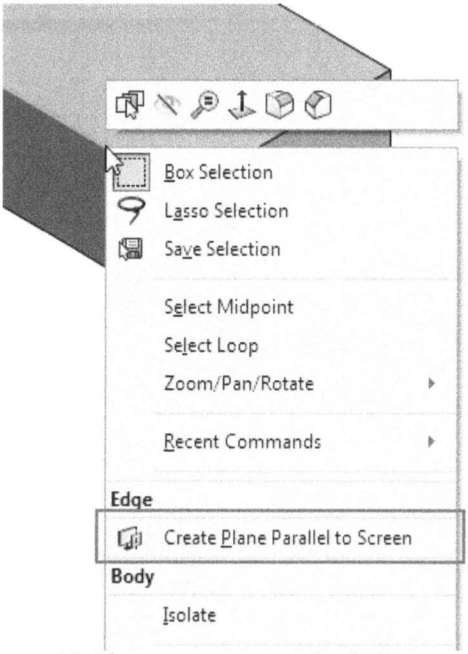

Figure-40. Shortcut menu on right clicking on a face

- Click on the **Create Plane Parallel to Screen** option from the shortcut menu. A plane parallel to screen will be created; refer to Figure-41.

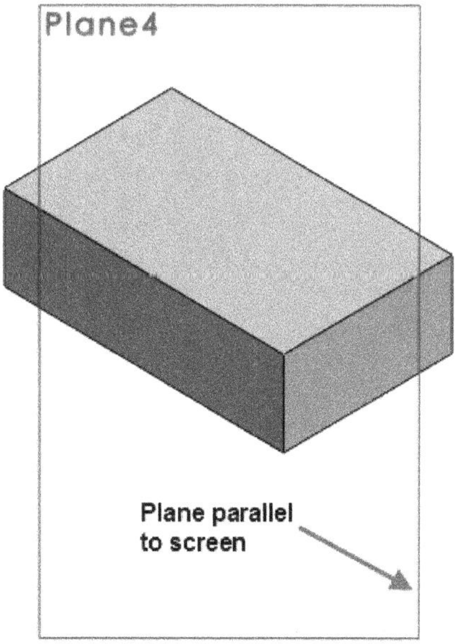

Figure-41. Plane parallel to screen

Axis

The **Axis** tool is used to create reference axes. An axis is useful in creating revolve features or to create planes at angle. To procedure to create axis by using the **Axis** tool is given next.

- Click on the **Axis** tool from the **Reference** drop-down. The **Axis PropertyManager** will display as shown in Figure-42.
- Select the desired button from the **PropertyManager**. The buttons in this **PropertyManager** are explained next.

Figure-42. Axis PropertyManager

One Line/Edge/Axis

Select this button if you want to create an axis coincident to the selected line/edge/axis. After selecting this button, click on the line/edge/axis. The axis will be created coincident to the selected line/edge/axis; refer to Figure-43.

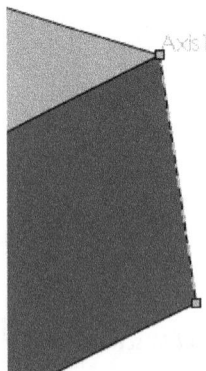

Figure-43. Axis created on edge

Two Planes

Select the **Two Planes** button if you want to create axis at the intersection of the two selected planes/faces. After selecting this button, click on the two intersecting. The axis will be created at the intersection; refer to Figure-44.

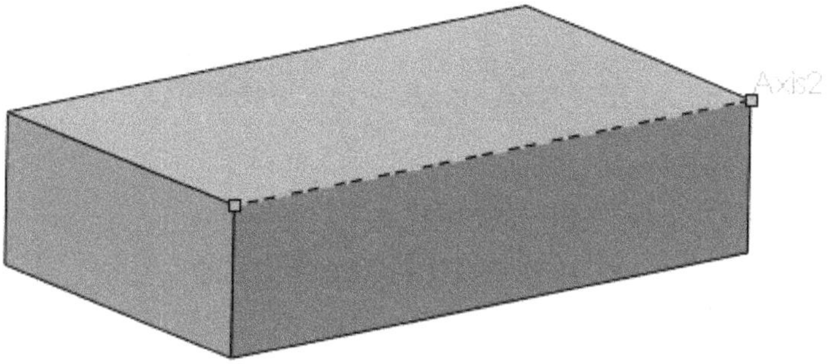

Figure-44. Axis at intersection of planes

Two Points/Vertices

Select the **Two Points/Vertices** button if you want to create axis passing through the selected two points/vertices; refer to Figure-45.

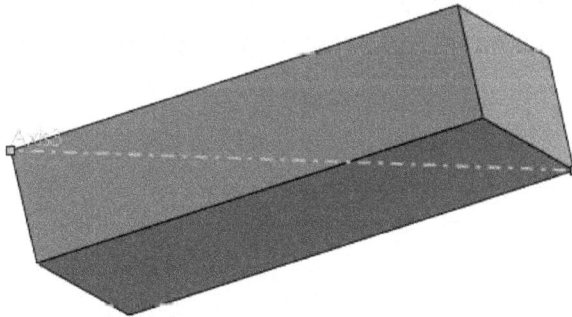

Figure-45. Axis passing through two points or vertices

Cylindrical/Conical Face

Select the **Cylindrical/Conical Face** button and select a cylindrical/conical face. An axis passing through center of cylindrical/conical face will be created; refer to Figure-46.

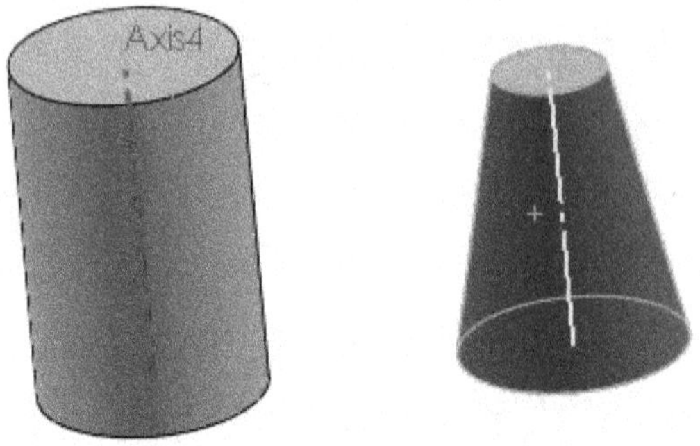

Figure-46. Axis through cylinder and conical

Point and Face/Plane

Select the **Point and Face/Plane** button if you want to create an axis passing through the selected point and perpendicular to the selected face/plane.

Coordinate System

The **Coordinate System** tool is used to create reference coordinate system. The steps to create coordinate system are explained next.

• Click on the **Coordinate System** tool from the **Reference** dropdown. The **Coordinate System PropertyManager** will display as shown in Figure-47.

Figure-47. Coordinate System PropertyManager

- Click on the point where you want to place the coordinate system.
- Click in the box for which you want to specify direction reference and select the reference like plane, axis and so on. Figure-48 shows a coordinate system created on the face.

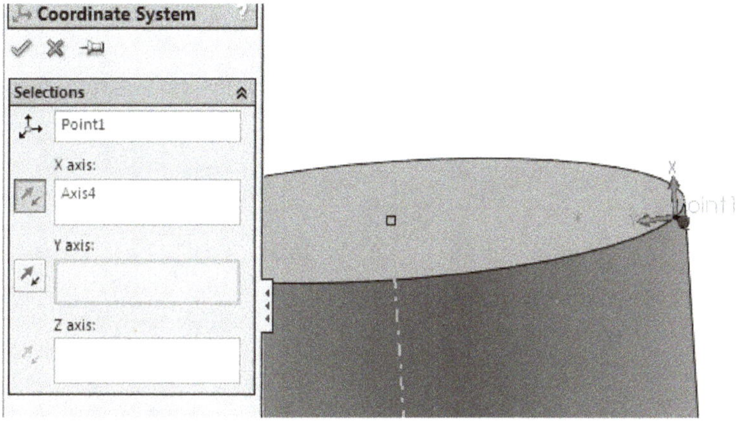

Figure-48. Coordinate system creation

Point

The **Point** tool is used to create reference points on the model. The steps to create points are given next.

- Click on the **Point** tool from the **Reference** drop-down. The **Point PropertyManager** will display as shown in Figure-49.
- Select the desired button to specify the type of point you want to create. In this case, we have selected **Center of Face** button.
- Select the reference (face of the model in this case). Preview of the point will display; refer to Figure-50.

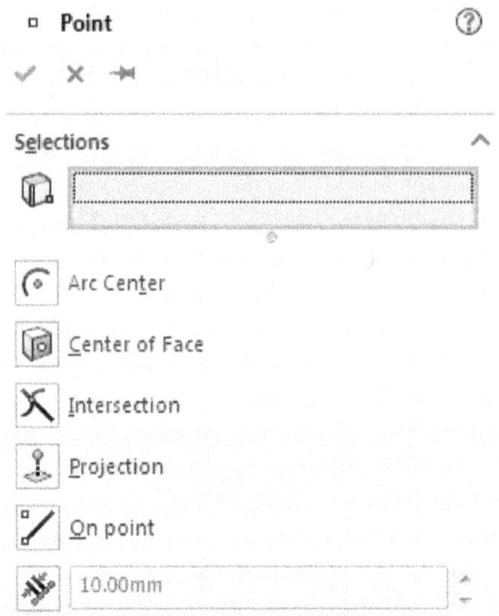

Figure-49. Point PropertyManager

Figure-50. Preview of point

- Click on the **OK** button to create the point.

You can create array of points along a curve by selecting  button.

Center of Mass

The **Center of Mass** tool is used to display the center of mass of the model. The coordinates of center of mass are generally required in some calculations related to inertia of the objects. Identification of center of mass is also helpful in checking the stability of object in constraint free environment. To display the center of mass, click on the **Center of Mass** tool from the **Reference** drop-down and the center of mass will display in the viewport; refer to Figure-51.

Figure-51. Center of mass of cylinder

LOFTED BOSS/BASE TOOL

The **Lofted Boss/Base** tool is used to create a solid volume joining two or more sketches created on different planes; refer to Figure-52. The procedure to create lofted features is given next.

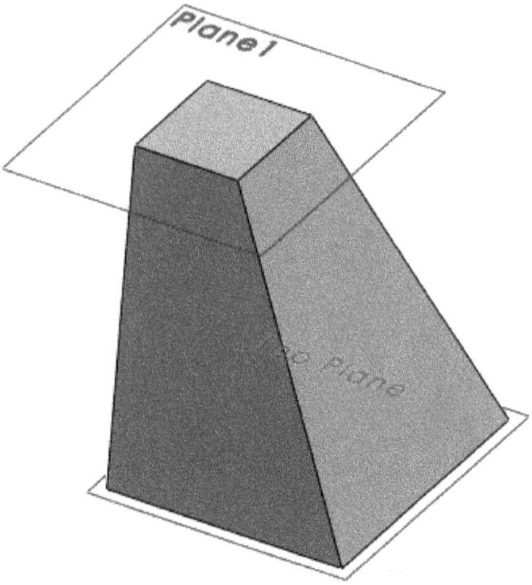

Figure-52. Lofted feature example

- Click on the **Lofted Boss/Base** tool from the **Ribbon**. The **Loft PropertyManager** will display as shown in Figure-53.

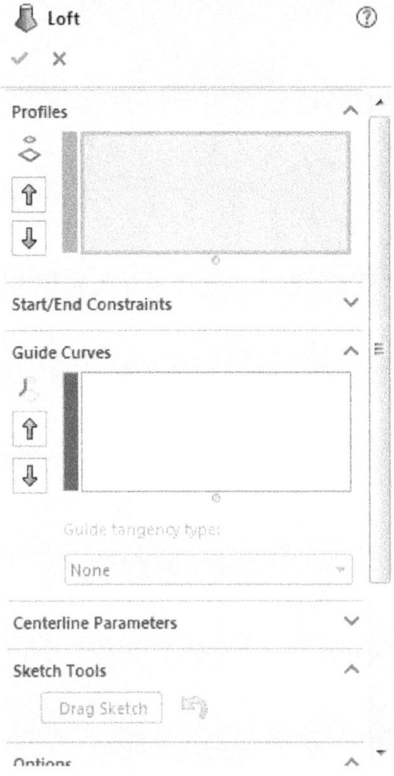

Figure-53. Loft PropertyManager

- By default, **Profiles** selection box is active and you are asked to select sketches for lofted feature.
- Click one by one on the sketches created at different planes. Note that you need to select the sketches in the order by which they can be joined to each other successively. The preview of the lofted feature will display as shown in Figure-54.

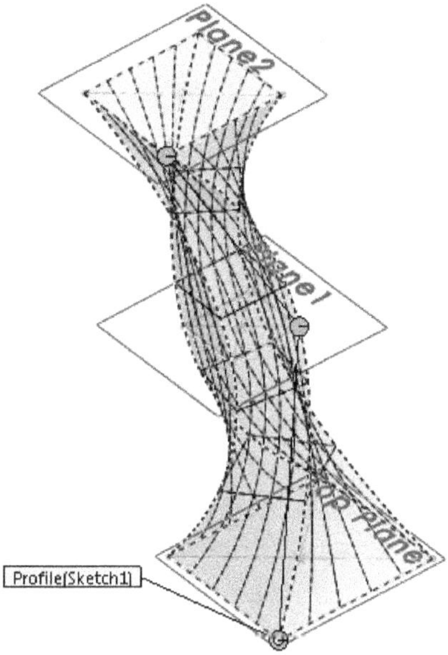

Figure-54. Preview of the lofted feature

- Drag the green handle to align edges of the lofted feature. After aligning the edges, the above figure will be displayed as shown in Figure-55.

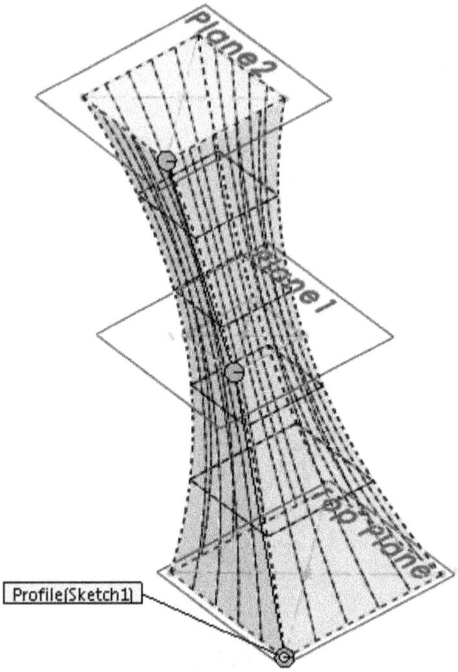

Figure-55. Lofted feature after aligning edges

- If you want to change the starting or end conditions of the loft feature then expand the **Start/End Constraints** rollout and select the desired option from the **Start Constraint** and **End Constraint** drop-down. Figure-56 shows the preview of model after changing the constraints.
- If you have guide curve/curves, then click in the Guide Curves box and select them to refine the shape of lofted feature; refer to Figure-57.

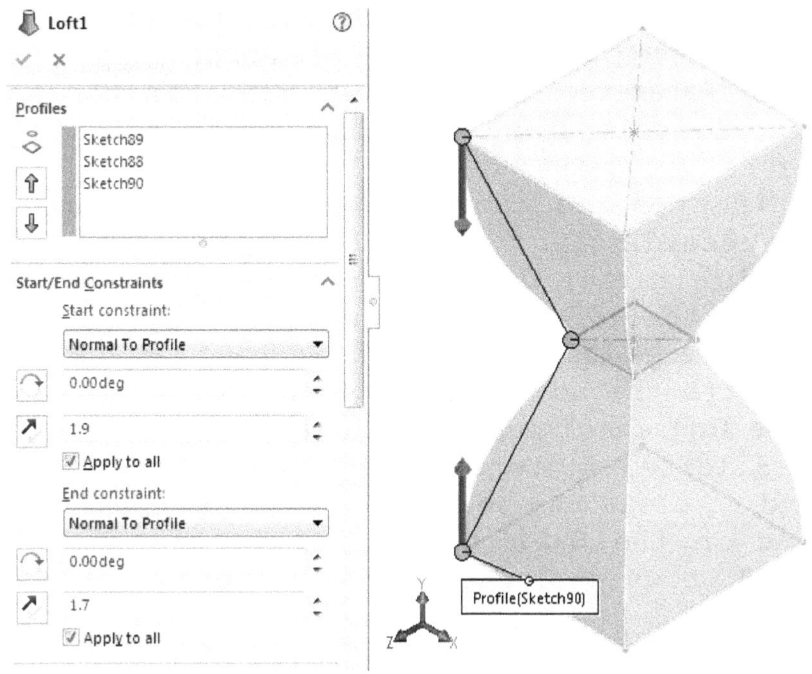

Figure-56. Preview after changing constraints

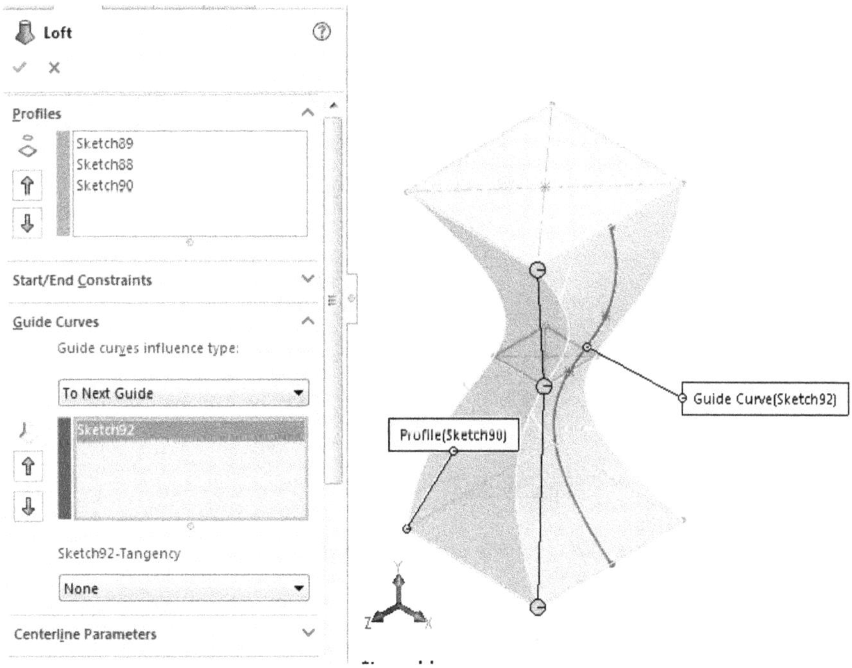

Figure-57. Preview of loft after selecting guide curves

- If you have a curve passing through center by which you want to control the shape of feature then expand the **Centerline Parameters** rollout and click in the selection box of this rollout. You are asked to select a center line or center curve
- Select the desired curve, the lofted feature will be modified accordingly.
- Next, click in the **Options** rollout to change the basic options of the lofted feature.
- Select the **Merge tangent faces** check box to merge the lofted feature with the adjoining tangent features.
- The **Close loft** check box is selected if you want to create a closed lofted feature. Note that to use this option, you need to clear the selection of guide curves by right-clicking in the **Guide Curves** box to use this feature. Figure-58 shows a closed lofted feature.

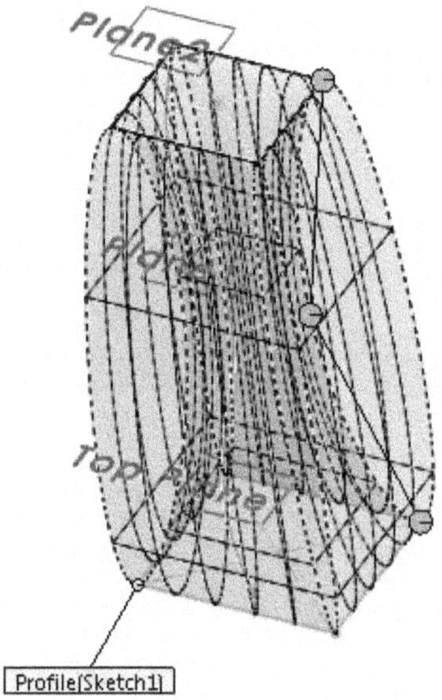

Figure-58. Preview of closed lofted feature

- Clear the **Merge Result** check box if you want to create it an individual object and don't want to join it with other solids in the model.

Note that the **Merge Result** check box will be available for each feature creation tool like **Extrude** and **Revolve**, after you have created the first solid/surface feature in the viewport. As discussed earlier, you can create a thin feature by using the options in the **Thin Feature** rollout.

BOUNDARY BOSS/BASE TOOL

The **Boundary Boss/Base** tool is used to create a solid volume by joining curves in different directions. On selecting this tool, the **Boundary PropertyManager** is displayed as shown in Figure-59. The tool works in the same way as the **Lofted Boss/Base** tool do. But, by using this tool, you can simultaneously analyze the surface of the feature being created and make it smoother or rougher as per requirement. Figure-60 shows the preview of the boundary boss/base feature with the mesh preview, Zebra stripes and Curvature combs.

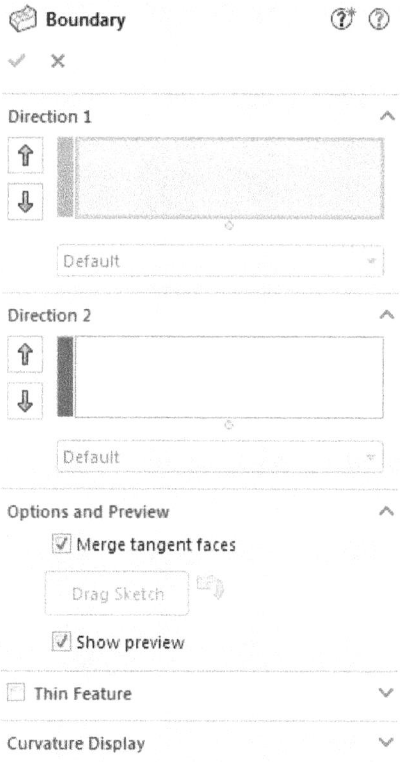

Figure-59. Boundary PropertyManager

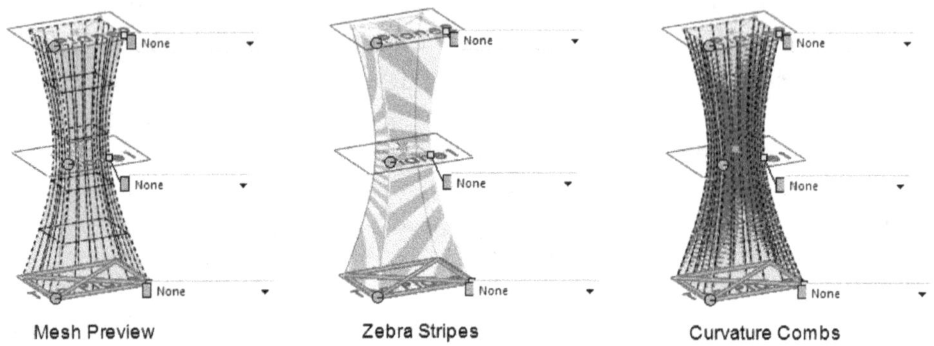

Mesh Preview　　　　Zebra Stripes　　　　Curvature Combs

Figure-60. Previews of Boundary Boss

REMOVING MATERIAL FROM SOLID OBJECTS

Till this point, you have created base/boss features by using various feature creation tools. But in Engineering, machining an object means removing material from it. So, it is equally important to know the ways you can remove material from objects in SolidWorks. All the tools to remove material are available in the next column to the one discussed earlier; refer to Figure-61.

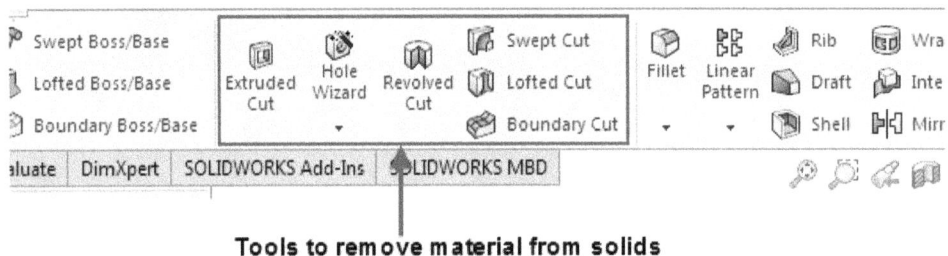

Tools to remove material from solids

Figure-61. Tools to remove material

The tools except the **Hole Wizard** work in the same way as their respective Boss/Base tool. All these tools are explained one by one as follows:

Extruded Cut

The **Extrude Cut** tool is used to remove material by extruding the sketch. The steps to use this tool are given next.

- Click on the **Extruded Cut** tool. The **Extrude PropertyManager** will display.

- Select the face from which you want to start removing the material. The sketch environment will activate.
- Click the sketch using which you want to remove the material.
- Click on the **Exit Sketch** button from the **Ribbon**. The preview of cut feature will display; refer to Figure-62.

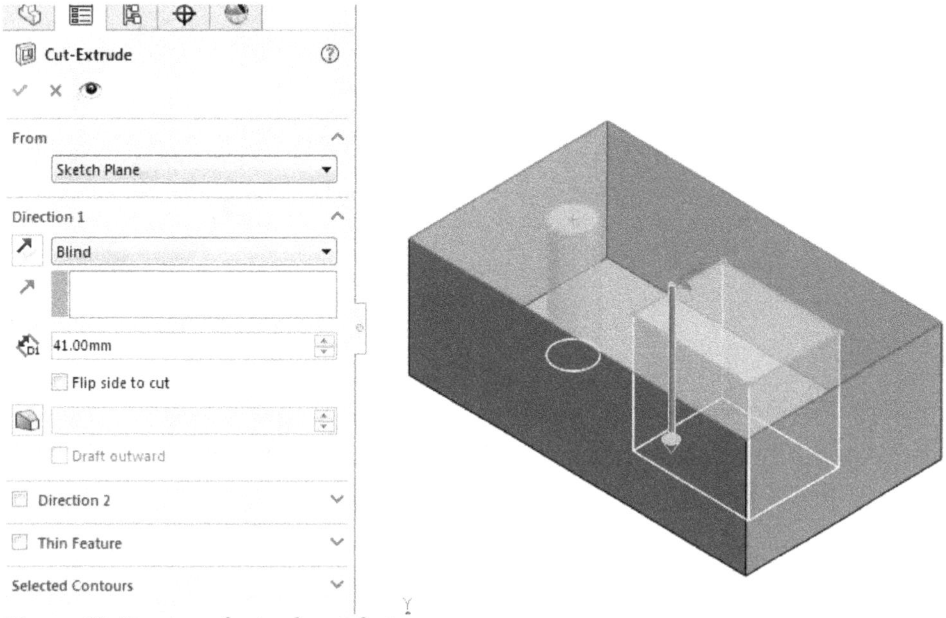

Figure-62. Preview of extrude cut feature

- Specify the depth of material removal and other parameters in the **PropertyManager** as discussed for **Extrude Boss/Base** tool and then Click on the **OK** button.

Revolved Cut

The **Revolve Cut** tool is used to remove material by revolving the sketch. The steps to use this tool are given next.

- Click on the **Revolved Cut** tool. The **Revolve PropertyManager** will display.
- Select the face from which you want to start removing the material. The sketch environment will be activated.
- Create the sketch of cut feature and a center line then click on the **Exit Sketch** button. Preview of revolved cut feature will display; refer to Figure-63.

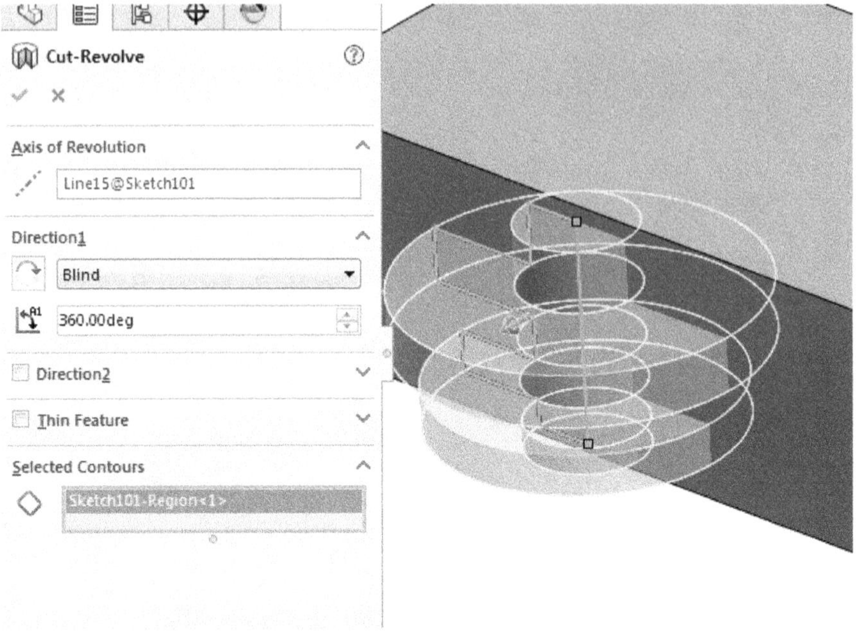

Figure-63. Preview of revolve cut feature

- Specify the parameters as discussed for **Revolved Boss/Base** tool and click on **OK** button to remove material.

Swept Cut

The **Swept Cut** tool is used to remove material by sweeping a section along the specified path. The steps to use this tool are given next.

- Click on the **Swept Cut** tool. The **Cut-Sweep PropertyManager** will display; refer to Figure-64. By default, the **Sketch Profile** radio button is selected in the **Cut-Sweep PropertyManager** and you are asked to select a sketch section.
- Select a closed sketch section and then an open sketch for path.
- Click on the **OK** button from the **PropertyManager** to create the swept cut; refer to Figure-65.

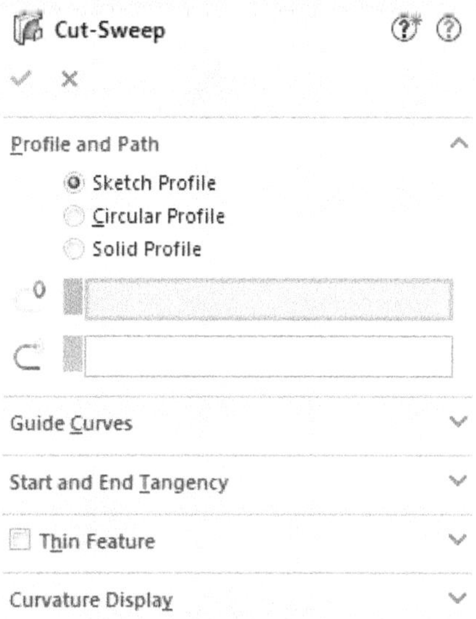

Figure-64. Cut Sweep PropertyManager

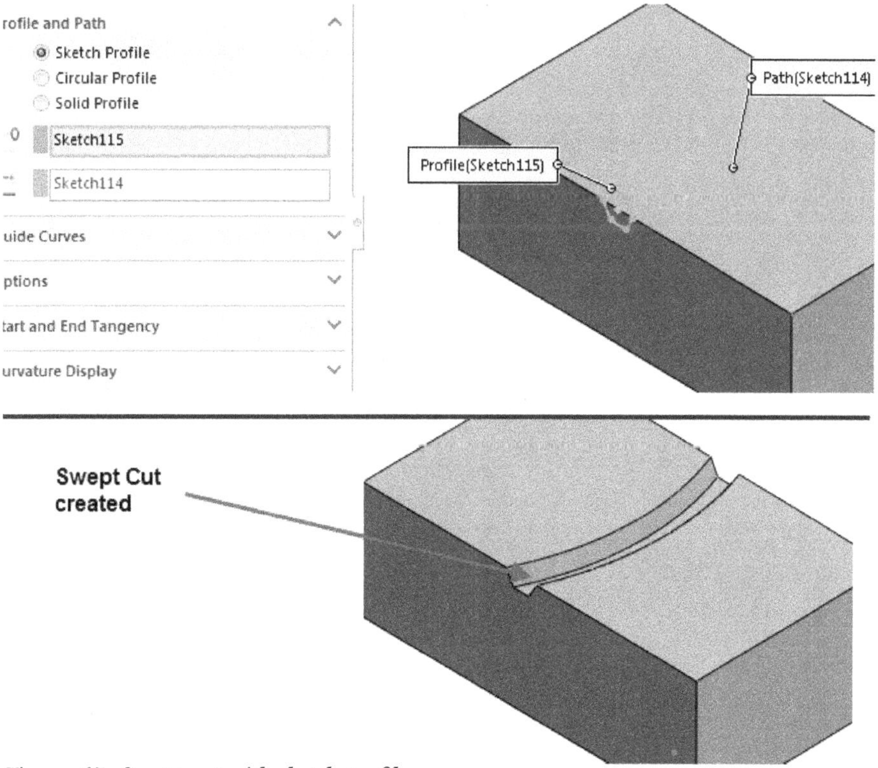

Figure-65. Swept cut with sketch profile

Swept Cut with Circular Profile

- Select the **Circular Profile** radio button from the **Cut-Sweep PropertyManager**. You are asked to select a path for the swept cut.
- Select an open sketch for the path. Preview of the cut will be displayed; refer to Figure-66.

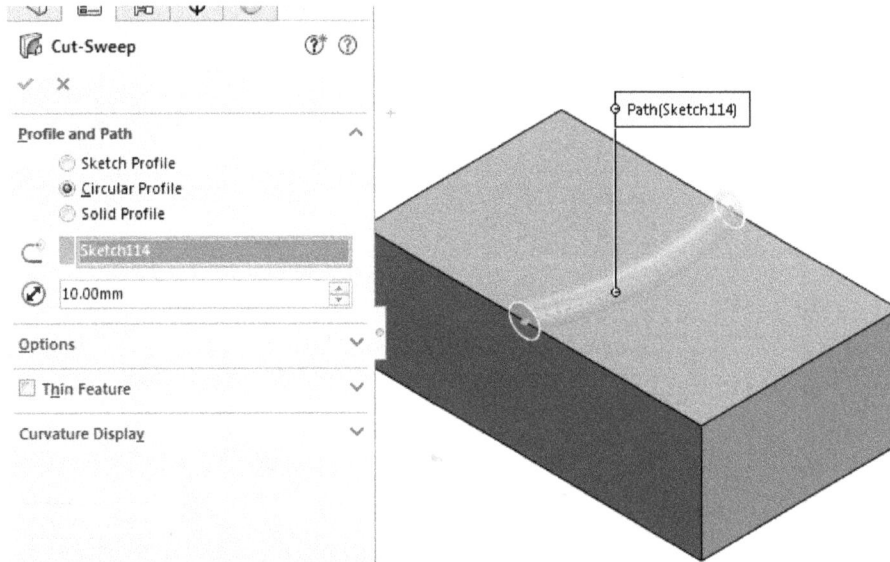

Figure-66. Swept cut with circular profile

- Specify the desired value for diameter of circular profile and click on the **OK** button from the **PropertyManager** to create the feature.

Swept Cut with Solid Profile

- Select the **Solid Profile** radio button from the **PropertyManager**. You are asked to select a solid revolve or cylindrical extrude feature as tool body. Note that the solid should be created by using only arc and line, and should not be merged with other bodies in model.
- Select the solid body and then select the path. Note that path should start from or within the tool body; refer to Figure-67.

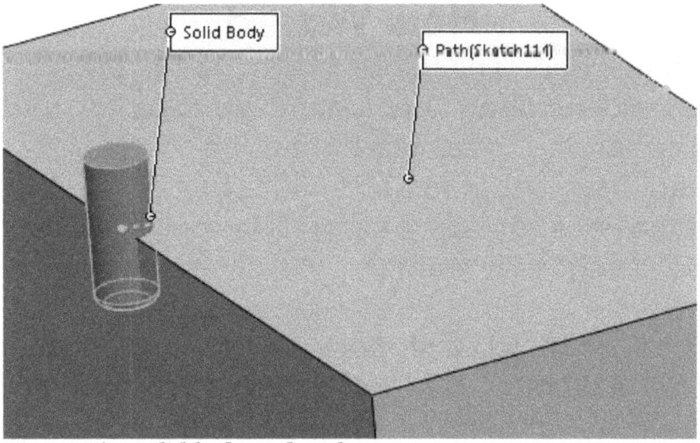

Figure-67. Solid body and path

- Click on the **OK** button from the **PropertyManager** to create the swept cut; refer to Figure-68. This option can be useful in checking the tool impression on a solid body before making any CAM programming.

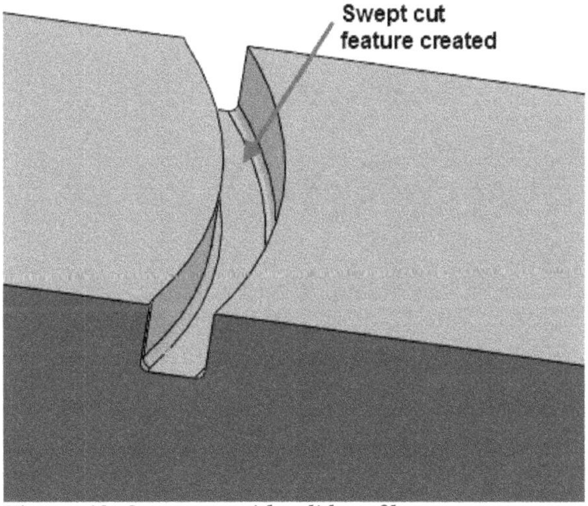

Figure-68. Swept cut with solid profile

The other tools like Lofted Cut and Boundary Cut work in the same way as their respective Boss/Base tool. I would be very happy if you practice those tools by yourself. If you get any doubt, please let me know at cadcamcaeworks@gmail.com

Hole Wizard

The **Hole Wizard** tool is used to create holes that comply
with the real machining tools. SolidWorks has library of
standard holes and slots that can be created in the solid
model. You can use this standard library or you can create
a customized hole/slot by using the tool. The procedure to
use this tool is given next.

- Click on the **Hole Wizard** tool from the **Hole Wizard** drop-
  down in the **Ribbon**. The **Hole Specification PropertyManager**
  will display as shown in Figure-69.

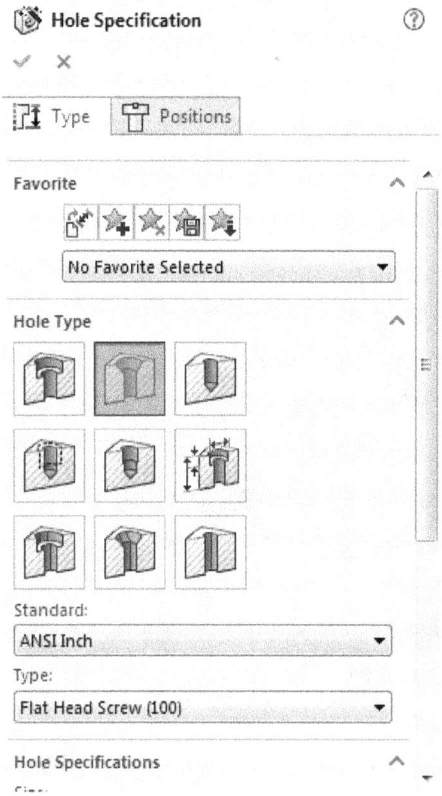

Figure-69. Hole Specification PropertyManager

- The **Favorite** rollout is used to store and reuse specific
  type of holes that you need again and again. Refer to
  Figure-70.

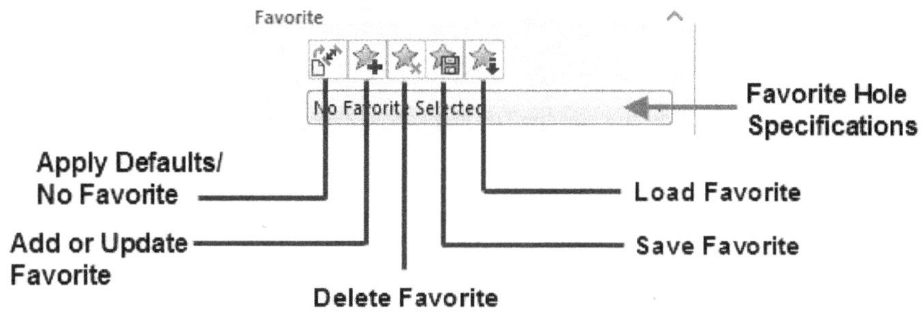

Figure-70. Favorite rollout

- Click on the desired type of hole from the **Hole Type** rollout. The parameters of the selected hole type will display in the **PropertyManager.**
- Select the desire hole standard and hole type from the **Standard** and **Type** drop-downs, respectively. The sizes of holes related to selected hole standard and type will display in the **Size** drop-down in the **Hole Specification** rollout.
- Click on the **Show custom sizing** check box, if you want to customize the hole.
- Enter the desired parameters and select the options as per your need from the other rollouts.
- Click on the **Add or Update Favorite** button from the **Favorite** rollout if you want to use this hole many times in the model.
- Now, click on the **Positions** tab to specify the position of the hole. The **PropertyManager** will display as shown in Figure-71.

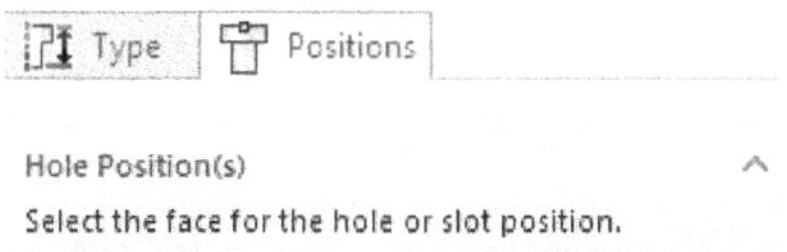

Figure-71. Positions tab of Hole Specification PropertyManager

- Click on the face of the solid model where you want to place the hole. Click to specify the position of the hole. Note that later we can position it with dimensions.
- Click on the **OK** button to create the hole.
- On creating the hole, a node of it will be added in the **FeatureManager Design Tree.** Expand this node; refer to Figure-72.

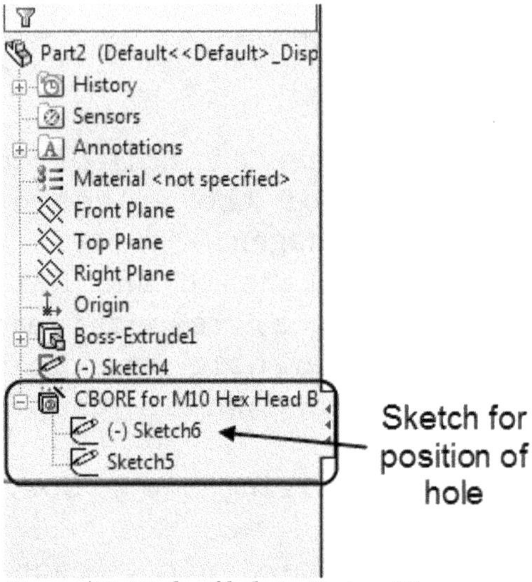

Figure-72. Node of hole in FeatureManager Design Tree

- Right-click on the sketch for position of the hole and select the **Edit Feature** button from the tool box; refer to Figure-73.

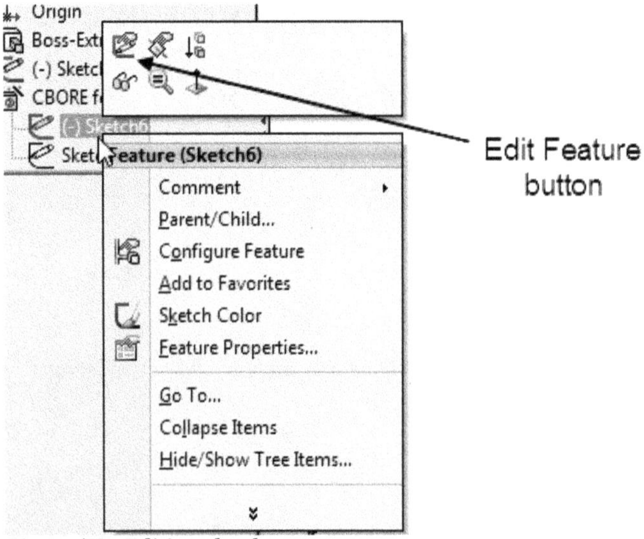

Figure-73. Editing sketch

- The sketching environment will be displayed.
- Here, specify the position of point by dimensioning. Note that here you can create multiple points by using the **Point** tool and the holes will be created at all the points you created. Refer to Figure-74.

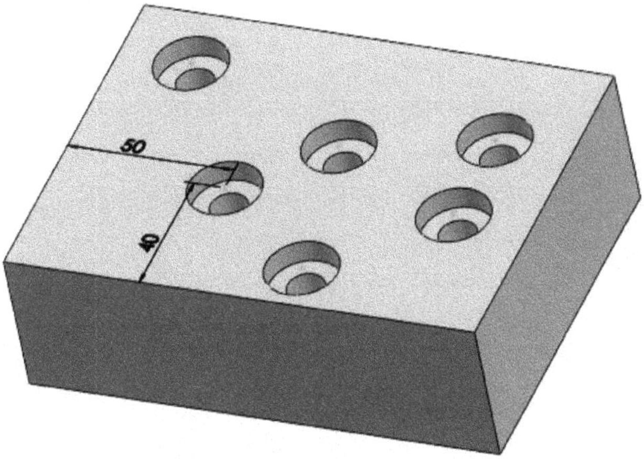

Figure-74. Holes created at sketched points

Thread

The **Thread** tool is used to cut helical thread on cylindrical faces. Using this tool, you can save the custom threads in library. The procedure to use this tool is discussed next.

- Click on the **Thread** tool from the **Hole Wizard** drop-down in the **Features** tab of the **Ribbon**; refer to Figure-75. The **Thread PropertyManager** will be displayed; refer to Figure-76.

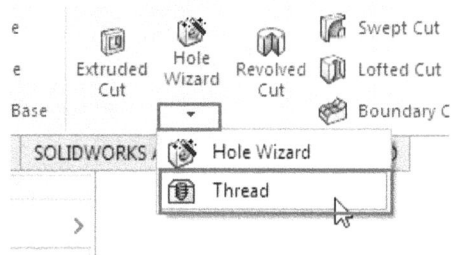

Figure-75. Thread tool

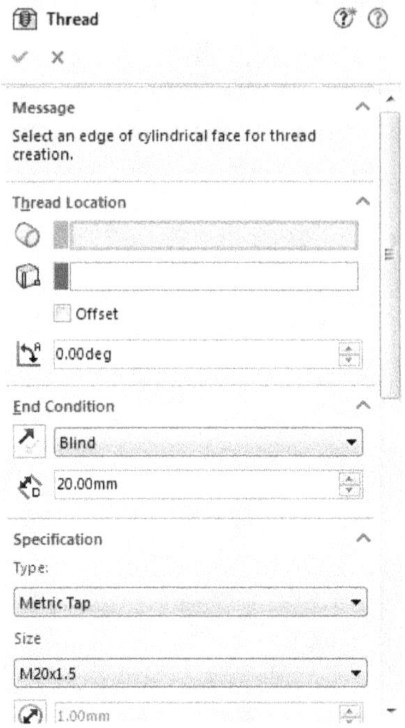

Figure-76. Thread PropertyManager

- Select the round edge of the cylindrical face of hole/ boss feature. Preview of the thread will be displayed; refer to Figure-77.

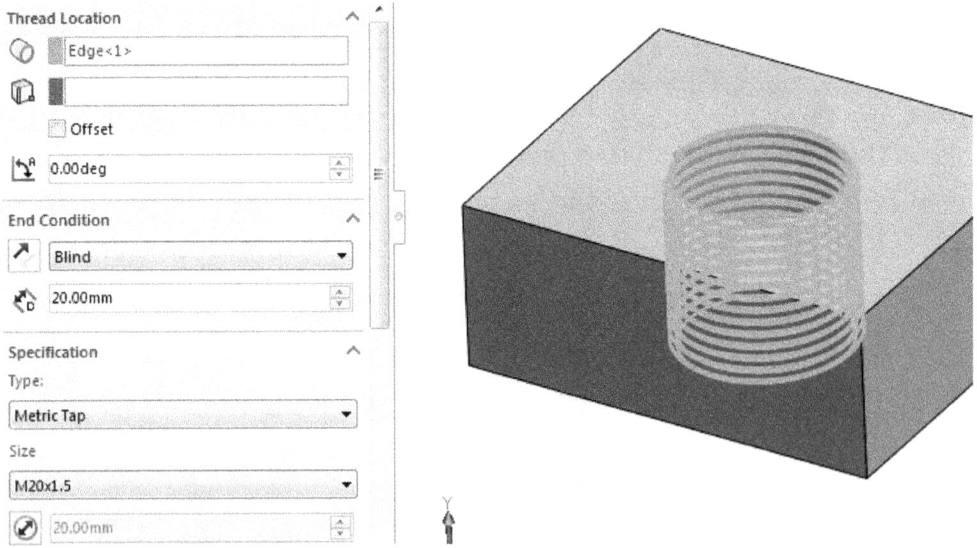

Figure-77. Preview of thread

- Select the starting location of the thread after clicking in the **Start Location** selection box. [icon] [icon] _____ Note that this option is optional and you may not require it in your operation.
- Select the **Offset** check box if you want to start the thread at an offset distance from the starting location. After selecting this check box, an edit box becomes active below it where you can specify the value of offset; refer to Figure-78.

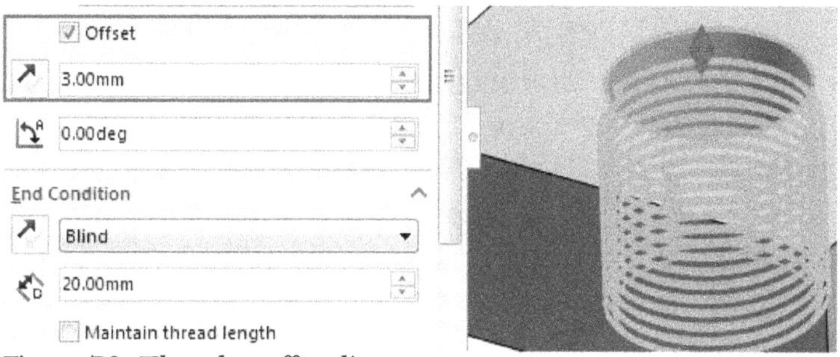

Figure-78. Thread at offset distance

- From the **End Condition** rollout, specify the depth of the thread. By default, the **Blind** option is selected in the drop-down and you need to specify the depth of thread in the edit box below it.
- Select the desired threading tool from the **Type** drop-down in the **Specification** rollout of the **PropertyManager**; refer to Figure-79.

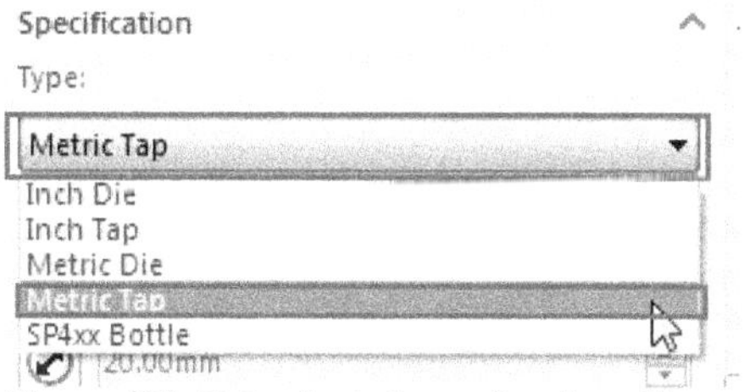

Figure-79. Type drop-down for threading tools

- Select the thread size from the **Size** drop-down below the **Type** drop-down in the **Specification** rollout. You want to specify custom size of thread then click on the **Override Diameter** and **Override Pitch** buttons below the drop-down and specify the desired values; refer to Figure-80.

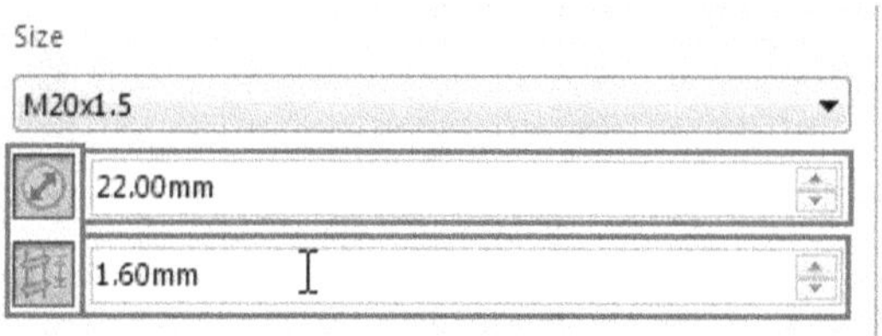

Figure-80. Custom size of thread

- Select the desired method from the **Thread Method** area of the rollout. There are two radio buttons available in this area; **Cut thread** and **Extrude thread**. Select the **Cut thread** radio button if you are creating thread inside a hole like in nut. Select the **Extrude thread** radio button if you want to create thread on a boss feature like bolt.

- You can cut/extrude the mirror image of current thread specifications by using the **Mirror Profile** check box. On selecting this check box, the **Mirror horizontally** and **Mirror vertically** radio buttons will become active. Select the **Mirror horizontally** radio button if you want to use horizontal mirror image of the current thread profile or select the **Mirror vertically** radio button if you want to use the vertical mirror image of the current thread profile; refer to Figure-81.

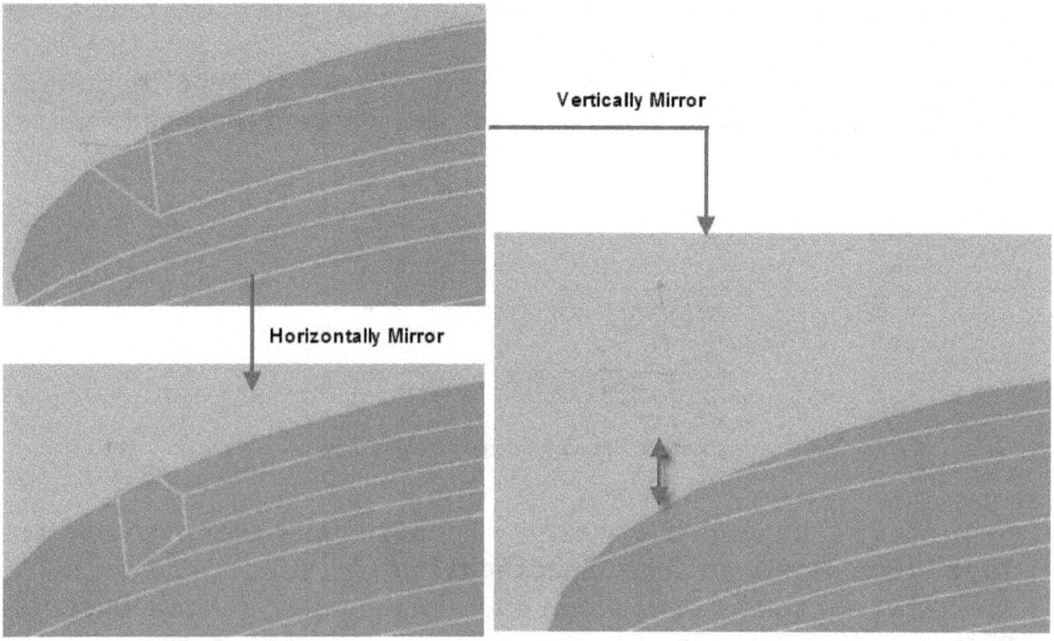

Figure-81. Mirroring thread profile

- Specify the thread angle using the **Rotation Angle** edit box.
- From the **Thread Options** rollout, you can select the thread as left handed or right handed.
- Check the preview of threads by using the desired radio button from the **Preview Options** rollout in the **PropertyManager**.
- Click on the **OK** button from the **PropertyManager** to create the thread; refer to Figure-82.

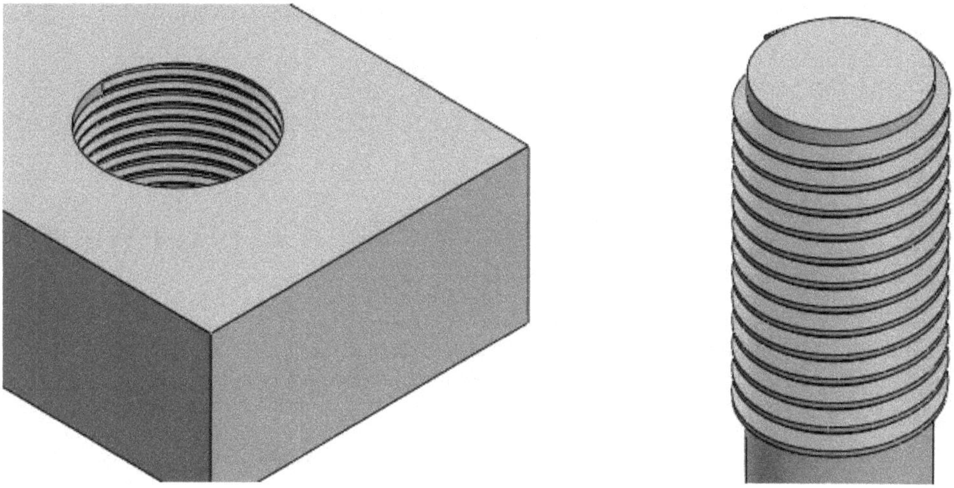

Figure-82. Threaded solids

Till this chapter, we have learned creation of Solid objects and removing material from them. In the next chapter, we will learn the modifying operations that can be performed on solid models. Also, we will practice all the tools we have discussed till this point.

SELF ASSESSMENT

Q1. A sketch which exists in more than one plane is called _____ .

Q2. Using the _____ key from keyboard, you can change the plane while creating the 3D sketch.

Q3. We can create a box using rectangle as sketch by using the _____ tool.

Q4. We can create a cylinder using the rectangle as sketch by using the _____ tool.

Q5. The _____ tool is used to create a solid volume by moving a sketch along the selected path.

Q6. We can create sweep feature only at one side if the section lies in between the path. (T/F)

Q7. We can create both guide curves and path in same sketch for creating sweep feature. (T/F)

Q8. Using the **Profile Twist** options of the **Swept Boss/Base** tool, you can create springs. (T/F)

Q9. For Circular Profile Sweep feature, the profile of section is by default circle and we are required to specify the diameter of that. (T/F)

Q10. By default,_____ (hint: number) planes are available in SolidWorks.

Q11. How many number of maximum references can be provided for creating a reference plane?
a. 2
b. 3
c. 4
d. 5

Q12. After invoking the **Plane PropertyManager** if you select an existing plane then which of the following button is selected by default in the **PropertyManager**?

a. Parallel
b. Perpendicular
c. Coincident
d. Offset distance

Q13. Which of the following combination can be used to create plane at an angle?

a. An Edge and connected planar face
b. An Edge and intersecting plane
c. Two vertices and flat face
d. All of the above.

Q14. Which of the following cannot be selected as origin for creating coordinate system?

a. Vertices of solid
b. Edge of solid
c. Mid Point of an edge
d. Center line of revolve feature

Q15. Which of the following option is available for **Swept cut** tool but not for **Swept Boss/Base** tool?

a. **Circular Profile** radio button
b. **Solid Profile** radio button
c. **Guide Curves** rollout
d. **Mesh preview** check box

FOR STUDENT NOTES

FOR STUDENT NOTES

FOR STUDENT NOTES

Solid Editing
and Practical

Chapter 5

Topics Covered

The major topics covered in this chapter are:

- *Fillet/Chamfer tool.*
- *Pattern tools.*
- *Rib tool.*
- *Draft tool.*
- *Shell tool.*
- *Wrap tool, Intersect tool, and Mirror tool.*
- *Practical and Practice*

In the previous chapter, we have learned to create solid models and remove material from them. In this chapter, we will learn to edit the models. Like the other tools, SolidWorks has packed all the editing tools into one column. These tools are available in the column next to the column intended for removing material; refer to Figure-1. The tools in this column are explained next.

Figure-1. Solid Editing tools

FILLET

The **Fillet** tool is used to apply radius at the edges. This tool works in the same way as the **Sketch Fillet** do. It is recommended that you apply the fillets after creating all the featured required in model, if possible. The procedure to use this tool is given next.

Constant Size Fillet

- Click on the **Fillet** tool from the **Fillet** drop-down in the **Ribbon**. The **Fillet PropertyManager** will display as shown in Figure-2.
- Select the edge on which you want to apply the fillet. In place of selecting edge, you can select the two adjoining faces. The fillet will be created at the intersection of the two faces; refer to Figure-3.
- In the **Fillet Type** rollout **Constant size** button is selected.
- Specify desired radius for fillet in the **Radius** edit box in **Fillet Parameters** rollout of the **PropertyManager**.
- Select the **Full preview** radio button to check the preview of the fillet.
- You can select the desired profile fillet by using the **Profile** drop-down in **Fillet Parameters** rollout. Preview of fillet on selecting different profile options is given in Figure-4.

Figure-2. Fillet PropertyManager

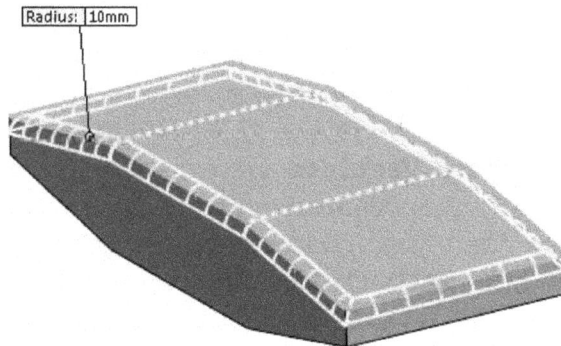

Figure-3. Constant size fillet

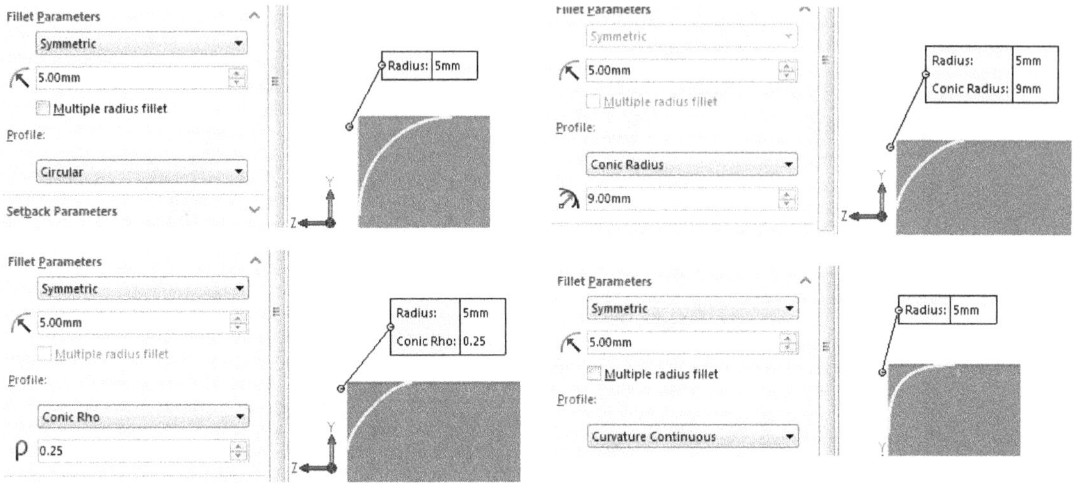

Figure-4. Fillet with different profiles

- You can specify different radius for both sides of round with respect to selected edge by using the **Asymmetric** option from the drop-down in the **Fillet Parameters** rollout.
- You can set the setback parameters at the vertices by using the options in the **Setback Parameters** rollout; refer to Figure-5.

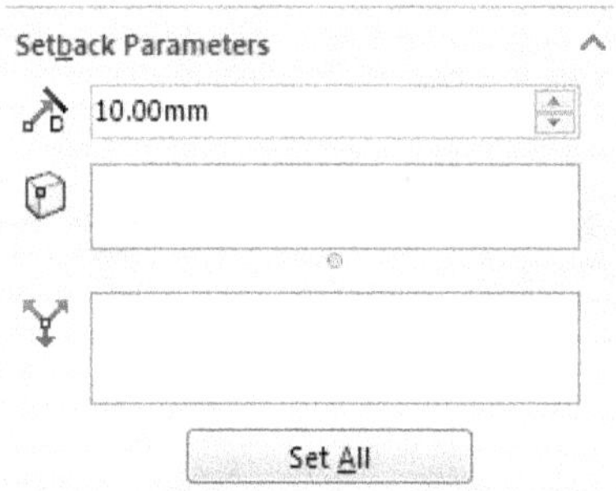

Figure-5. Setback Parameters rollout

- Click in the **Setback Vertices** selection box and click on the vertex for which you want to specify setback parameters. Fillet radius at the different joining edges will be displayed in the **Setback Distances** box.
- Select the desired direction from the **Setback Distances** box and specify the value of distance by using the **Distance** edit box in the **Setback Parameters** rollout.
- Click on the **OK** button from the **PropertyManager** to create the fillet.

Variable Radius Fillet

You can create fillet with varying radius by selecting the **Variable size** button. On selecting this button, the **Variable Radius Parameters** rollout is added in the **Fillet PropertyManager**; refer to Figure-6.

- Set the number of point for changing the radius by using the **Number of Instances** spinner in the rollout.
- Click in the **Attached Radii** box for start and end point in the viewport and specify the starting and ending radius.
- Click on the desired point in the preview of fillet and specify the desired radius; refer to Figure-7.
- You can change the profile for round by using the **Profile** drop-down in the rollout.

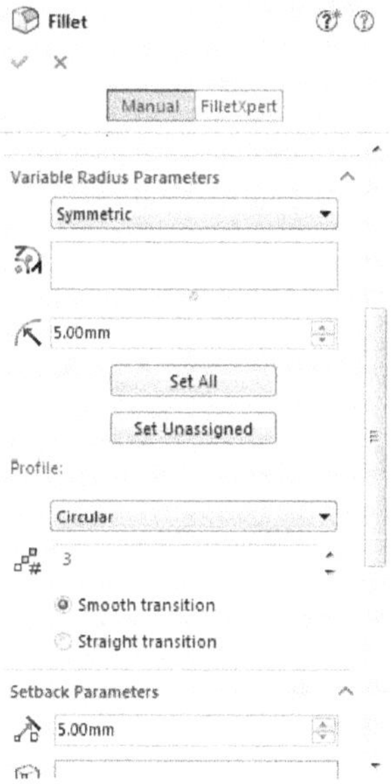

Figure-6. Variable Radius Parameters rollout

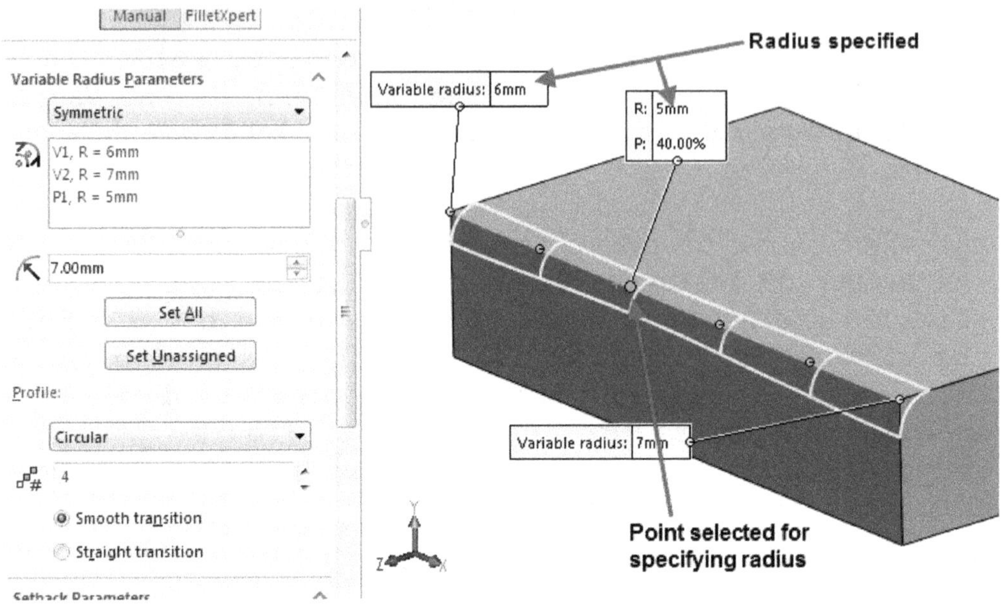

Figure-7. Creating variable radius fillet

Face fillet

- Select the **Face fillet** button from the **Fillet Type** rollout to create fillet at the joining edge of two faces.
- On selecting this radio button, the **Items to Fillet** rollout is added in the **Fillet PropertyManager**.
- Click in the first box and select the first face/faces.
- Next, click in the second box and select the second face/faces. Figure-8 shows preview of the fillet.

Full round fillet

Select the **Full round fillet** button to create a fillet where three faces meet each other. Using this option, you can create dome feature; refer to Figure-9.

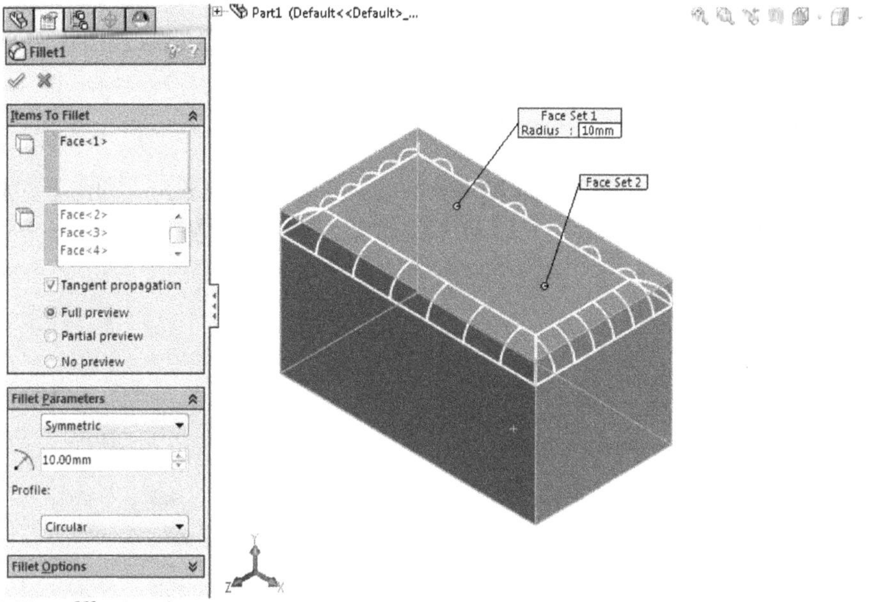

Figure-8. Face fillet preview

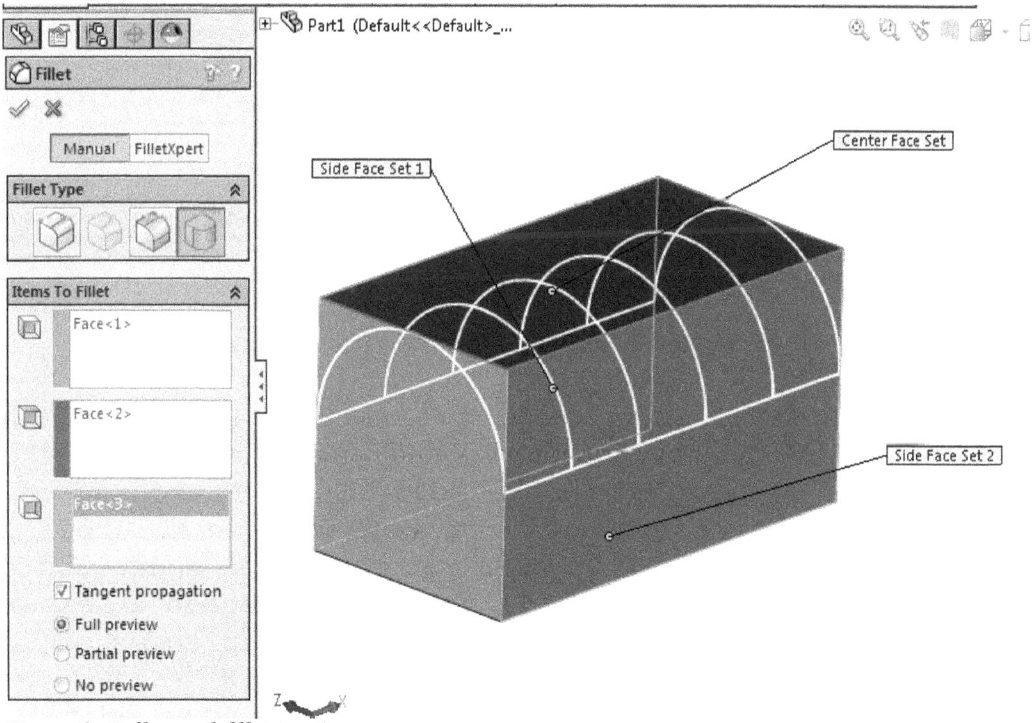

Figure-9. Full round fillet preview

FilletXpert

The **FilletXpert** tool is used to apply different type of fillets in one single mode. The procedure to use this tool is discussed next.

- Click on the **FilletXpert** button from the **Fillet PropertyManager**. The **FilletXpert PropertyManager** will be displayed; refer to Figure-10.

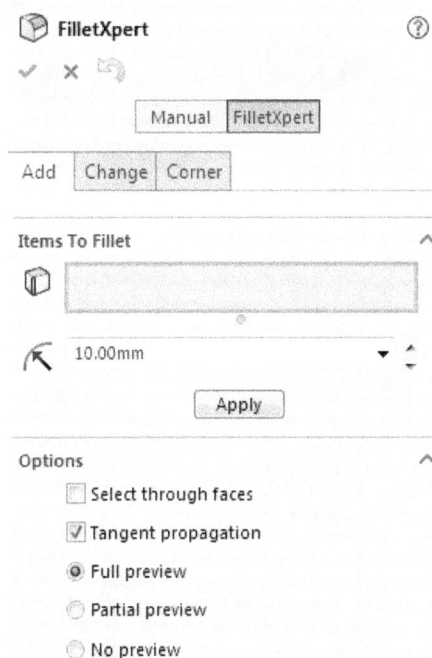

Figure-10. FilletXpert PropertyManager

- Select the face, edge, loop, or feature. Preview of the fillet will be displayed. If you have selected an edge then fillet will be applied to the edge. If you have select a face then all the edges of the face will be filleted. If you have selected a loop then all the edges connected to the loop will be filleted. If you have selected a feature then fillet will be applied to all the sharp edges of the feature.
- Set the desired value of fillet radius in the edit box below **Items To Fillet** selection box.
- Click on the **Apply** button to apply fillet and start with a new set of fillet.

CHAMFER

The **Chamfer** tool is used to bevel the sharp edges of the model. This tool works in the same way as the **Sketch Chamfer** do. The procedure to create chamfer by using this tool is given next.

- Click on the **Chamfer** tool from the **Fillet** drop-down. The **Chamfer PropertyManager** will display as shown in Figure-11.
- Select the edges from the model on which you want to apply chamfer.
- By default, the **Angle Distance** radio button is selected in the **PropertyManager**. Specify the angle and distance parameters in the respective boxes.
- If you want to specify distances for both sides of chamfer then select **Distance distance** radio button and specify the parameters.
- If you want to create chamfer at corners then select the **Vertex** radio button and select the corner of the model.

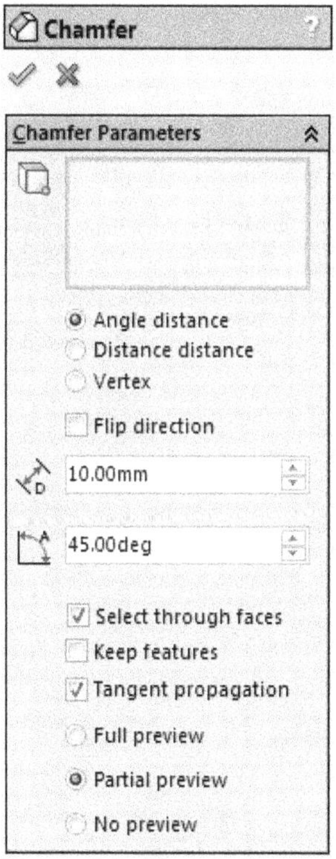

Figure-11. Chamfer Property-tyManager

Figure-12 shows chamfer created at edges and vertexes.

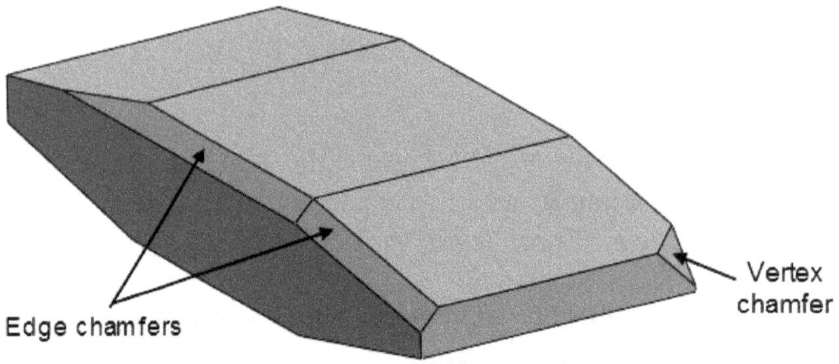

Figure-12. Chamfers created on edges and vertex

LINEAR PATTERN

The **Linear Pattern** tool is used to create linear pattern of solid features in the Modeling environment. This tool is similar to the sketch linear patterns. The procedure to create linear pattern is given next.

*   Select all the features that you want to pattern and select the **Linear Pattern** tool from the **Linear Pattern** drop-down. The **Linear Pattern PropertyManager** will display as shown in Figure-13.

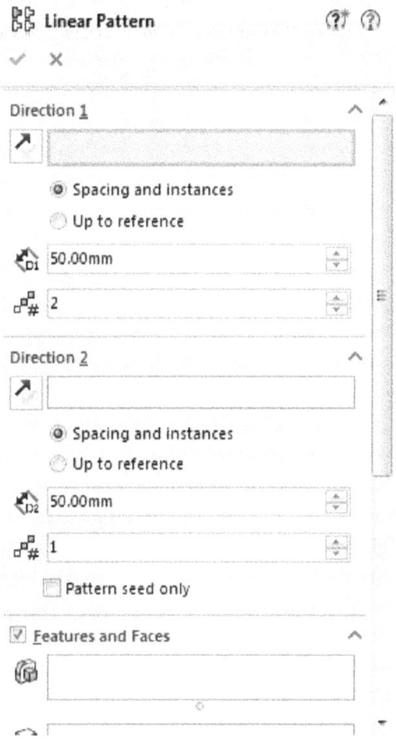

Figure-13. Linear Pattern PropertyManager

- Select the direction reference for Direction 1 (like edge, face, plane, and so on) and specify the related parameters.
- Similarly, specify the reference for Direction 2 and specify the related parameters; refer to Figure-14.

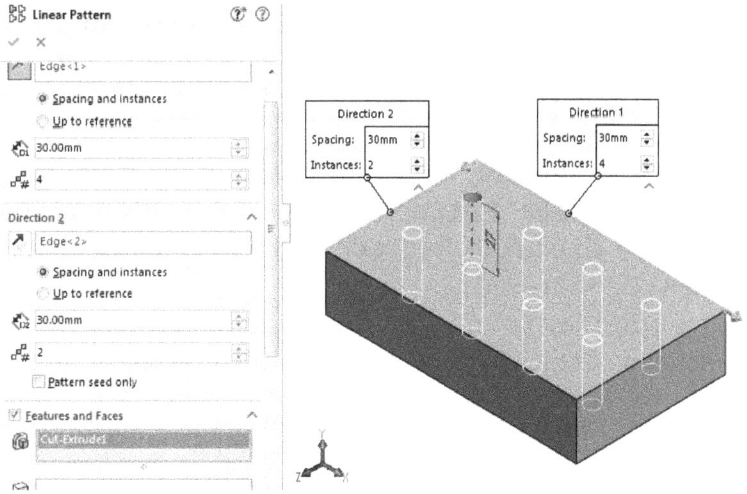

Figure-14. Preview of Linear pattern

- If you want to skip any instance in the pattern then expand the **Instances to Skip** rollout and select the pink dots in the preview to skip them; refer to Figure-15.

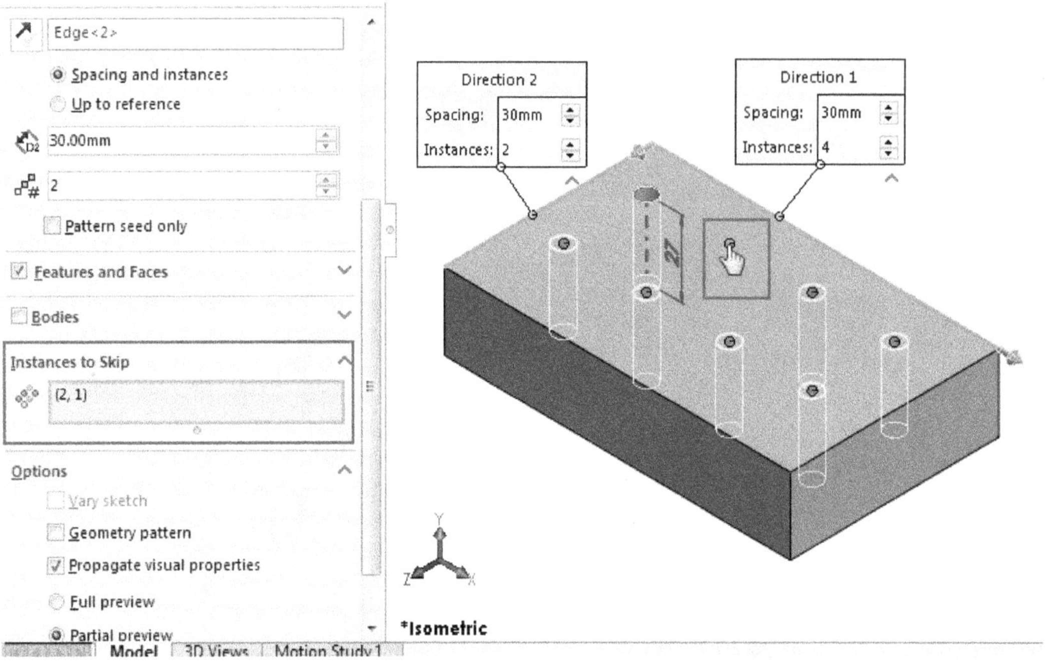

Figure-15. Skipping instances in pattern

- Click on the **OK** button from the **PropertyManager** to create the pattern.

CIRCULAR PATTERN

The **Circular Pattern** tool is used to create circular pattern of solid features in the Modeling environment. This tool is similar to the sketch circular pattern. The procedure to create circular pattern is given next.

- Select all the features that you want to pattern and select the **Circular Pattern** tool from the **Linear Pattern** drop-down. The **Circular Pattern PropertyManager** will display as shown in Figure-16.
- Select the edge/axis/circular face(axis of circular face will be automatically selected) about which you want to create the pattern.
- Specify the required parameters like angle between two instances, number of instances, and so on. The preview of pattern will display; refer to Figure-17.

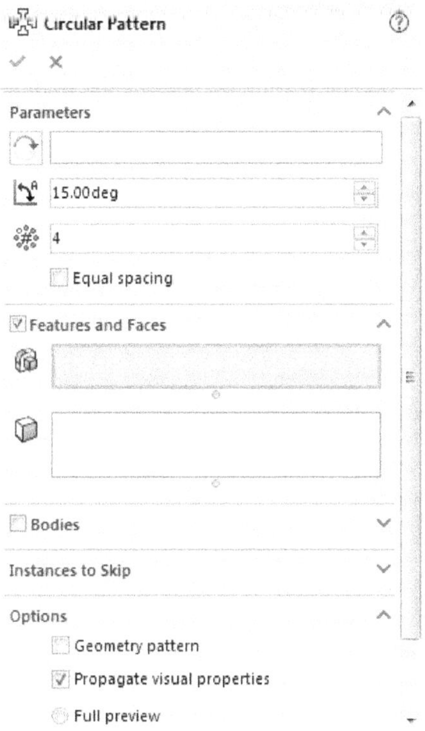

Figure-16. Circular Pattern PropertyManager

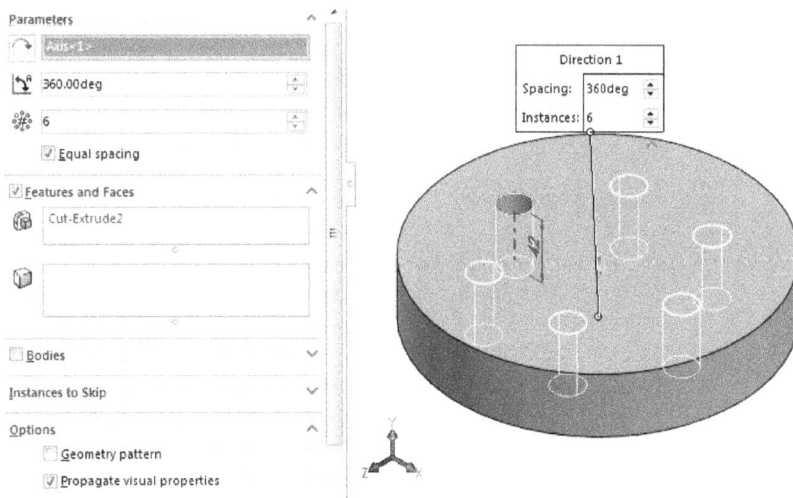

Figure-17. Preview of circular pattern

CURVE DRIVEN PATTERN

The **Curve Driven Pattern** tool is used to create multiple instances of a solid features along the selected path. The procedure to create curve driven pattern is given next.

- Select all the features that you want to pattern and select the **Curve Driven Pattern** tool from the **Linear Pattern** drop-down. The **Curve Driven Pattern PropertyManager** will display as shown in Figure-18.

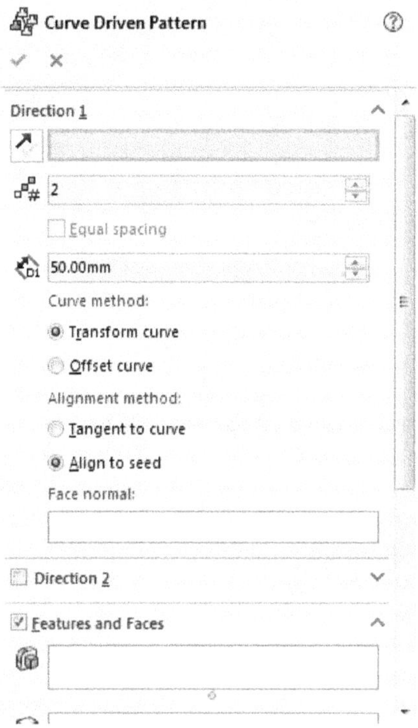

Figure-18. Curve Driven Pattern
PropertyManager

- Select the curve along which you want to create the pattern.
- Specify the number of instances and other parameters. Preview of pattern will display; refer to Figure-19.

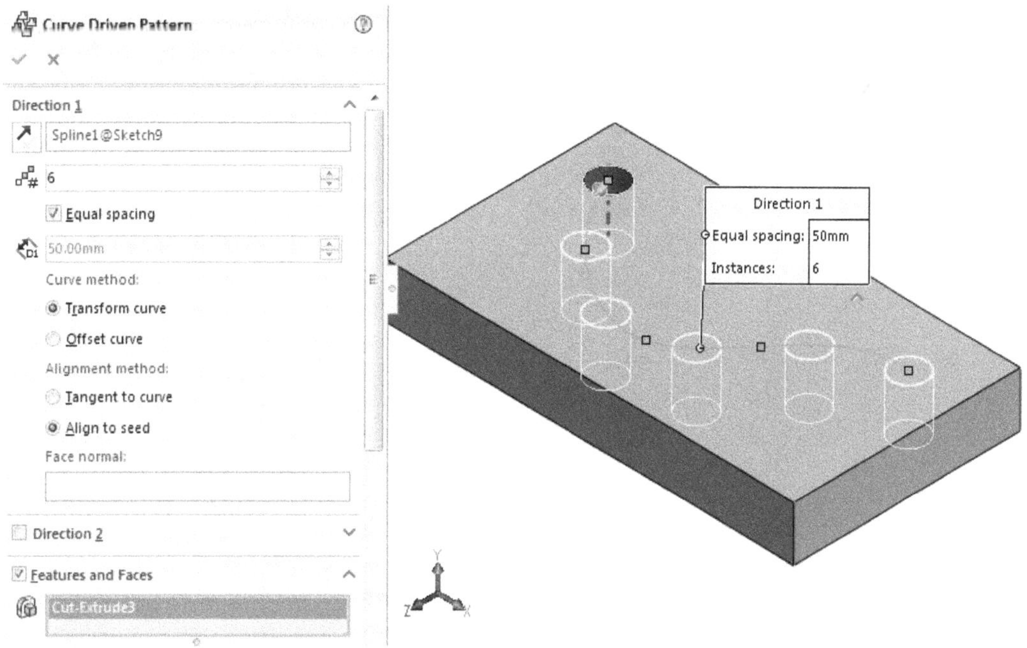

Figure-19. Preview of curve driven pattern

SKETCH DRIVEN PATTERN

The **Sketch Driven Pattern** tool is used to create multiple instances of a solid features as per the points specified in the selected sketch. The procedure to create sketch driven pattern is given next.

- Select all the features that you want to pattern and select the **Sketch Driven Pattern** tool from the **Linear Pattern** drop-down. The **Sketch Driven Pattern PropertyManager** will display as shown in Figure-20.

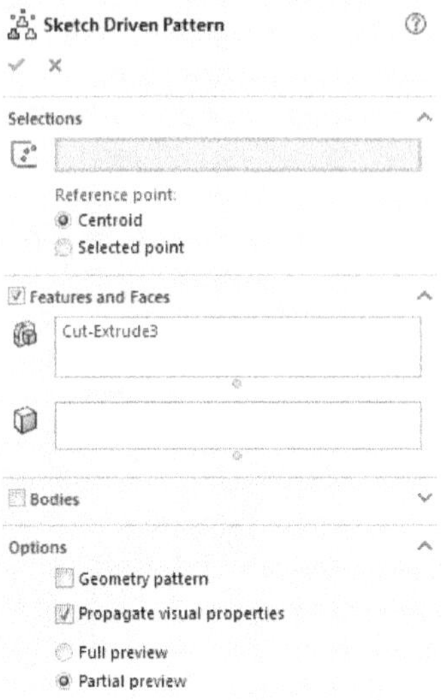

*Figure-20. Sketch Driven Pattern
PropertyManager*

- Select the sketch as reference for pattern. Preview of the pattern will be displayed; refer to Figure-21.
- Specify the other parameters as required and then click on the **OK** button from the **PropertyManager** to create the pattern.

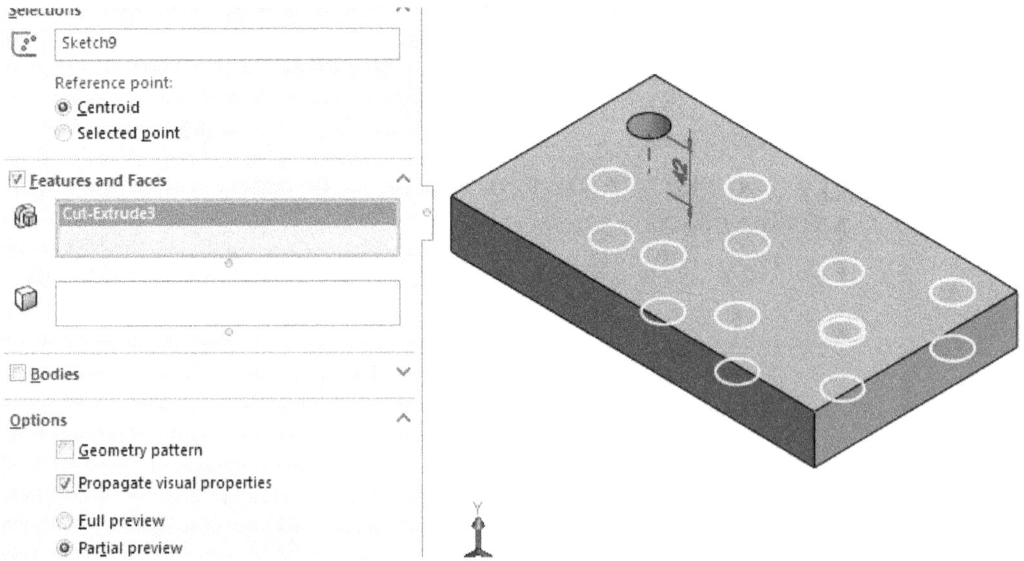

Figure-21. Preview of the sketch driven pattern

TABLE DRIVEN PATTERN

The **Table Driven Pattern** tool is used to create multiple instances of a features as per the coordinates specified in the table. Note that you must have a coordinate system created in the model for referencing the coordinates of the table. The procedure to create table driven pattern is given next.

* Select all the features that you want to pattern and then click on the **Table Driven Pattern** tool from the **Linear Pattern** drop-down in the **Ribbon**. The **Table Driven Pattern** dialog box will be displayed; refer to Figure-22.

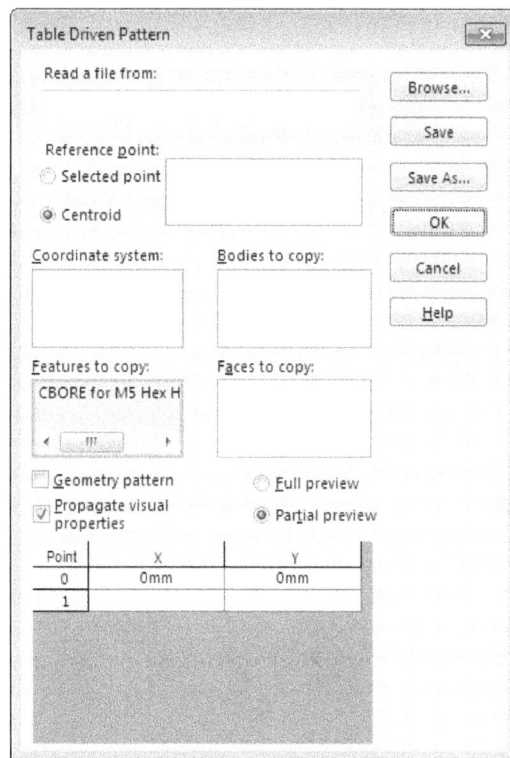

Figure-22. Table Driven Pattern dialog box

* Click in the **Coordinate system** selection box and select the coordinate system.
* Now, double-click in the cells of table displayed below in the table and specify the desired coordinates. Preview of the pattern will be displayed; refer to Figure-23.
* Specify the other parameters if required and then click on the **OK** button from the dialog box to create the pattern.

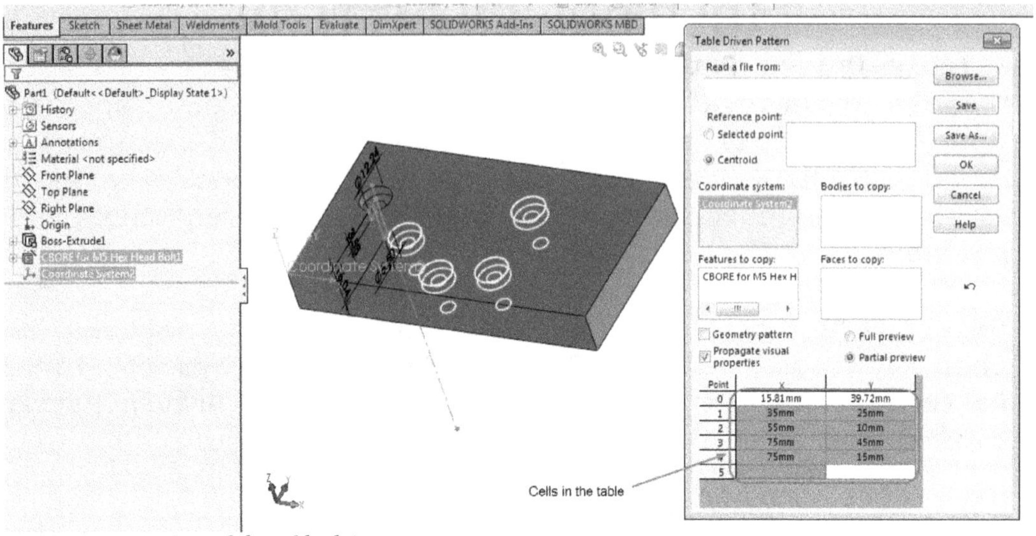

Figure-23. Preview of the table driven pattern

FILL PATTERN

The **Fill Pattern** tool is used to create multiple instances of a features by filling the selected bounded region. Note that you must have a closed loop sketch created in the model for referencing the boundary region. The procedure to create fill pattern is given next.

- Select all the features that you want to pattern and then click on the **Fill Pattern** tool from the **Linear Pattern** drop-down in the **Ribbon**. The **Fill Pattern PropertyManager** will be displayed; refer to Figure-24.
- Select the sketch for filling with instances of pattern feature. Preview of the pattern will be displayed; refer to Figure-25.
- Specify the desired parameters like spacing between instances, space between boundary line and instances, and so on.
- Click on the **OK** button from the **PropertyManager** to create the fill pattern.

Figure-24. Fill Pattern Property-Manager

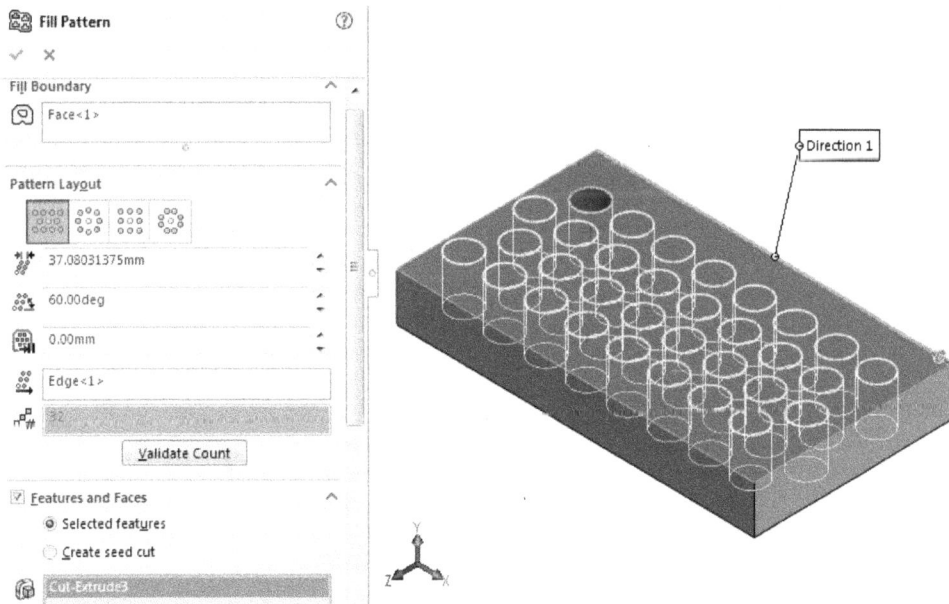

Figure-25. Preview of the fill pattern

MIRROR

The **Mirror** tool is used to create mirror copy of the features in the Modeling environment. This tool is similar to the sketch mirror. The procedure to create mirror is given next.

*   Select all the features that you want to mirror.
*   Click on the **Mirror** tool from the **Linear Pattern** drop-down or from the **Ribbon**. The **Mirror PropertyManager** will display as shown in Figure-26.
*   Select the plane or face about which you want to mirror the features. Preview of the mirror will be displayed; refer to Figure-27.
*   Click on the **OK** button from the **PropertyManager** to create the feature.

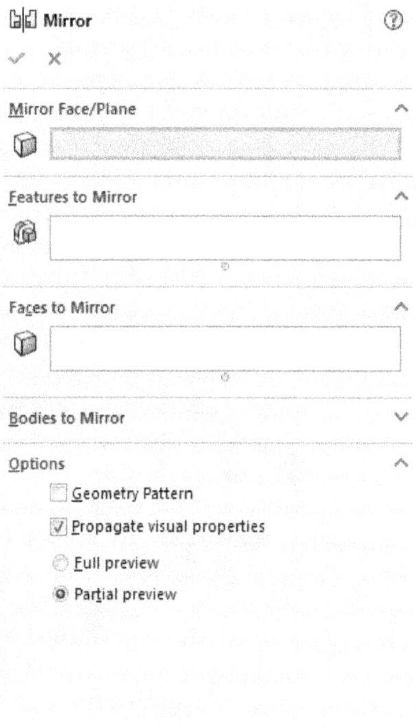

Figure-26. Mirror PropertyManager

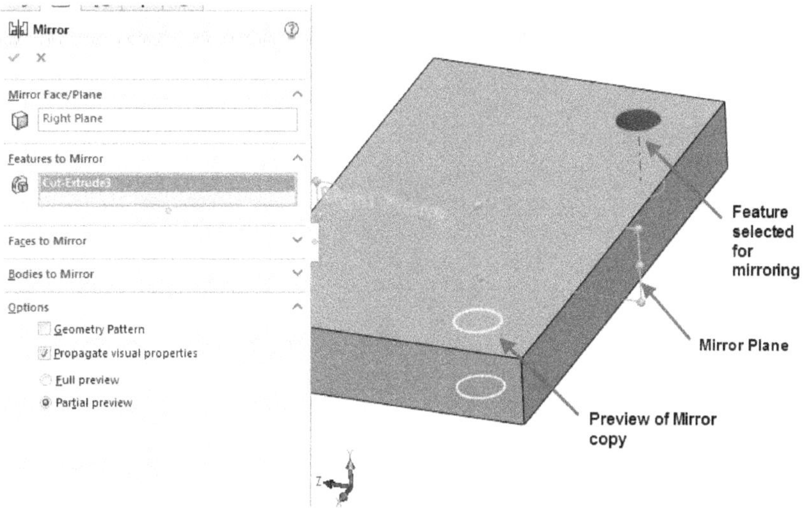

Figure-27. Preview of mirror

If you want to mirror a complete body with respect to mirror plane as in the above figure, then follow the steps given next.

- Click on the **Mirror** tool. The **Mirror PropertyManager** will display.
- Expand the **Bodies to Mirror** rollout and click on the bodies in the viewport that you want to mirror.
- Click in the **Mirror Face/Plane** box in the **Mirror Face/Plane** rollout of the **ProperyManager** and select the mirror plane. Preview of mirror will display; refer to Figure-28. Make sure that you clear the **Merge solids** check box before creating the mirror copy.
- Click on the **OK** button to create the mirror copy.

Figure-28. Preview of body mirror

RIB

The **Rib** tool is used to create support in the structures to increase their strength. You can find use of rib in various stands that are fastened to the wall or in the building columns. The procedure to create rib feature is given next.

- Click on the **Rib** tool from the **Ribbon**. You are asked to select a sketch or a sketching plane.

Note that the sketch should be created in such a way that its projection is within solid faces of the model; refer to Figure-29. For creating such sketch, you might need to create reference planes.

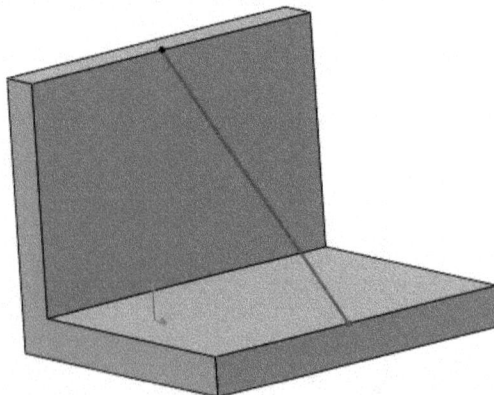

Figure-29. Sketch for rib

- Create a sketch or select the existing one. Preview of the rib feature will display. Also, the options in the **PropertyManager** will be modified as per selection; refer to Figure-30.

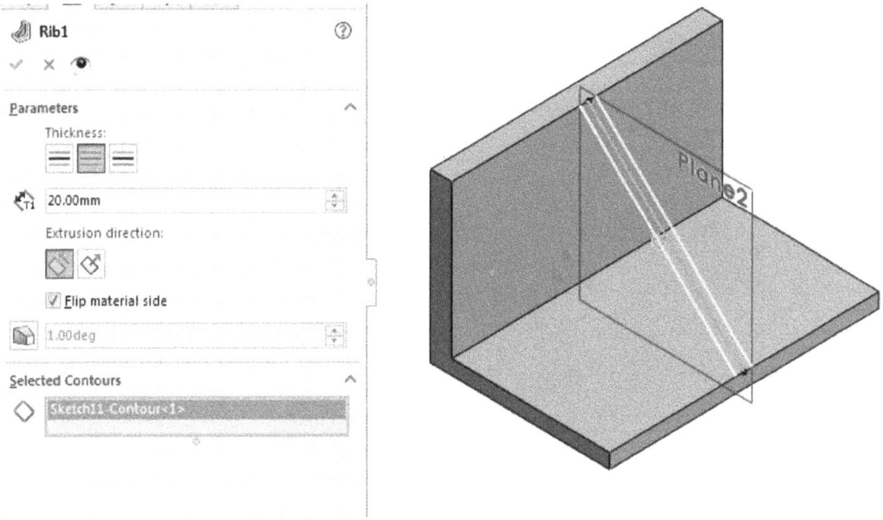

Figure-30. Preview of rib feature

- Specify the thickness of rib feature and the draft angle in the corresponding spinners.
- You can flip the direction of rib by using the buttons for **Extrusion direction** in the **PropertyManager**. Note that the direction should be in such a way that the rib feature terminates by solid faces; refer to Figure-30.
- You can also change the thickness side by using the three buttons given for **Thickness**.
- Click on the **OK** button to create the rib feature.

DRAFT

The **Draft** tool is used to apply taper to the faces of a solid model. This tool is mainly useful when you are designing components for molding or casting. **Draft** tool applies taper on the faces and this taper allows easy and safe ejection of part from the dies. The procedure to use this tool is given next.

- Click on the **Draft** tool from the **Ribbon**. The **DraftXpert PropertyManager** will display; refer to Figure-31.

- Click on the **Manual** button at the top in the **PropertyManager**. The **Draft PropertyManager** will display; refer to Figure-32.

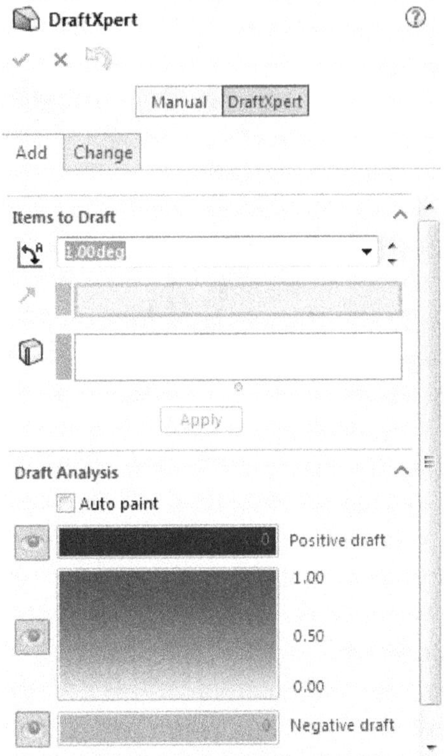

Figure-31. DraftXpert PropertyManager

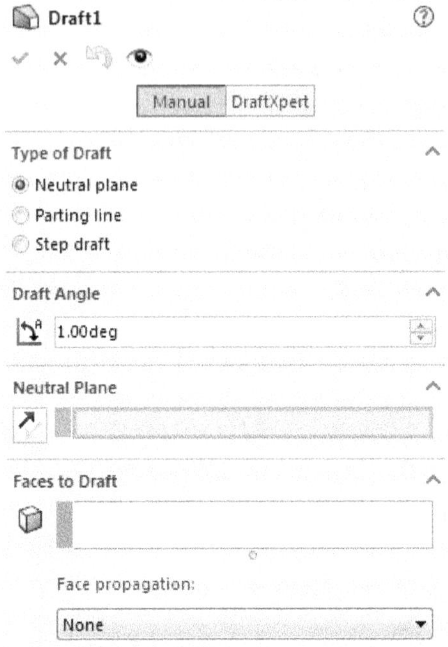

Figure-32. Draft PropertyManager

- Make sure that the **Neutral Plane** radio button is selected and then select a face with respect to which you want to measure all the draft angles. You can flip the direction of draft by using the **Flip** button adjacent to the **Neutral Plane** box.
- On selecting the face, you are asked to select the walls on which you want to apply the draft.
- Change the draft angle by using the spinner in the **Draft Angle** rollout.
- Click on the **Detail Preview** button to check the preview of the draft. Click again on the button to exit the preview; refer to Figure-33.
- Click on **OK** button to create the draft.

We will learn about other options later in the chapter **Mold Tools**.

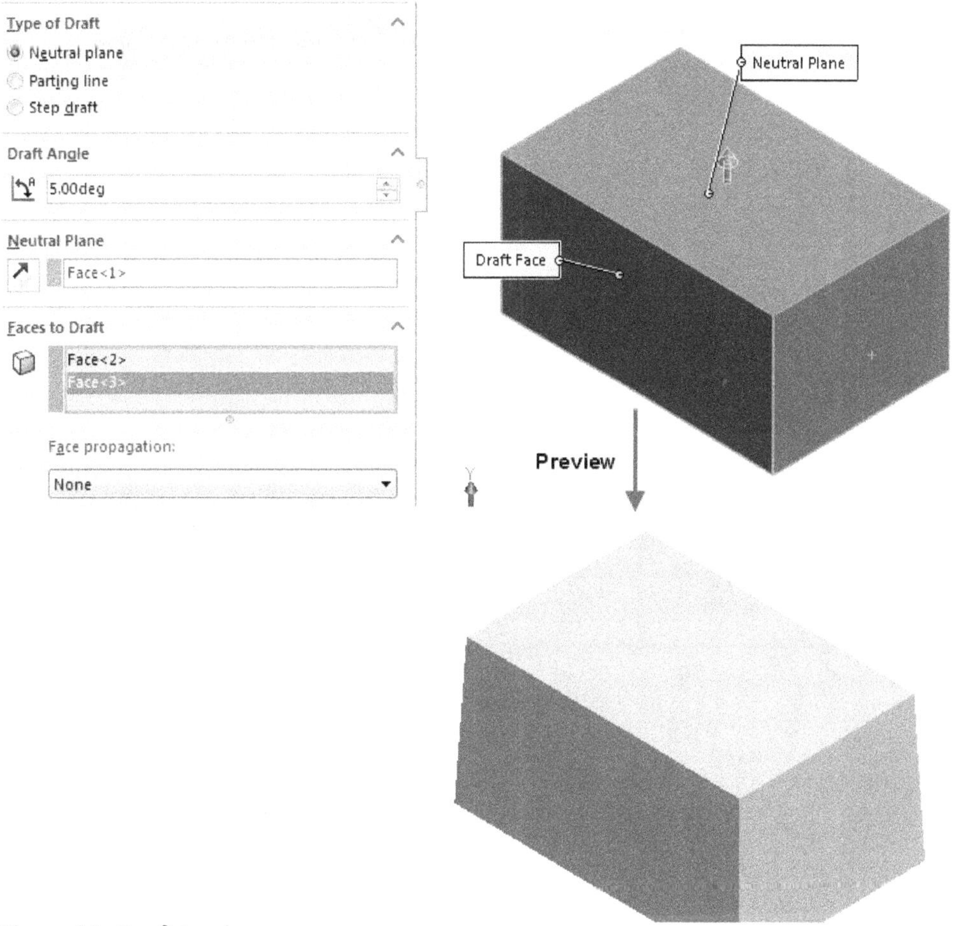

Figure-33. Draft preview

SHELL

The **Shell** tool is used to make a solid part hollow and remove one or more faces. The procedure to use this tool is given next.

- Click on the **Shell** tool from the **Ribbon**. The **Shell PropertyManager** will display; refer to Figure-34.
- Specify the desired thickness in the spinner.
- Click on the **Show preview** check box to display the preview.
- Select a face that you want to remove.
- You can flip the direction of shell by clicking on the **Shell outward** check box.
- Click on the **OK** button to create the feature.

Figure-35 shows the preview of the shell feature and output of shell.

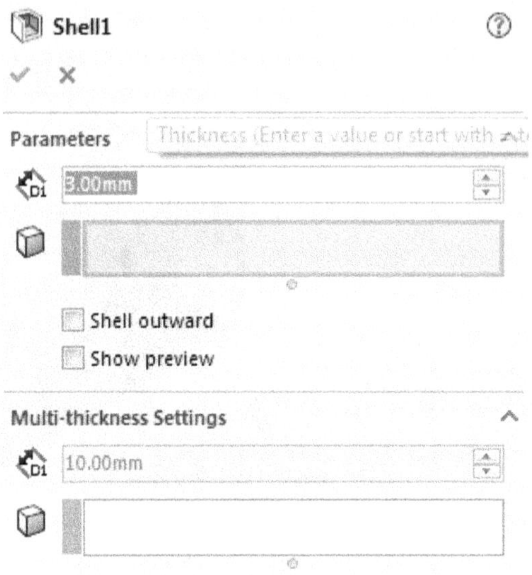

Figure-34. Shell PropertyManager

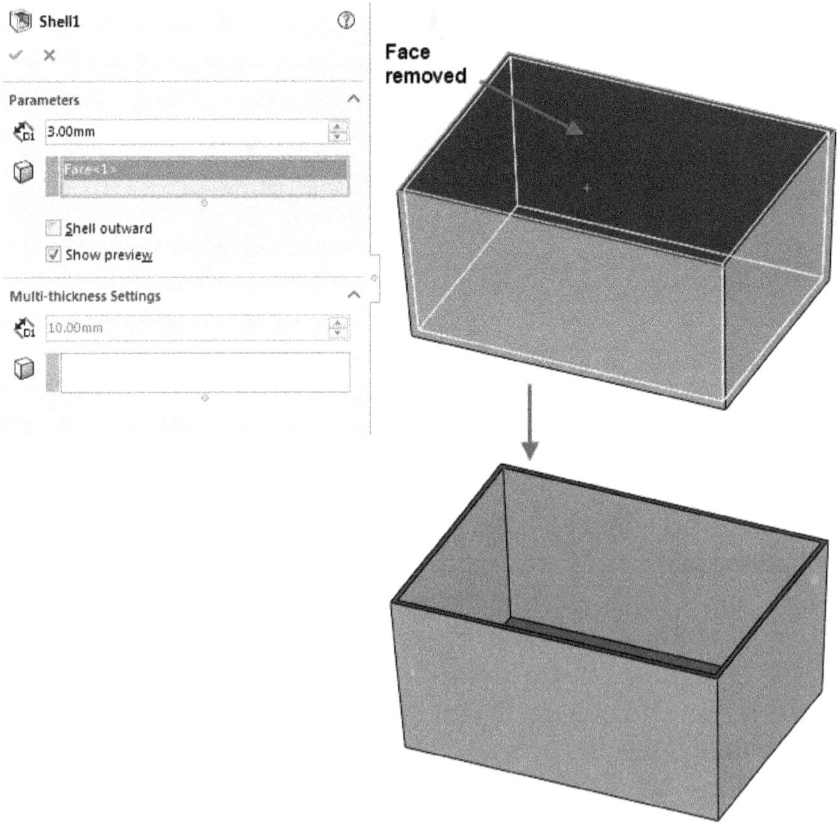

Figure-35. Preview and output of shell

WRAP

The **Wrap** tool is used to wrap text or any curve/curves on the cylindrical faces. Before using this tool, create a sketch that you want to wrap on a plane parallel to the wrapping face; refer to Figure-36. The steps to create wrap are given next.

Figure-36. Sketch for wrap

• Select the sketch and then click on the **Wrap** tool. The **Wrap PropertyManager** will display as shown in Figure-37.

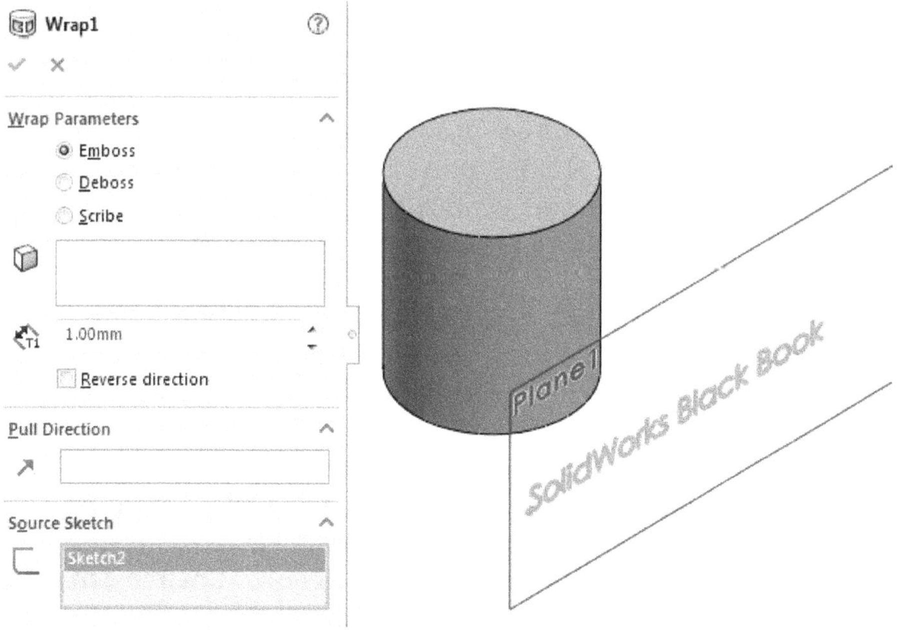

Figure-37. Wrap PropertyManager and sketch

- Select the desired radio button for embossing, engraving (debossing), or scribing.
- Select the cylindrical face on which you want to emboss or engrave the sketch. Preview will display.
- Set the thickness value in the spinner.
- Click **OK** button to create the feature. Figure-38 shows a sketch embossed on the cylindrical face.

Figure-38. Output of wrap

INTERSECT

The **Intersect** tool is used to separate the volume create by intersection of solids, surfaces or a combination of both. This tool can be very helpful for creating mold tools manually. The steps to use this tool are given next.

- Click on the **Intersect** tool from the **Ribbon**. The **Intersect PropertyManager** will display as shown in Figure-39.
- Select the surfaces/solids from which you want to extract the intersecting volume.
- There are two type of regions formed while intersecting solids; Intersecting regions and internal regions formed during intersection. There are two radio buttons in the **PropertyManager** to find out these regions; **Create intersecting regions** and **Create internal regions**. You can select the

Create both radio button to include both regions in the list.

- Click on the **Intersect** button from the **PropertyManager**, the preview will be displayed in the viewport. Also, the region that you can exclude will be displayed in the **Regions to Exclude** rollout; refer to Figure-40.

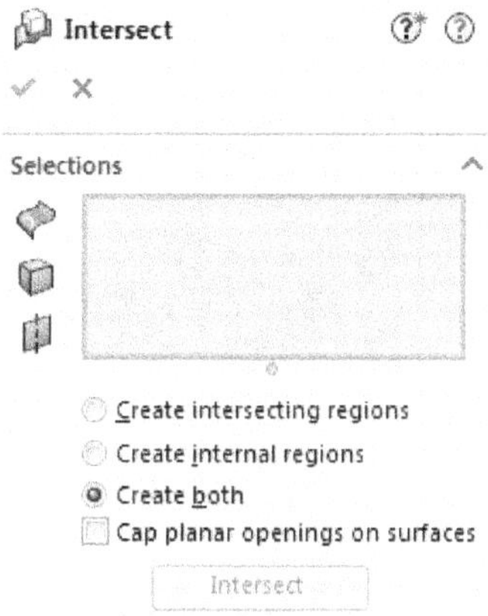

Figure-39. Intersect PropertyManager

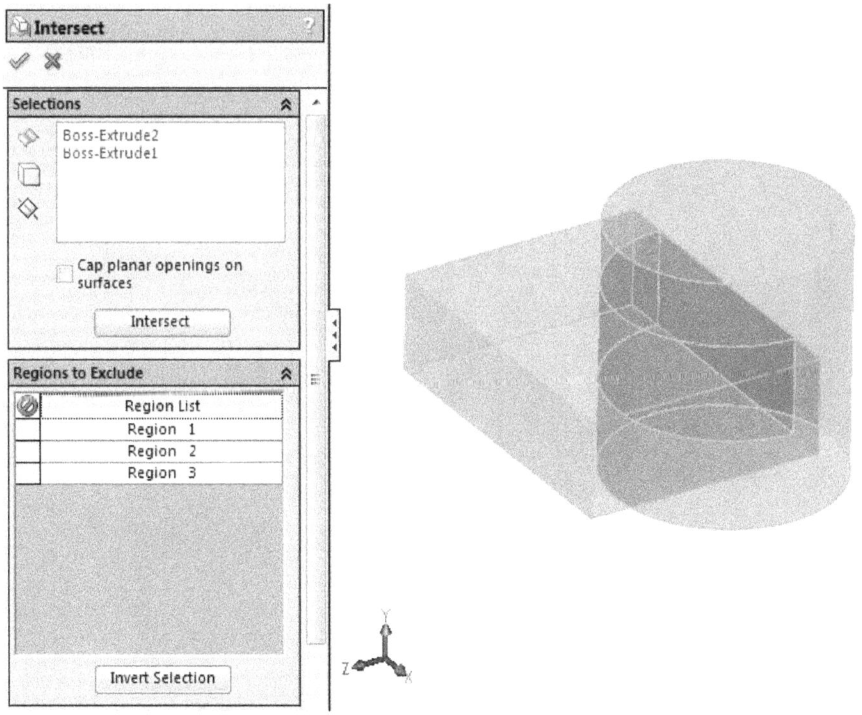

Figure-40. Preview of intersection

- Click on the check boxes of feature that you want to exclude from the **Regions to Exclude** rollout or select the portion of model from the viewport.
- Click on the **OK** button from create the intersection portion. Figure-41 shows the output of the intersection.

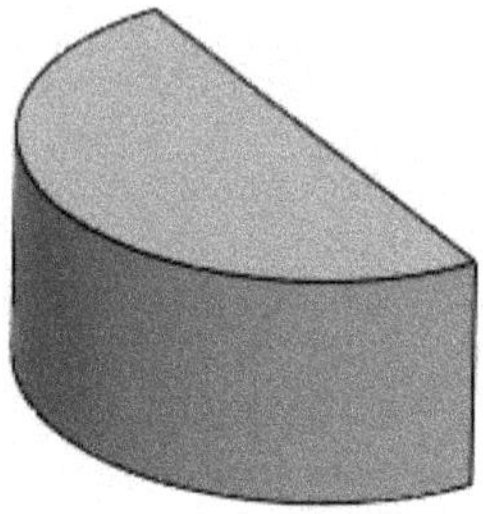

Figure-41. Output of intersection

Practical 1

Create the model(isometric view) as shown in Figure-42. The views of the model with dimensions are given in Figure-43.

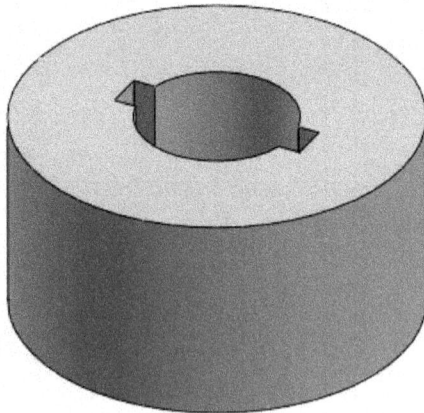

Figure-42. Practical1 model

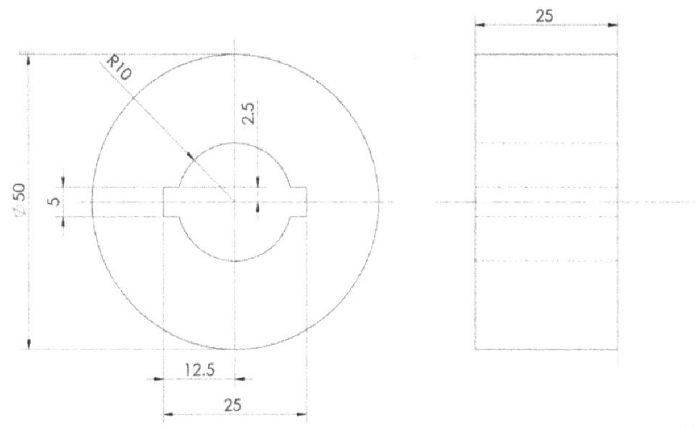

Figure-43. Views for Practical 1

Before we start working on the Practical, it is important to understand two terms; first angle projection and third angle projection. These are the standards of placing views in the engineering drawing. The views placed in the above figure are using third angle projection. In first angle projection, the top view of model is placed below the front view and right side view is placed at left of the front view. You will learn more about projection in chapter related to drafting.

Starting SolidWorks Modeling environment and creating Extrude feature

- Double-click on the SolidWorks icon from desktop if you have not started SolidWorks.
- Click on the **Features** tab from the **Ribbon** if not selected.

From the isometric view as well as from the other views, we can judge that this model can be easily created by extruding the sketch.

- Click on the **Extrude Boss/Base** tool. The **Extrude PropertyManager** will display.
- Select the **Top** plane from the **FeatureManager Design Tree** or from the viewport.

We will draw the sketch on top plane to get the isometric view as shown in Figure-42.

- Click on the **Normal To** button from the **View Orientation** drop-down in the **Heads-up View toolbar** if the plane is not parallel to screen.
- Click on the **Circle** tool from the **Ribbon**, select the **Diameter Dimensions** check box.
- Click at the coordinate system to place center of the circle.
- Drag the cursor and enter the value as **50** in the **Dimension box**.
- Again click at the coordinate system and draw circle of diameter **20**.
- Click on **OK** button from the **Circle PropertyManager**.
- Select the **Center Rectangle** tool from the **Rectangle** drop-down in the **Ribbon** and click at the center of the circles.
- Drag the cursor and specify the dimension of rectangle as **5** and **25** for height and width respectively. Click **OK** from the **PropertyManager**. The sketch after performing the above steps will display as shown in Figure-44.

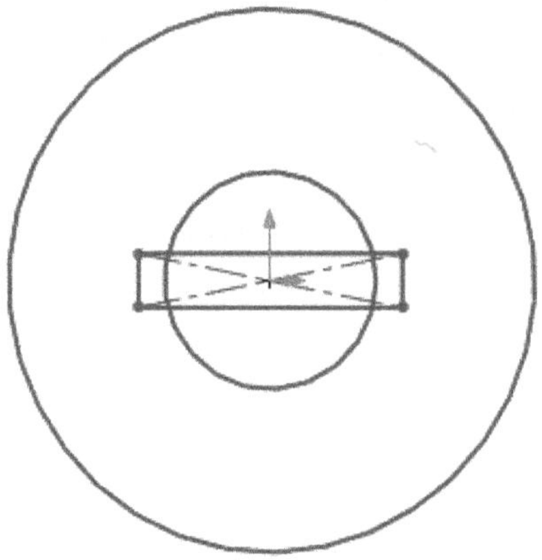

Figure-44. Sketch after creating circles and rectangle

- Select the **Trim Entities** tool from the **Ribbon** and trim the entities in such a way that the sketch is displayed as shown in Figure-45.

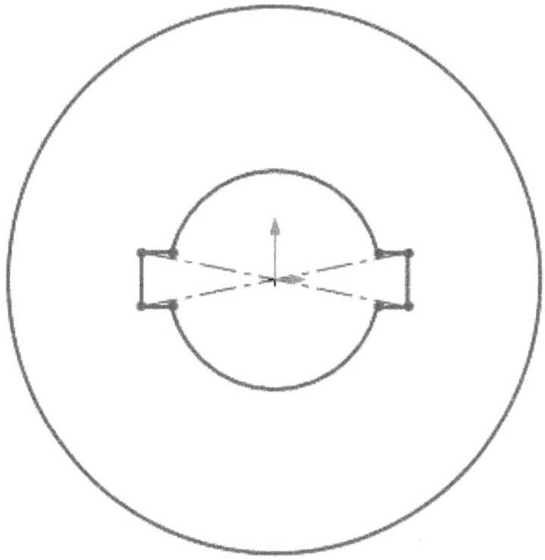

Figure-45. Sketch after trimming

- Click on the **Smart Dimension** tool and dimension the sketch. Refer to Figure-46.

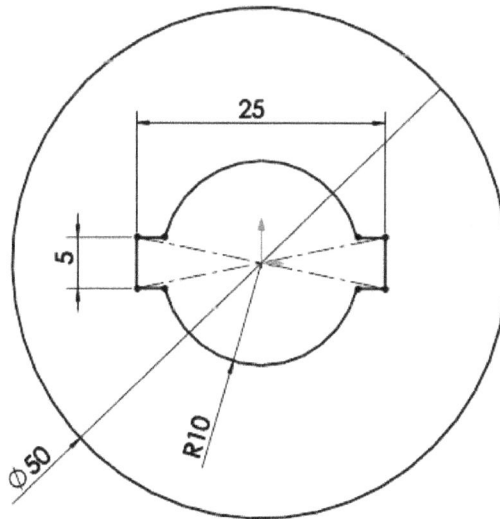

Figure-46. Dimensioned sketch

- Click on the **Exit Sketch** button. Preview of the Extrude feature will display as shown in Figure-47.

Figure-47. Preview of extrude

- Specify the height of extrusion as **25** and then click on the **OK** button from the **PropertyManager**. The model will be created as in Figure-42.

Practical 2

Create the model(isometric view) as shown in Figure-48. The dimensions and views are given in Figure-49.

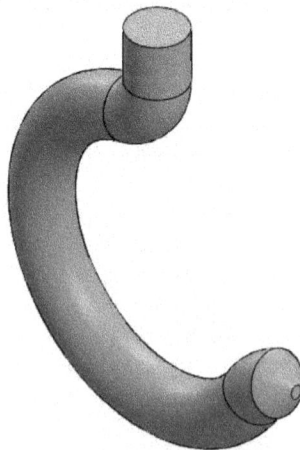

Figure-48. Model for Practical2

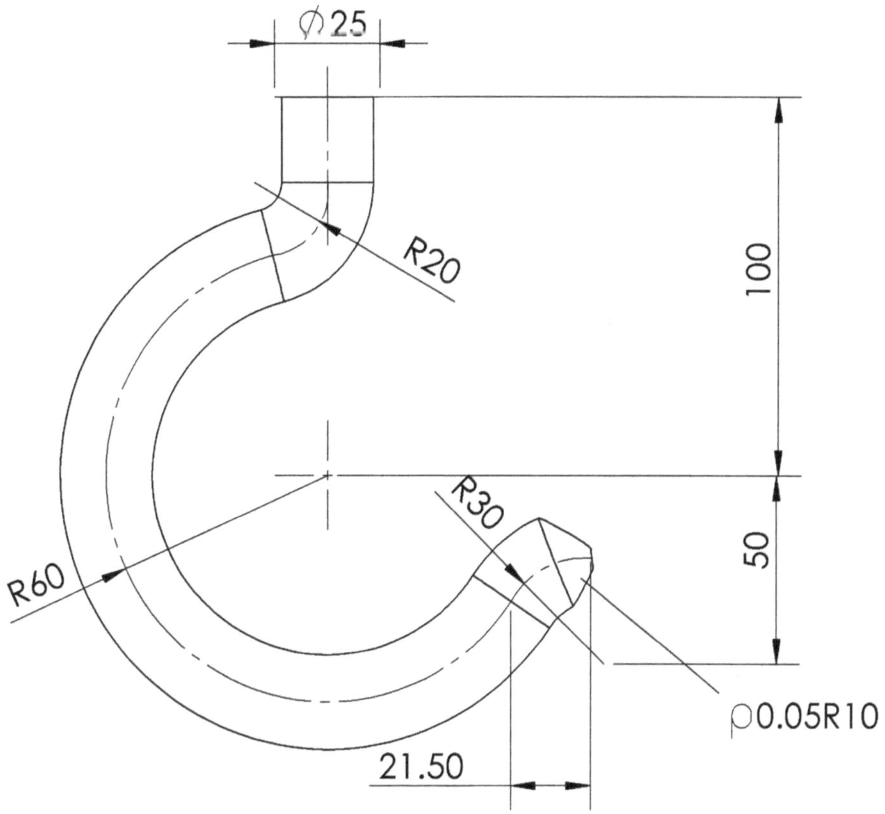

Figure-49. Practical2 drawing view

Start the SolidWorks if not started and open the modeling environment as explained in previous practical.

You can find out from the model that this hook can be easily created with the help of **Swept Boss/Base** tool. But before we use that tool, we must have sketches for path as well as section. The steps to create them are as follows:

Creating Sketches for Hook

- Start a new sketch on Front Plane.
- Click on the **Circle** tool and create the circle of diameter **120** taking coordinate system as center.
- Draw a straight line starting from top quadrant point and having length of approximately **40.**
- Draw a three point arc in the bottom area of the circle; refer to Figure-50. (It should look like the one given in figure. Accuracy is no required.)

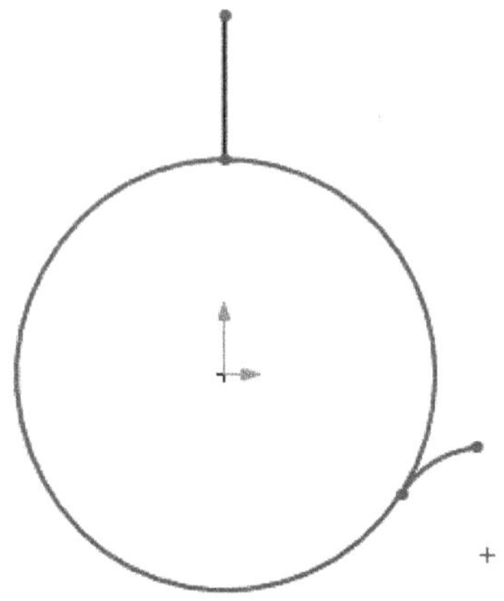

Figure-50. Sketch after creating three point arc

- Click on the **Trim Entities** tool and trim the portion between the straight line and the arc; refer to Figure-51.

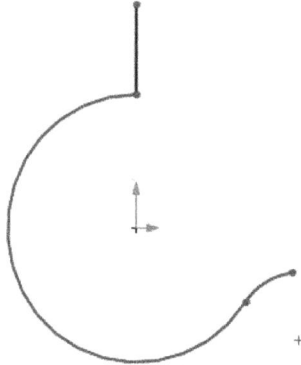

Figure-51. Sketch after trimming circle

- Click on the **Sketch Fillet** tool and apply fillet between the straight line and circle. Specify the fillet radius as **20**.
- Now, dimension the sketch as shown in Figure-52.

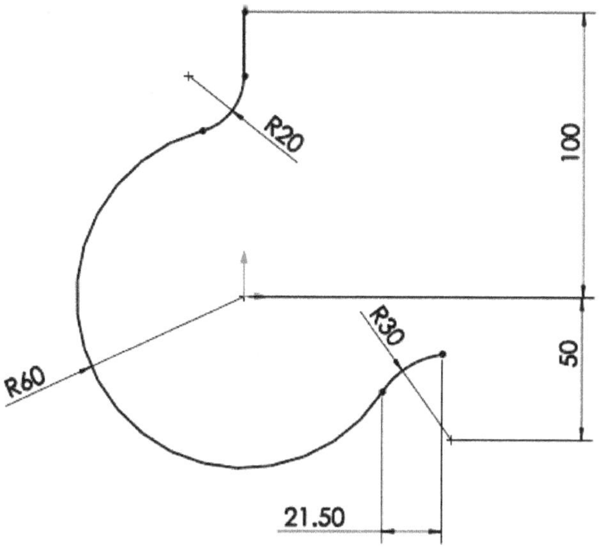

Figure-52. Dimension sketch of hook

- Exit the sketch by selecting **Exit Sketch** button from the **Ribbon**.

Above we have created the sketch of the path. Now, we will create the section sketch.

- Click on the **Plane** tool from the **Reference** drop-down in the **Features** tab of the **Ribbon**. The **Plane PropertyManager** will display as shown in Figure-53.
- Click on the top point of the line and select the **Top Plane**. Preview of the plane will display; refer to Figure-54.
- Click on the **OK** button from the **Plane PropertyManager** to create the plane.
- Select the newly created plane and select the **Sketch** button from the **Sketch** tab of the **Ribbon**. The sketching environment will display.

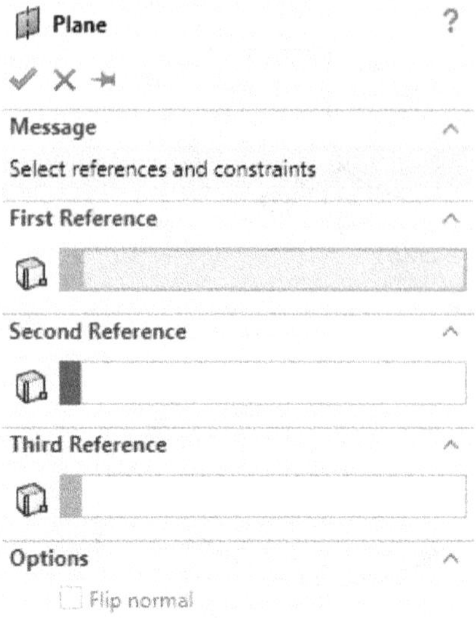

Figure-53. Plane Property-Manager

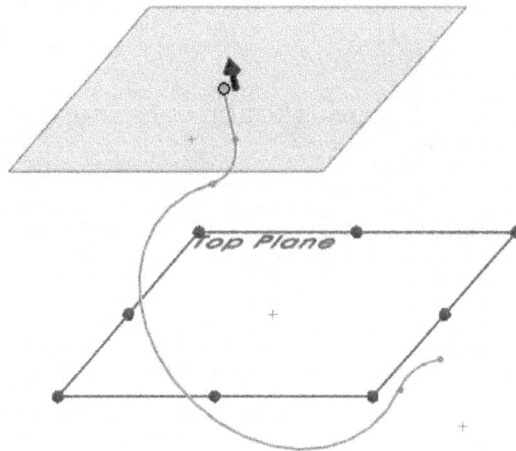

Figure-54. Preview of plane

- Select the **Normal To** button from the **View Orientation** box. The sketching plane will become parallel to the viewport.
- Create a circle of diameter **25** taking end point of line as center; refer to Figure-55.

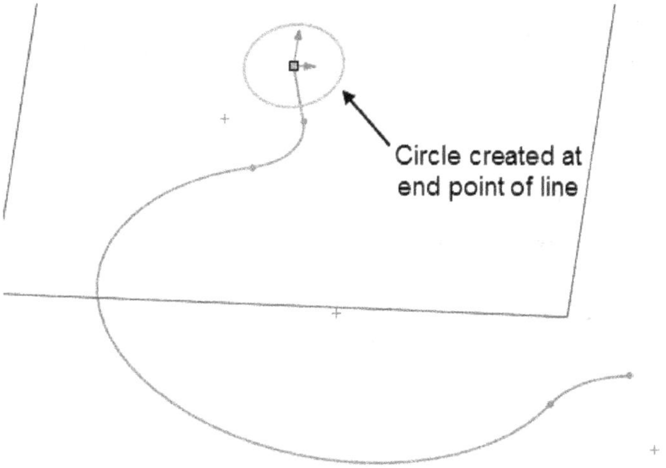

Figure-55. Circle to be created

Creating the Swept Boss/Base feature

Now, we have all the sketches to create the hook. The steps to create the swept boss/base feature using these sketches is given next.

- Click on the **Swept Boss/Base** tool from the **Ribbon**. The **Sweep PropertyManager** will display.
- Select the circle created for section and then select sketch created for the profile.
- Click on the **OK** button from the **PropertyManager**. The swept base feature will be created; refer to Figure-56.

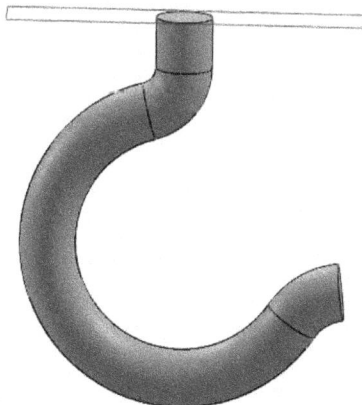

Figure-56. Swept feature created

Applying Conic fillet at the end

As we all know hooks doesn't end with sharp edges. So, we need to apply fillet at the end. Steps to do so are given next.

- Click on the **Fillet** tool from the **Fillet** drop-down in the **Ribbon**. The **Fillet PropertyManager** will display.
- Select the edge of the end and specify the parameters in the **PropertyManager** as shown in Figure-57.
- Click on the **OK** button from the **PropertyManager** to create the feature. The model will be displayed as shown in Figure-58.

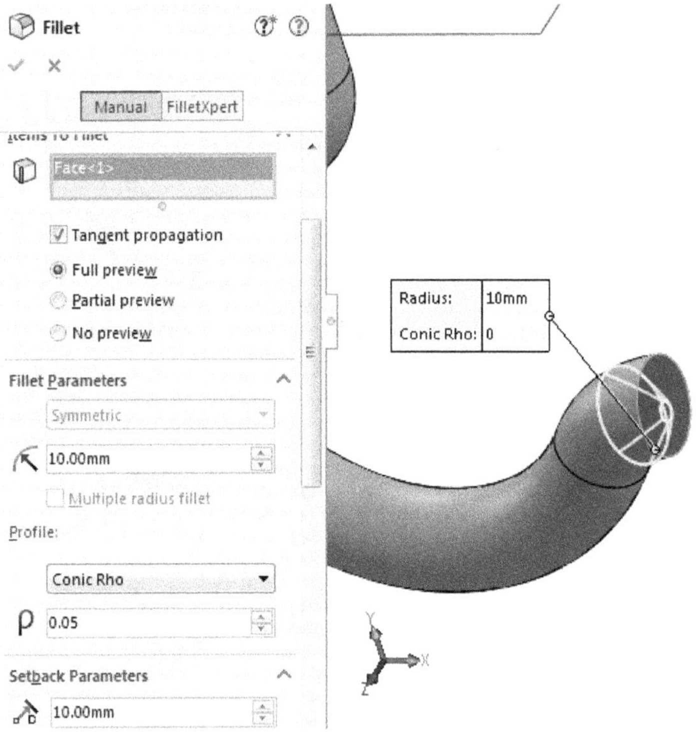

Figure-57. Fillet at the end edge

Figure-58. Final model

Practical 3

Create the model(isometric view) as shown in Figure-59. The dimensions of the model are given in Figure-60.

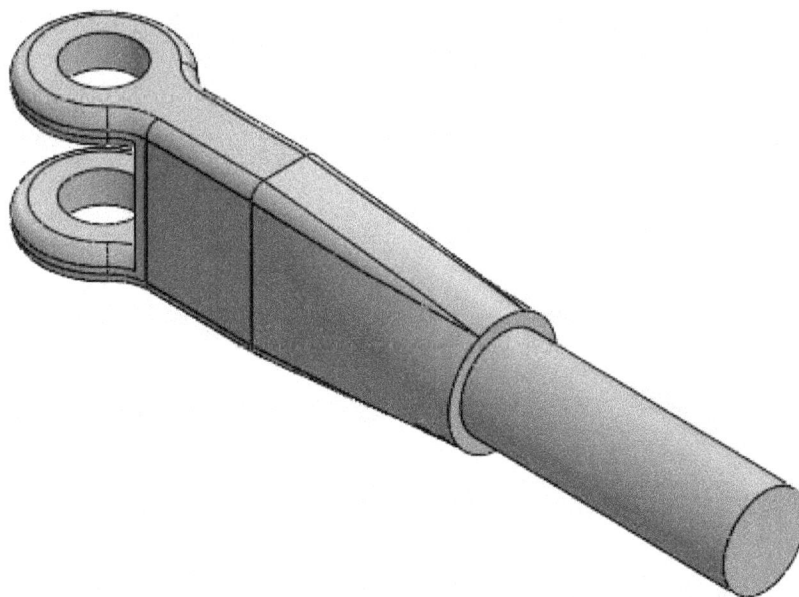

Figure-59. Model for Practical3

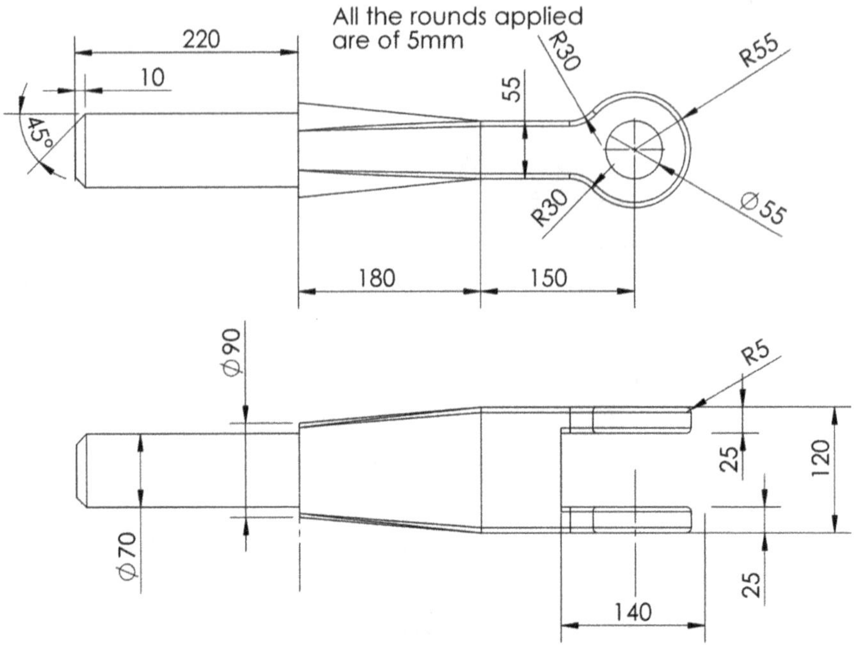

Figure-60. Practical3 drawing views

Creating first extrude feature

- Click on the **Extrude Boss/Base** tool from the **Ribbon**. The **Boss-Extrude PropertyManager** will display.
- Select the Top Plane from the viewport and create the sketch as shown in Figure-61.

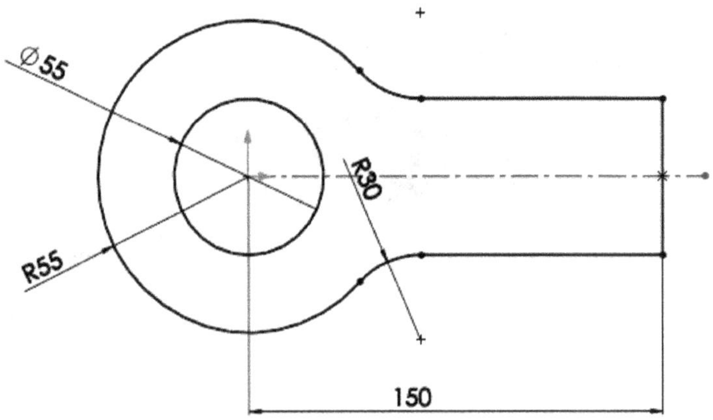

Figure-61. Sketch created at top plane

- Exit the sketch and extrude it to the height of **120**.

Creating loft feature

Before creating loft feature, we must have at least two sketch sections. First, we will create a plane at a distance of **180** from the vertical flat face of the model and then, we will create sketches and loft feature successively.

- Click on the **Plane** tool from the **Reference** drop-down and select the vertical flat face of the model.
- Specify the offset distance as **180**; refer to Figure-62.
- Click on **OK** button to create the plane.

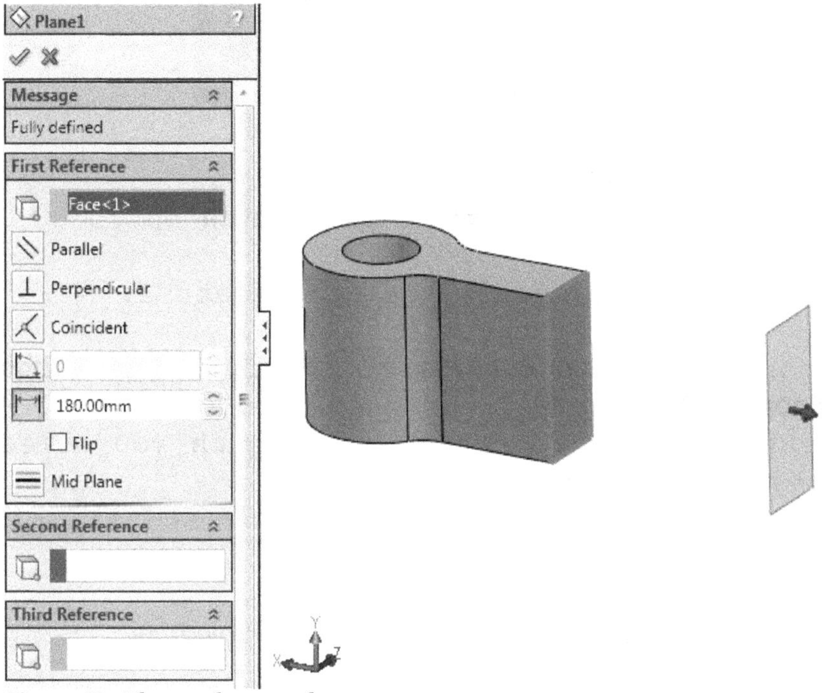

Figure-62. Plane to be created

- Select this plane and create the sketch as shown in Figure-63.

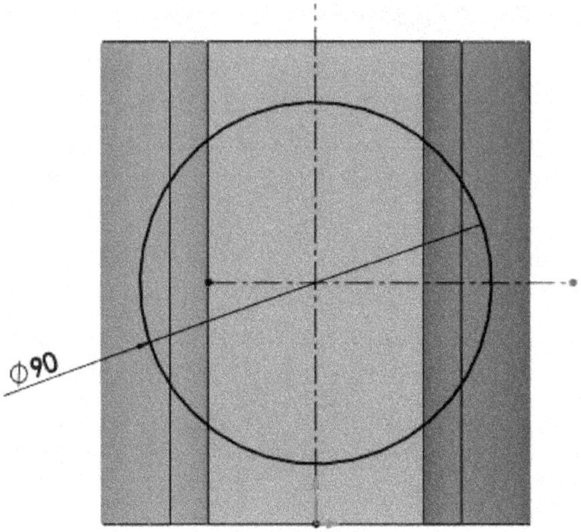

Figure-63. Sketch to created

- Exit the sketch by clicking **Exit Sketch** button.

Now, we are ready to create the Loft feature.

- Click on the **Lofted Boss/Base** tool from the **Ribbon**. The **Loft PropertyManager** will be displayed.
- Select the flat face of the model which you selected to create plane; refer to Figure-64.

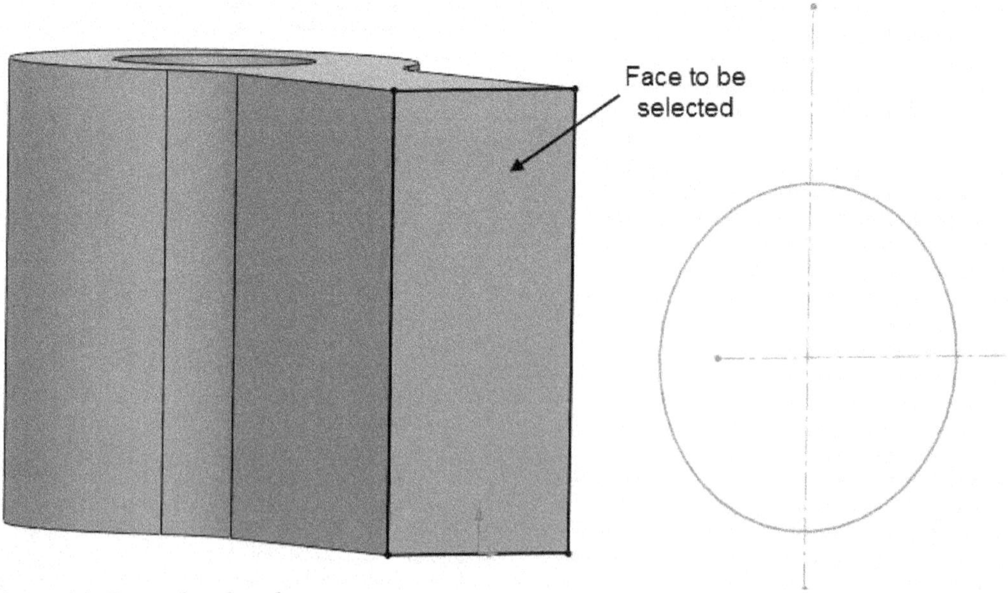

Face to be selected

Figure-64. Face to be selected

- Select the sketch created. Preview of the loft feature will be displayed; refer to Figure-64.
- Drag the green handles on the curves to manipulate the shape of lofted feature if you get it twisted.
- Click on the **OK** button from the **PropertyManager**.

Figure-65. Lofted feature to be created

Creating extrude feature

- Select the flat face of the lofted feature and click on the **Extruded Boss/Base** tool. The sketching environment will open.
- Create a circle of diameter **70** at the center and exit the sketch. Preview of extruded feature will display.
- Set the extrude depth as **220**. The preview of the model will display as shown in Figure-66.

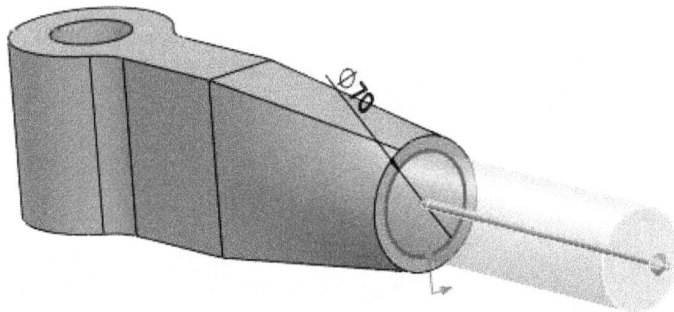

Figure-66. Preview of model after extrusion

- Click **OK** from the **PropertyManager**.

Now, we need to create the cut feature to give front shape to model.

Creating extrude cut feature

After looking at the model, we can find that the front cut can be easily created by using the **Extruded Cut** tool.

- Click on the **Extruded Cut** tool from the **Ribbon** and select the vertical plane passing through the model i.e. Front Plane. The sketching environment will be displayed.
- Create the sketch as shown in Figure-67.

Figure-67. Sketch for cut feature

- Exit the sketch. The preview of extruded cut will display.
- Select the **Mid Plane** option from the **End Condition** drop-down and specify the height as **200**.
- Click on the **OK** button. The cut feature will be created and the model will display as shown in Figure-68.

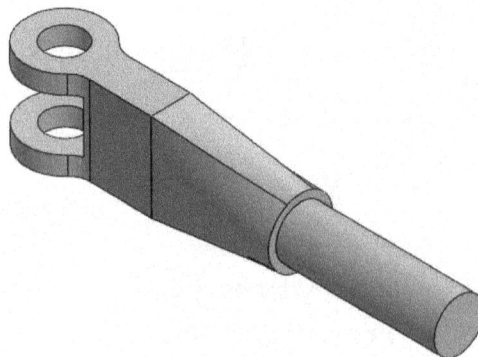

Figure-68. Model after cut feature

Creating Fillets and Chamfers

Click on the **Fillet** tool, specify the radius as **5,** and select all the edges on which you want to create the fillet. Similarly, click on the **Chamfer** tool, specify the parameters as per the drawing and select the edges to apply the chamfer. The model after applying the fillet and chamfers is displayed as shown in Figure-69.

Figure-69. Final Model for Practical3

PRACTICE 1

Create the model as shown in Figure-70. The dimensions are given in Figure-71.

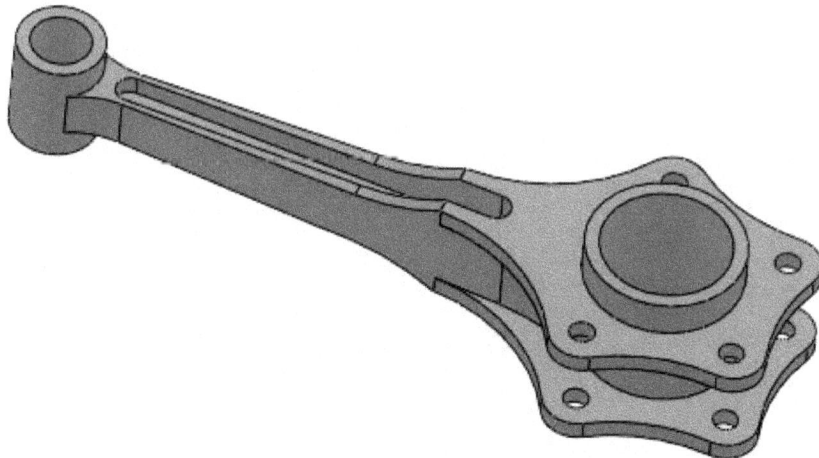

Figure-70. Practice1

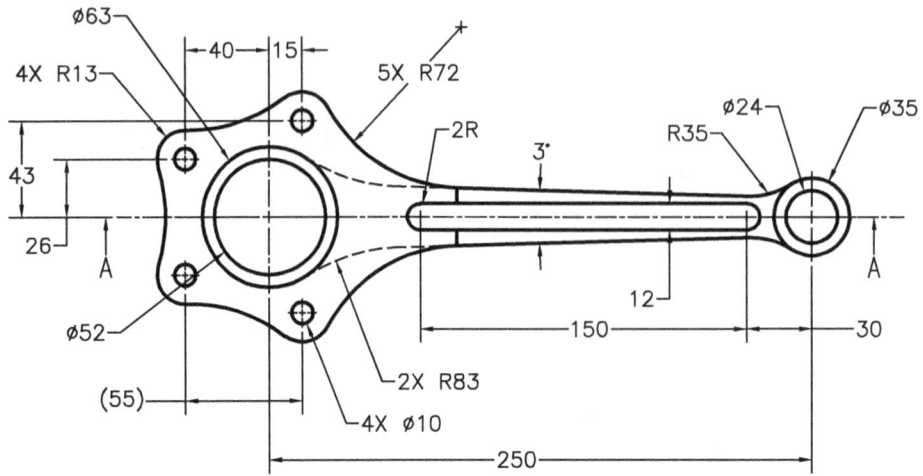

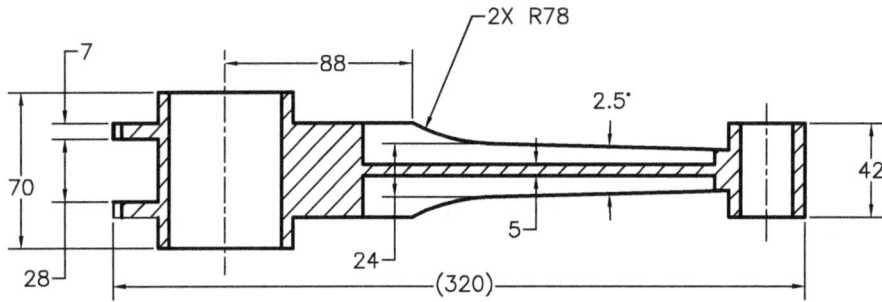

Figure-71. Dimensions of the Practice1 model

PRACTICE 2

Create the model using the drawings shown in Figure-72.

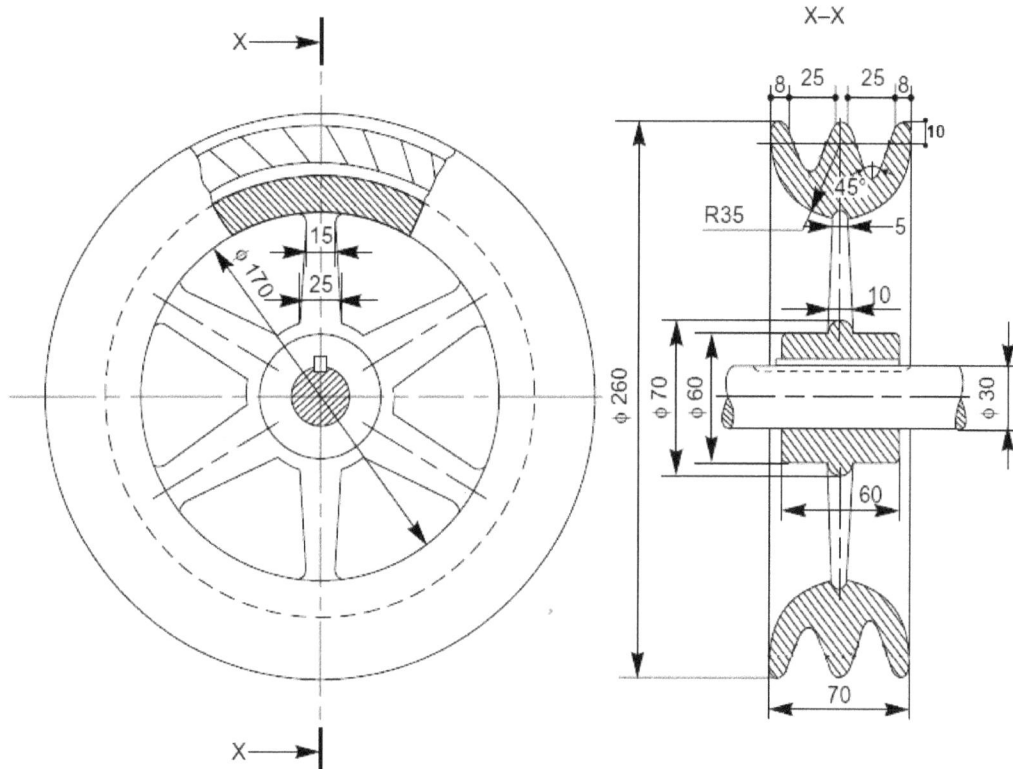

Figure-72. Rope Pulley

PRACTICE 3

Create the model as shown in Figure-73. Dimensions are given in Figure-74. Assume the missing dimensions.

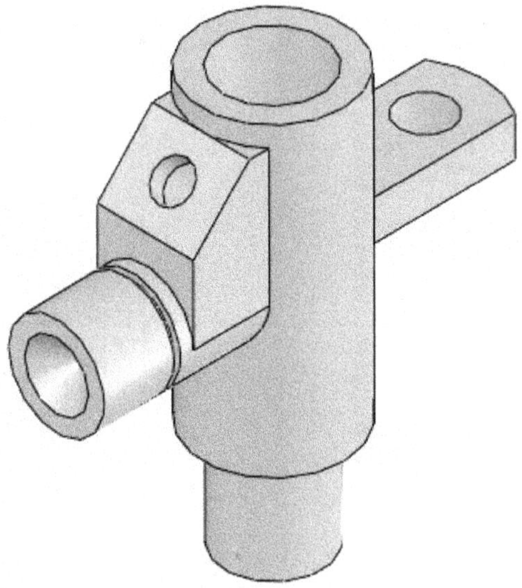

Figure-73. Practice3 model

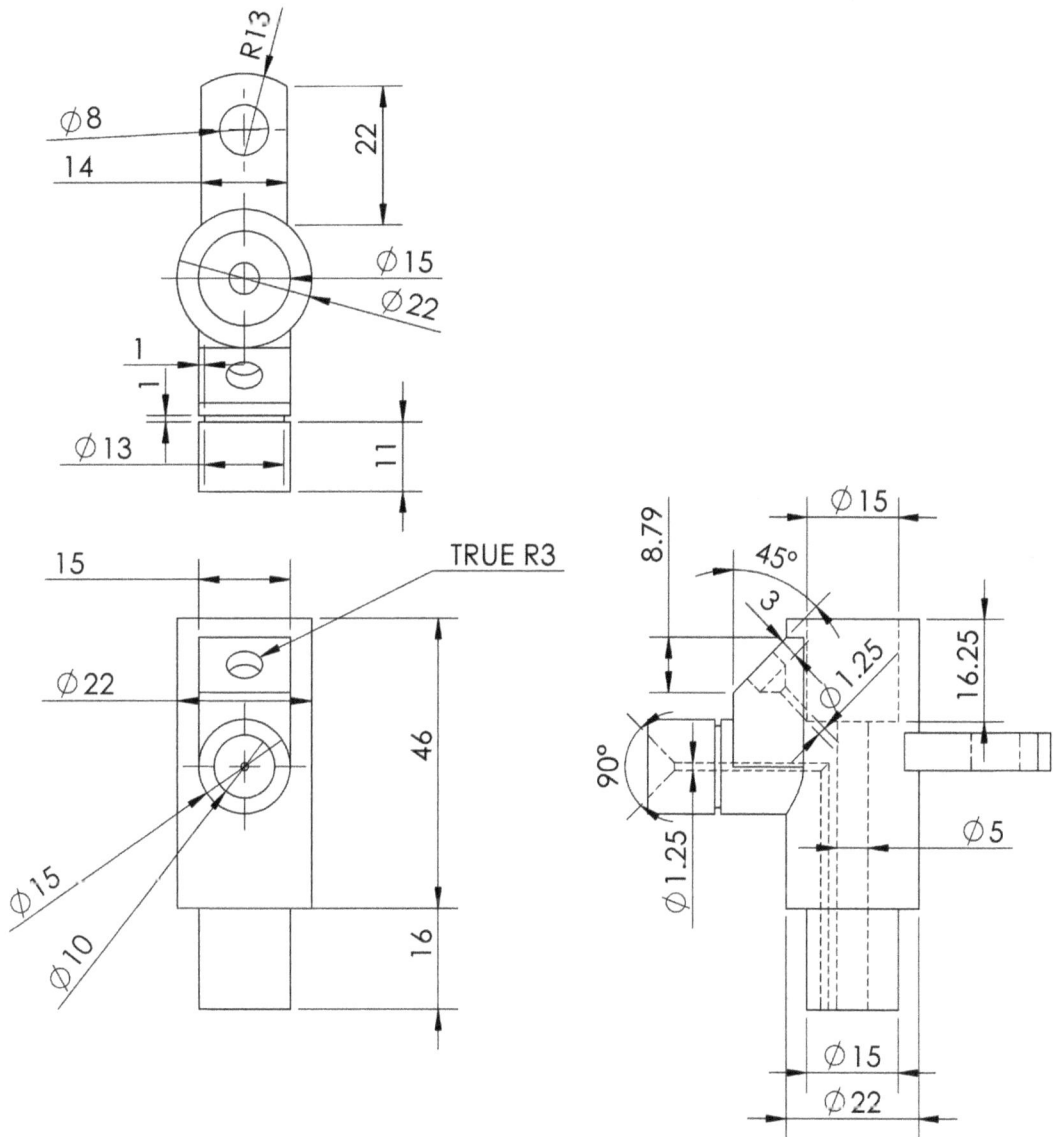

Figure-74. Practice3

PRACTICE 4

Create the model by using the dimensions given in Figure-75.

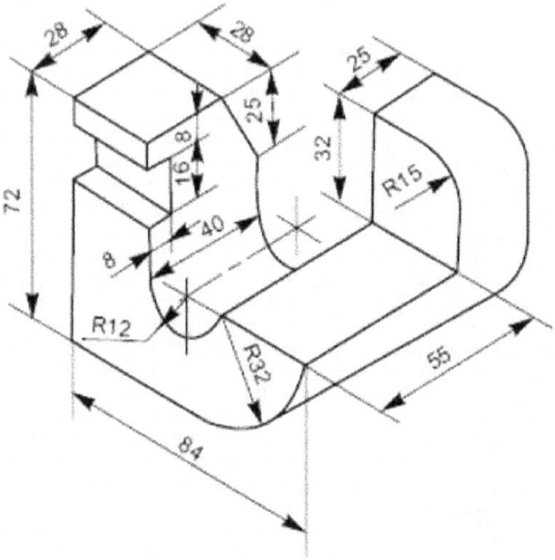

Figure-75. Practice4

PRACTICE 5

Create the model by using the dimensions given in Figure-76.

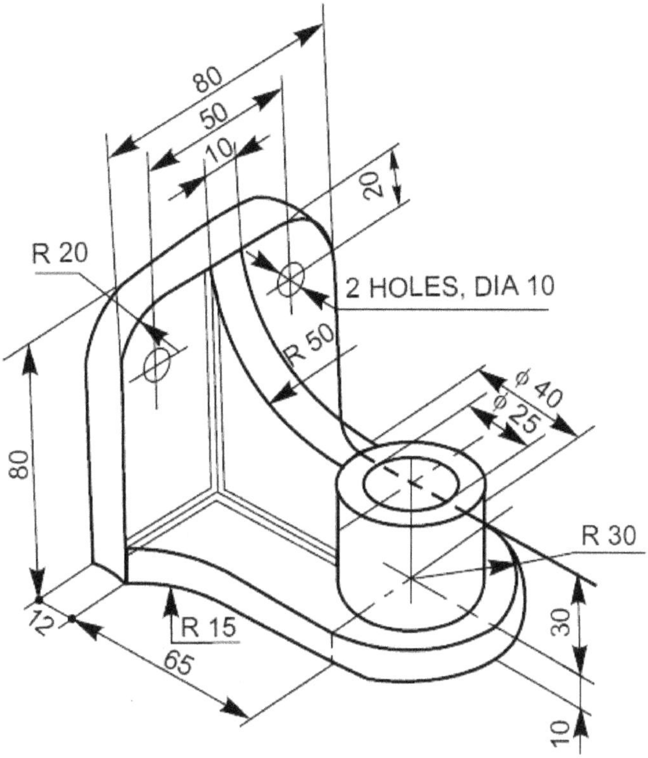

Figure-76. Practice5

PRACTICE 6

Create the model by using the dimensions given in Figure-77.

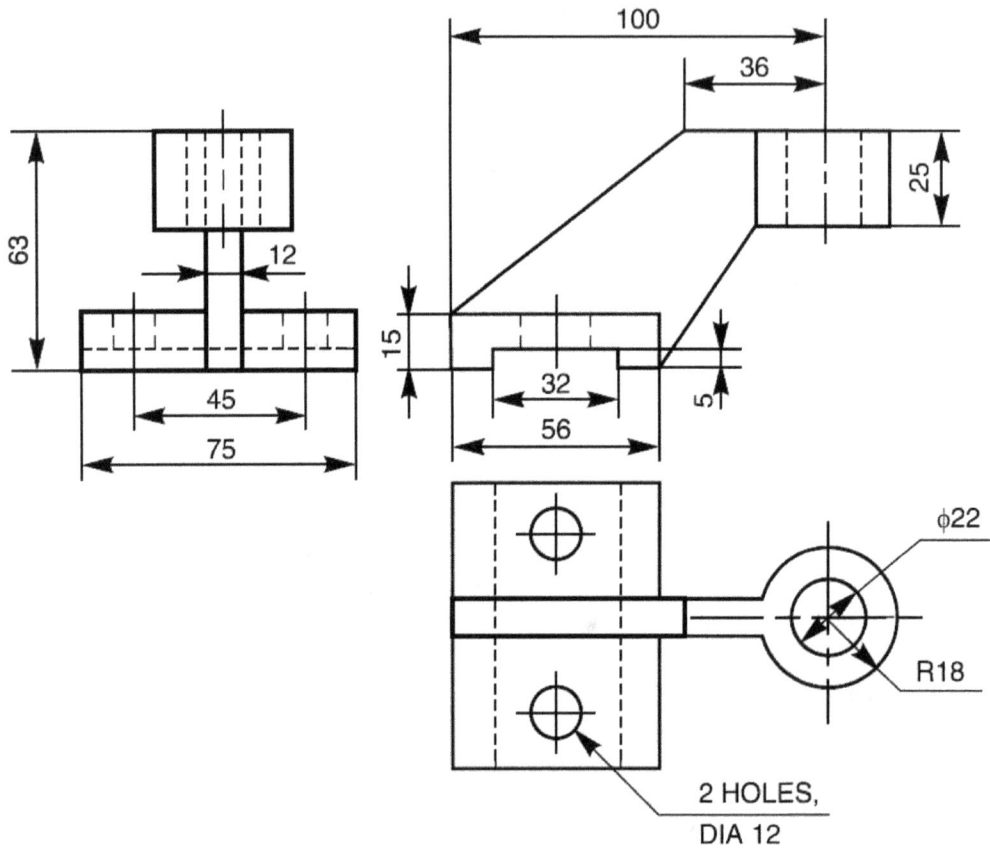

Figure-77. Practice6

SELF ASSESSMENT

Q1. The _____ tool is used to apply radius at the edges.

Q2. The _____ tool is used to apply different type of fillets in one single mode.

Q3. The _____ tool is used to bevel the edges of the model.

Q4. The _____ tool is used to create multiple instances of a solid features along the selected path.

Q5. The _____ tool is used to create multiple instances of a solid features as per the points specified in the selected sketch.

Q6. The _____ tool is used to create multiple instances of a features by filling the selected bounded region.

Q7. The _____ tool is used to create multiple instances of a features as per the coordinates specified in the table.

Q8. The _____ tool is used to create support in the structures to increase their strength.

Q9. The _____ tool is used to apply taper to the faces of a solid model.

Q10. The _____ tool is used to make a solid part hollow and remove one or more faces.

FOR STUDENT NOTES

Assembly and Motion Study

Chapter 6

Topics Covered

The major topics covered in this chapter are:

- *Inserting Components in Assembly.*
- *Assembly Constraints.*
- *Reference and Assembly features.*
- *Exploded View.*
- *Bill of Material.*
- *Motion Study.*

ASSEMBLY

In engineer's language, assembly is the combination of two or more components and these components are constrained to each other in a specified manner called assembly constraints. In SolidWorks, Assembly Design is a separate environment. To start the Assembly Design, click on the **New** button from the **Menu Bar**. The **New SolidWorks Document** dialog box will display; refer to Figure-1. Double-click on the **Assembly** button from the dialog box. The Assembly Design environment will display as shown in Figure-2. The tools related to assembly are available in the **Ribbon**. Note that in the left of the screen, the **Begin Assembly PropertyManager** is displayed. Using the options in this PropertyManager, you can start inserting components in the assembly.

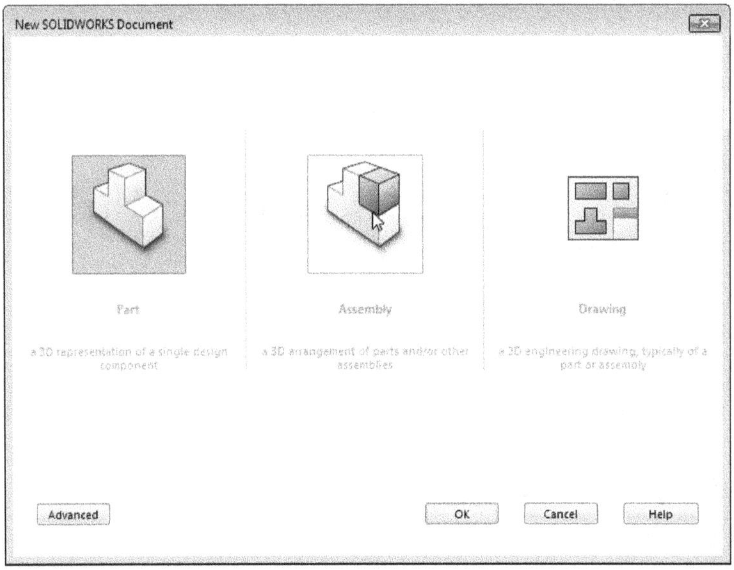

Figure-1. New SolidWorks Document dialog box

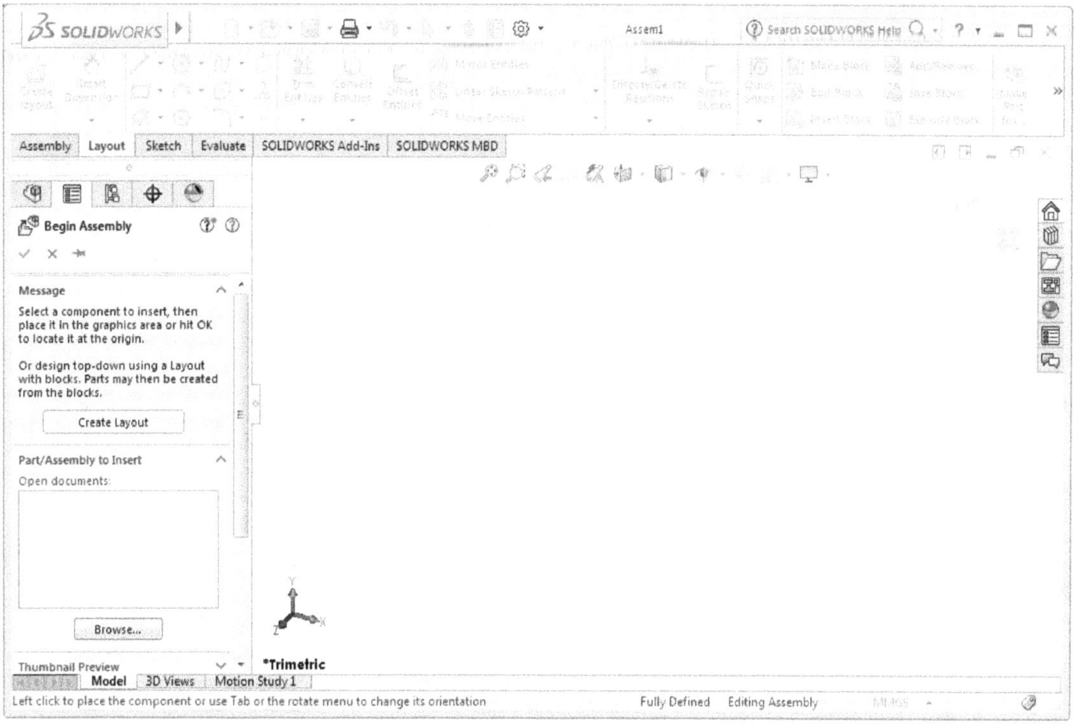

Figure-2. Assembly Design environment

INSERTING COMPONENTS IN ASSEMBLY

To insert components in the assembly follow the steps given next.

- Click on the **Browse** button from the **Begin Assembly PropertyManager** displayed at the left. The **Open** dialog box will display; refer to Figure-3.
- Browse to the location of your desired part file and double-click on the file to add it. The component will be attached to the cursor; refer to Figure-4. You can rotate the component by using the **Rotate** menu displayed.
- Click in the viewport to place the component. The component will be inserted in the assembly as a fixed part; refer to Figure-5. Note that the first component of assembly is inserted as fixed in SolidWorks and **(f)** is displayed next to its name in the **FeatureManager Design Tree**.

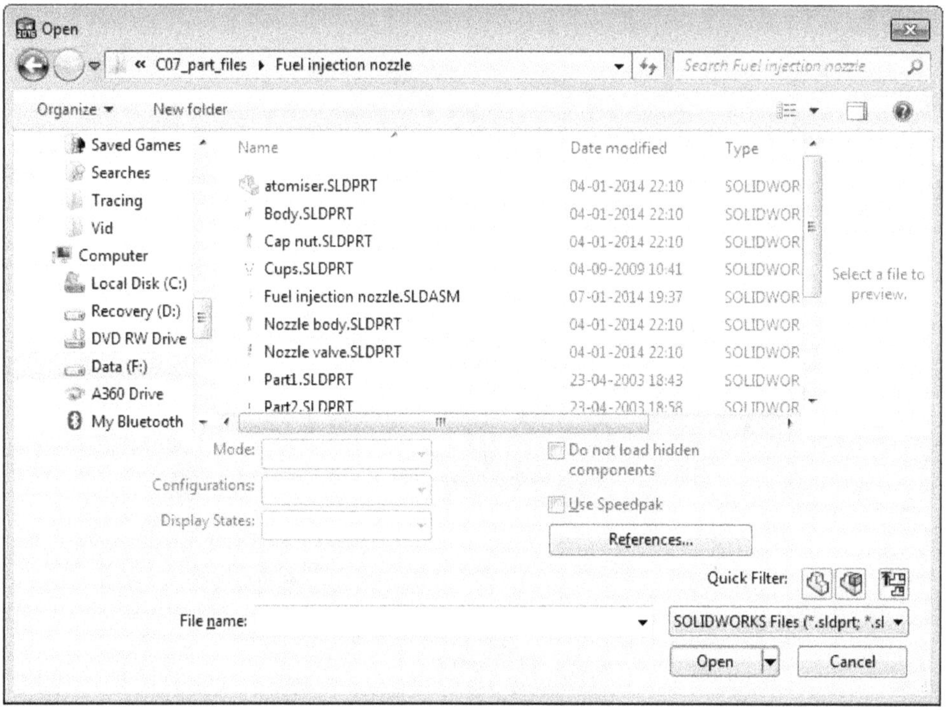

Figure-3. Open dialog box

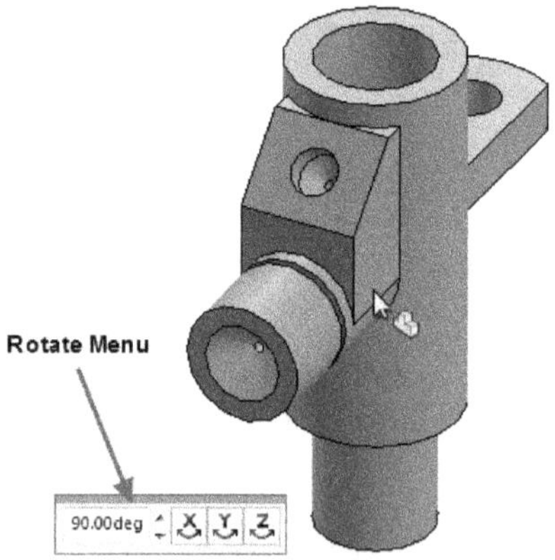

Figure-4. Component attached to cursor

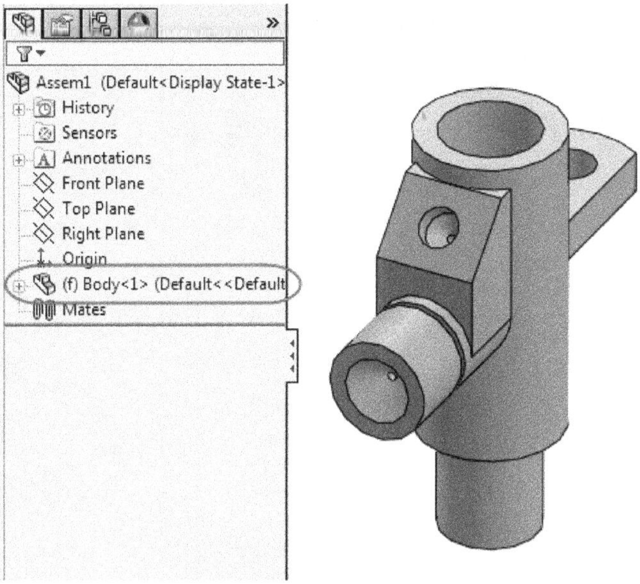

Figure-5. Component inserted

- To insert more components, click on the **Insert Components** button from the **Assembly** tab in the **Ribbon**. The **Insert Component PropertyManager** will display as shown in Figure-6.
- This **PropertyManager** is similar to the one displayed earlier. Click on the **Browse** button and open the desired part.
- Click anywhere in the viewport to place the part. The component will be placed as floating and can move anywhere by dragging. Also **(-)** mark will be added before its name in the **FeatureManager Design Tree** displayed at the left.
- You can use the **Rotate** menu to rotate the part before inserting.

- Similarly, you can insert more components as per your requirement in the assembly.

- If you want to reinsert a component then hold the CTRL key from keyboard and drag the desired component already existing in the display area. One more instance of component will be inserted; refer to Figure-7.

Figure-6. Insert Component PropertyManager

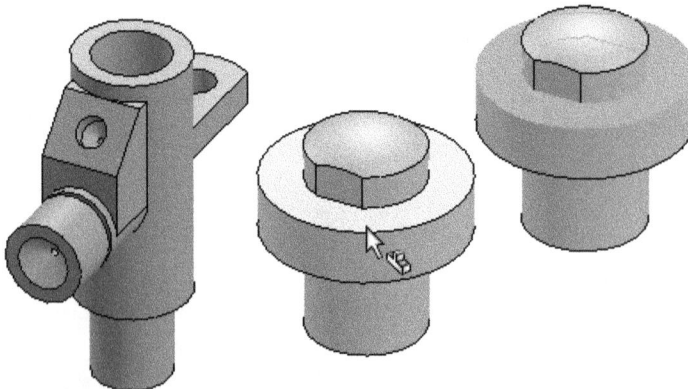

Figure-7. Reinserting a component

- In SolidWorks 2016, you can also choose the desired configuration of the part being inserted in assembly. To do so, click on the **Configuration** drop-down in the **Insert Component PropertyManager** and select the desired configuration; refer to Figure-8.

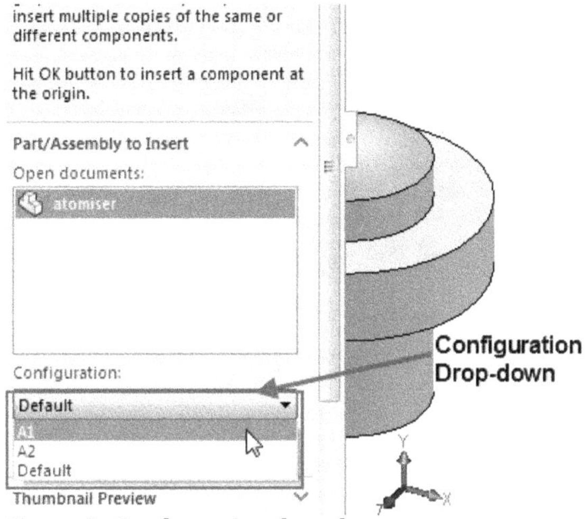

Figure-8. Configuration drop-down

After adding all the components in assembly, the next task is to apply proper constraints to them. The next section explains the use of assembly constraints (in SolidWorks **Mates**).

ASSEMBLY CONSTRAINTS (MATES)

The options to apply assembly constraints are available in the **Mate PropertyManager**. The use of options in this **PropertyManager** is explained next.

- After inserting the desired components, click on the **Mate** tool from the **Ribbon**. The **Mate PropertyManager** will display as shown in Figure-9.

- In SolidWorks 2016, you can make the first selection for mate as transparent for easy identification. To do so, scroll down in **Mate PropertyManager** and select the **Make first selection transparent** check box; refer to Figure-10.

Figure-9. Mate PropertyManager

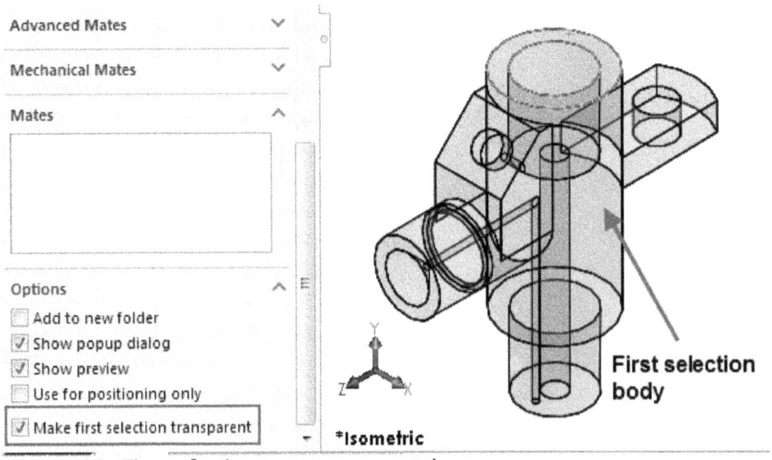

Figure-10. First selection transparency option

- If you want to position the components by using the mates but do not want to apply the mates then select the **Use for positioning only** check box from the **Options** rollout in the **PropertyManager**.

The buttons used for applying constraints are discussed next.

Coincident

The **Coincident** button is used to make two components coincide at the selected references. The steps to use this constraint are given next.

- Click on the **Coincident** button from the **PropertyManager**. You are asked to select two references for this mate.
- Select two axes/faces/planes/curves that you want to make coincident. The preview of constraint will display with a pop-up toolbar; refer to Figure-11.

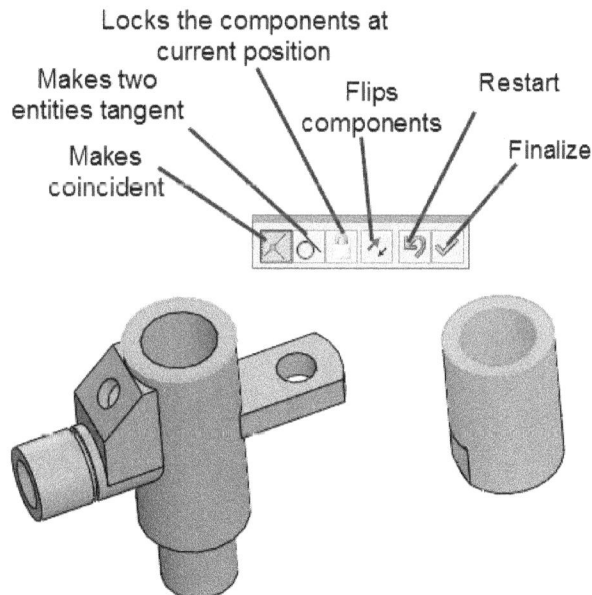

Figure-11. Coincident constraint preview

• Click on the **Flip** button to change the orientation of components; refer to Figure-12.

Figure-12. Components after flipping

Parallel

The **Parallel** button is used to make two components parallel with respect to each other. The steps to use this button are given next.

• Click on the **Parallel** button and select the two faces/ axes/planes. The components will be parallel with respect to these references. Refer to Figure-13.

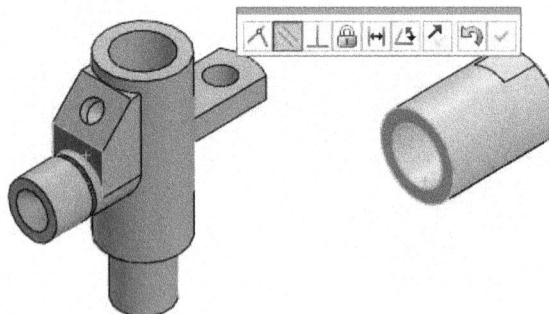

Figure-13. Parallel constraints

Note that the tools in the pop-up tool bar are also in the **PropertyManager** and work in the same way. The pop-up toolbar gives us facility to change the constraint type after the command is activated.

Perpendicular

The **Perpendicular** button is used to make two components perpendicular to each other. The steps are given next.

* Click on the **Perpendicular** button and select two faces/ axes/plane that you want to make perpendicular to each other. The preview will display; refer to Figure-14.

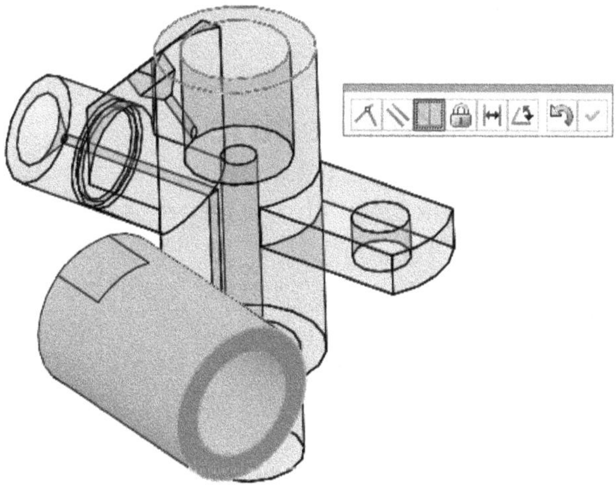

Figure-14. Perpendicular constraint

Tangent

The **Tangent** button is used to make two components tangent to each other. The steps are given next

* Click on the **Tangent** button and select two faces that you want to make tangent. The preview of tangent constraint will display as shown in Figure-15.

Figure-15. Tangent constraint

Concentric

The **Concentric** button is used to make two round components share the same center axis. The steps are given next.

- Click on the **Concentric** button and select two round faces that you want to make concentric. The preview of tangent constraint will display as shown Figure-16.

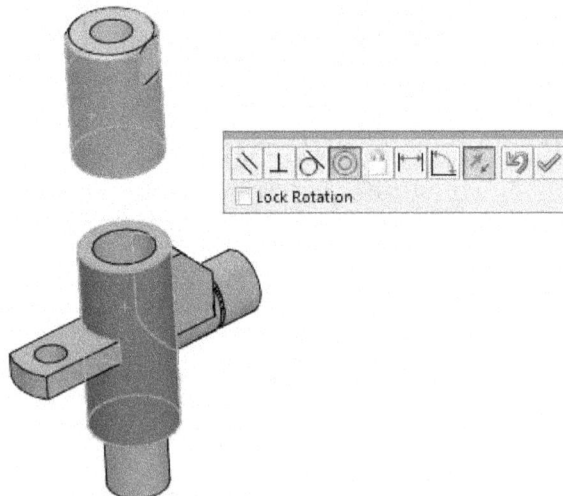

Figure-16. Concentric constraint

Note that you can lock the rotation of selected components by selecting the **Lock Rotation** check box in the **PropertyManager**/pop-up toolbar.

Lock

The **Lock** button is used to lock the component at its current position. The steps to do so are given next.

- Click on the **Lock** button and select the components you want to fix. The preview of components will display in blue color.
- Click **OK** to fix the component. Now, drag one of component to check the effect.

Note that you cannot move the first component of assembly by default. To move the first component, right click on its name in the **FeatureManager Design Tree** and select the **Float** option from the menu displayed; refer to Figure-17. Now, you will be able to move the component.

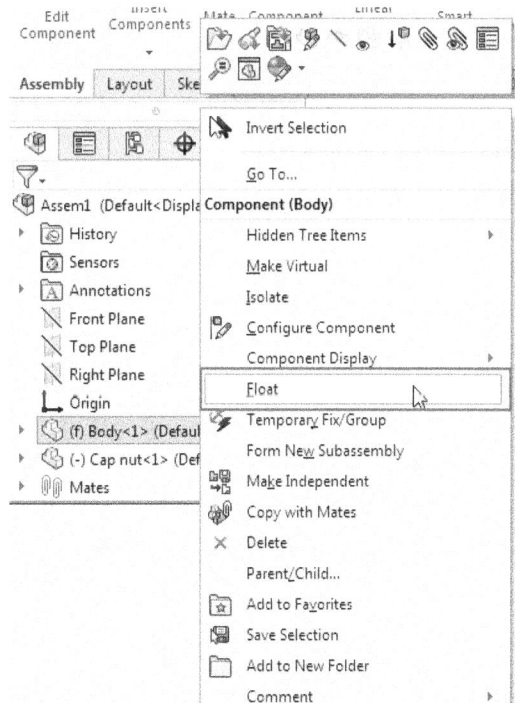

Figure-17. Float option in shortcut menu

Distance

The **Distance** button is used to set distance between two selected faces. The steps are given next.

• Click on the **Distance** button and select two flat faces that you want to use for setting distance. The preview of distance constraint will display; refer to Figure-18.

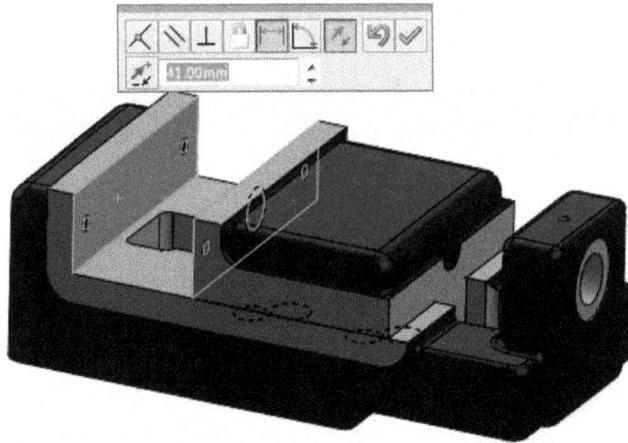

Figure-18. Distance constraints

• Set the desired distance in the pop-up toolbar and select the **Flip** button if required.

Angle

The **Angle** button is used to set angle between two selected faces. The steps are given next.

• Click on the **Angle** button and select two flat faces that you want to use for setting angle. The preview of angle constraint will display; refer to Figure-19.

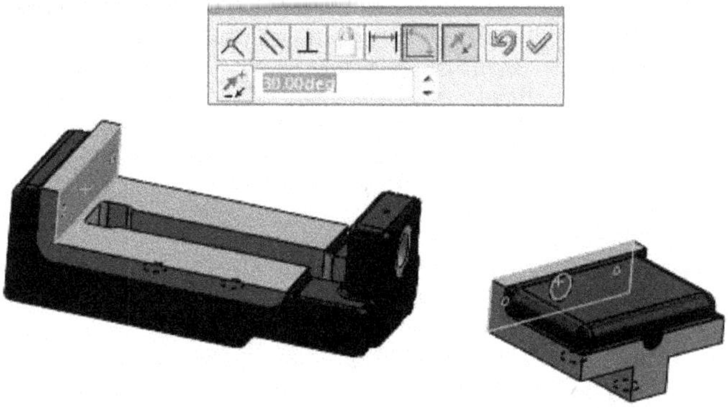

Figure-19. Angle constraint

- Set the desired angle in the pop-up toolbar and select the **Flip** button if required.

Note that if you have applied any wrong constraint and want to delete it then click on **+** sign next to **Mates** in the **FeatureManager Design Tree**, select the desired constraint from the list and press **DEL** button from keyboard.

Till this point, we have learned **Standard Mates** that are used for rigid assemblies. Now, we will move to advanced constraints (Mates) that play key role in motion study. Note that you might need to delete all the standard mates to apply advanced and mechanical mates for checking motion.

Expand the **Advanced Mates** rollout from the **FeatureManager** to display the advanced mates; refer to Figure-20. The mates in this rollout are explained next.

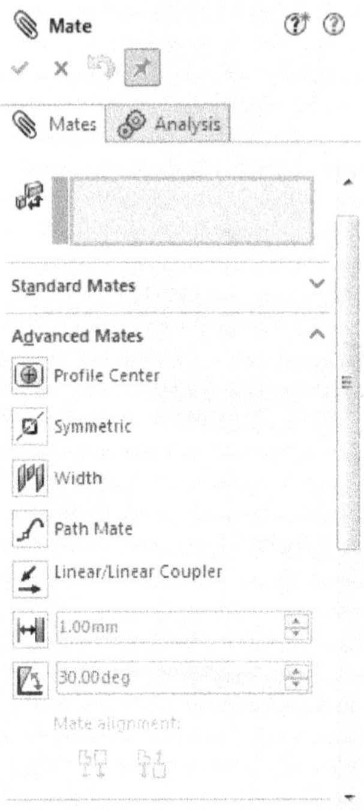

Figure-20. Advanced Mates

Profile Center

The **Profile Center** button is used to align the two components at a common center of the faces selected for mate. The steps to use this mate are given next.

- Click on the **Profile Center** button and select the flat faces of the components. The faces will be align at their profile centers; refer to Figure-21.
- Specify the desired distance between the faces by using the edit box displayed below the **Profile Center** button in the **PropertyManager**.

Figure-21. Applying Profile center mate

Symmetric

The **Symmetric** button is used to make faces of components symmetric with respect to a reference. In other words, the distance by which one component move will be the same as distance moved by other symmetric component. The steps are given next.

- Click on the **Symmetric** button and select two flat faces that you want to make symmetric.
- Click in the **Symmetry plane** selection box in the **Mate Selections** rollout and select the plane about which you want to make the components symmetric. The preview of applied mate will display; refer to Figure-22.

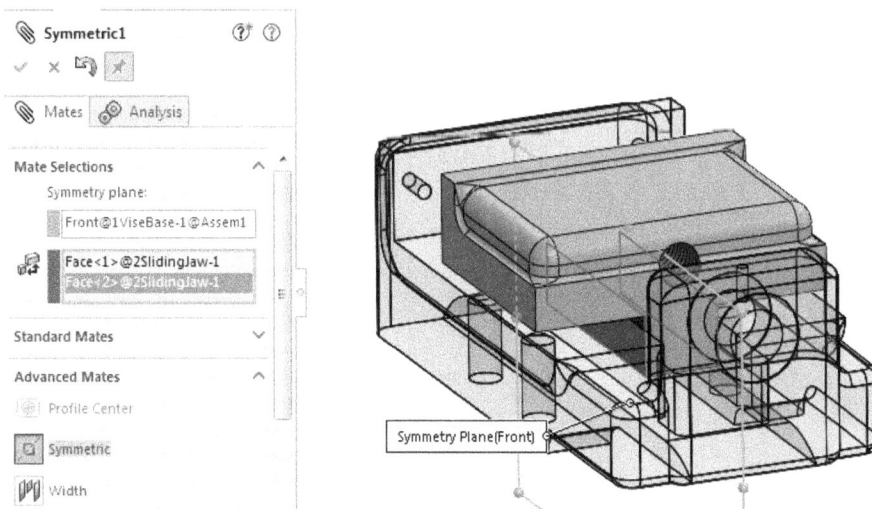

Figure-22. Symmetric constraint

Width

The **Width** button is used to fit a component in the selected width reference. The steps to use this mate are given next.

- Click on the **Width** button and select two flat faces that define the width reference.
- Select the two flat faces that define you components limit to fix them. The preview of mate will display; refer to Figure-23.

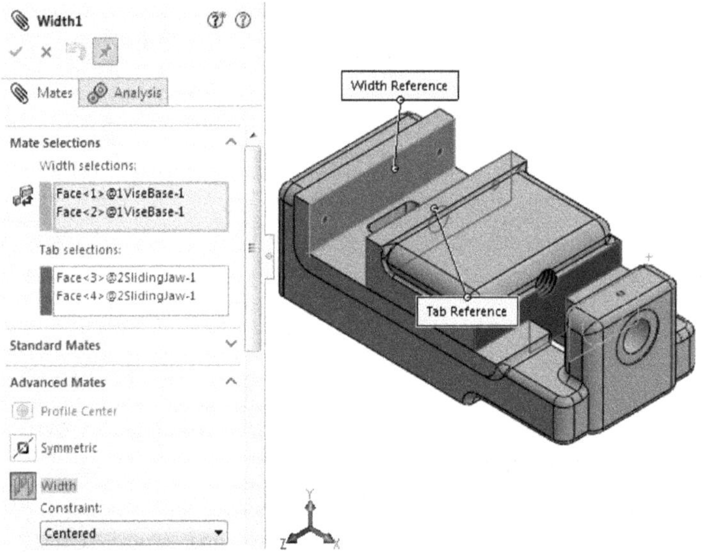

Figure-23. Width mate

Note that if the width of references is more than the limits of component then the component will be inserted in the middle of the width references.

Path Mate

The **Path Mate** button is used to make pointed component follow the specified path. The steps to use this mate are given next.

- Click on the **Path Mate** button and select the vertex of the component that you want to make follower.
- Select sketch of the path that you want to make as guide for the follower. The preview of path mate will display; refer to Figure-24.

Figure-24. Path mate constraint

- You can change the options related to movement of object by using the drop-down and edit boxes displayed below **Path Mate** in **PropertyManager**; refer to Figure-25.

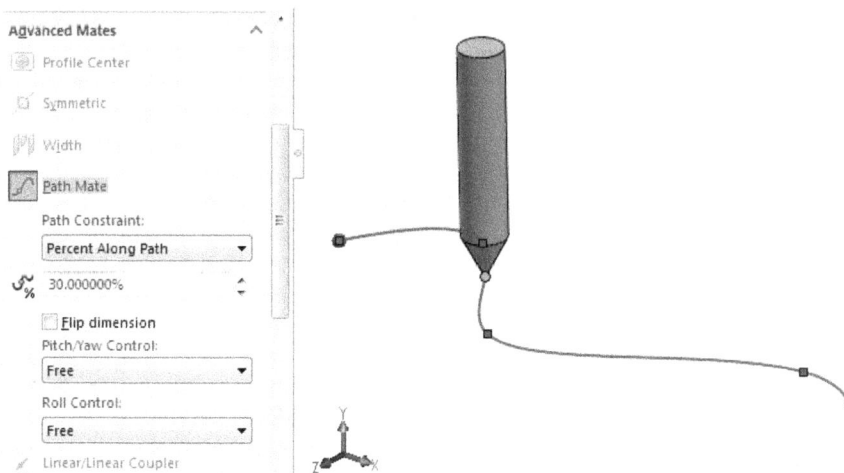

Figure-25. Options for path mate

Linear/Linear Coupler

The **Linear/Linear Coupler** button is used to make two components move with respect to each other by a specified ratio. The steps to use this mate are given next.

- Click on the **Linear/Linear Coupler** button and select the faces of the component between which you want to apply the mate. Preview of mate will display; refer to Figure-26.
- Set the desired ratio between the components. Click **OK** to apply the mate.
- To check the motion, drag one of the component in linear direction. The other component will move automatically.

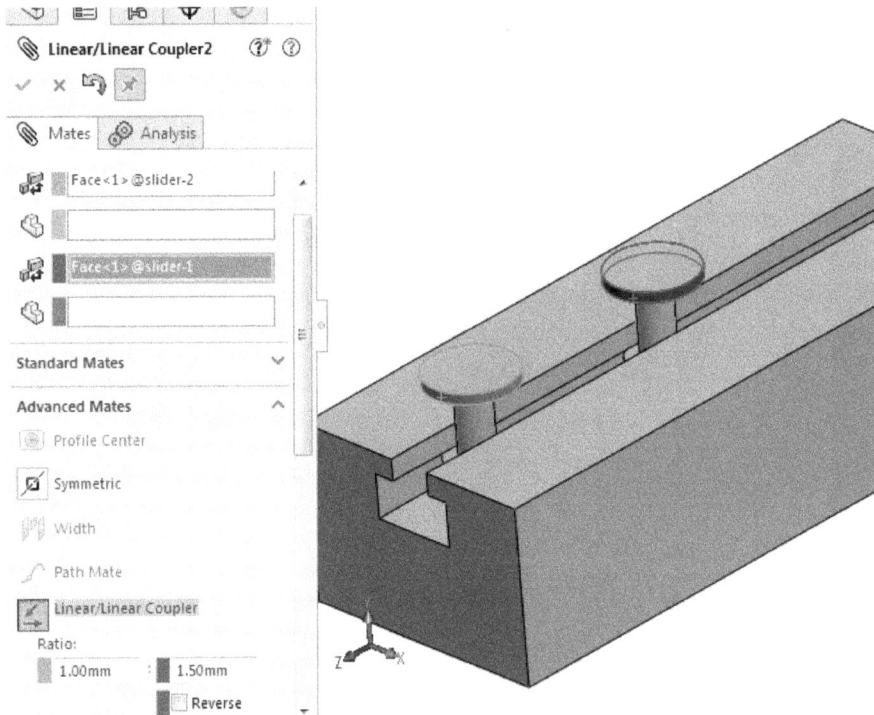

Figure-26. Linear coupler mate

Advanced Distance

The **Advanced Distance** button is used to apply maximum and minimum movement limit of a component. The steps to apply this mate are given next.

- Click on the **Distance** button from the **Advanced Mates** rollout. The options in the rollout will display as shown in Figure-27.
- Select the face/plane of component and then select the face/plane of reference to limit the movement. Preview of mate will display as shown in Figure-28.

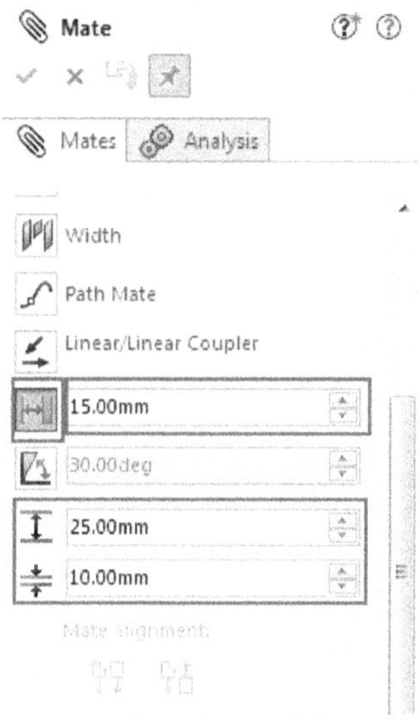

Figure-27. Advanced distance mate options

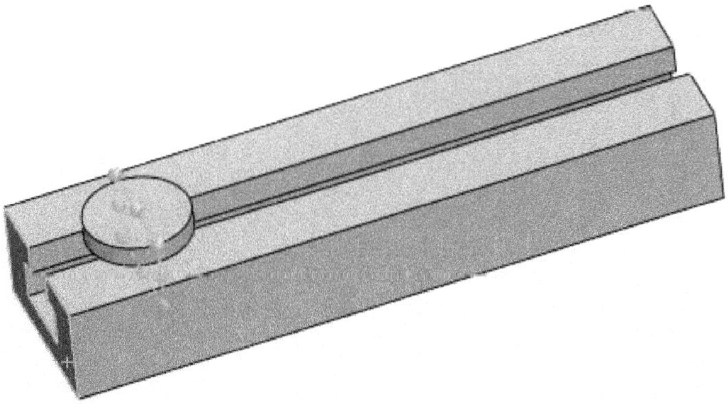

Figure-28. Planes selected for advanced distance

Advanced Angle

The **Advanced Angle** button is used to apply maximum and minimum rotation limit of a component. This mate works in the way similar to **Advanced Distance** mate.

Till this point, we have learned Standard Mates and Advanced Mates. There are also a few mates that represent mechanical motion in assembly. These constraints are grouped in a rollout named **Mechanical Mates**. To use these mates, expand the **Mechanical Mates** rollout; refer to Figure-29. The buttons in this rollout are explained next.

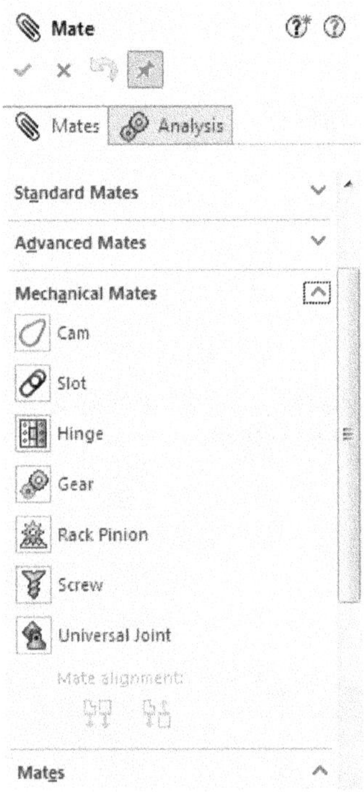

Figure-29. Mechanical Mates

Cam

The **Cam** button is used to create cam-follower mate between two entities. The steps given next explain the procedure.

- Click on the **Cam** button from the **Mechanical Mates** rollout. The selection boxes will be displayed as shown in Figure-30.

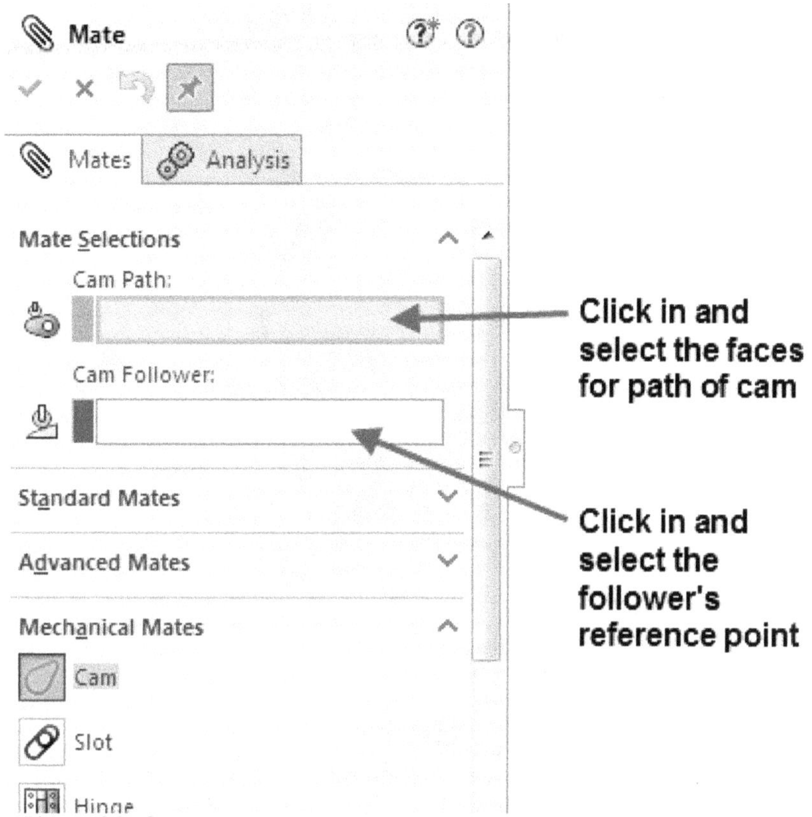

Figure-30. Cam Mate

- Follow the steps in the above figure. Figure-31 shows an example of cam mate.

Note that you need to limit the motion of follower so that it move only in vertical direction.

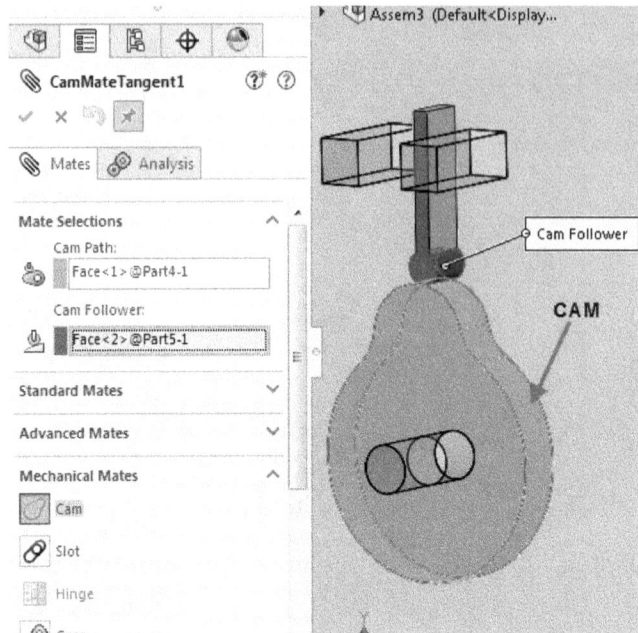

Figure-31. Example of cam mate

Slot

The **Slot** button is used to create slot-follower mate between two entities. After applying this mate, the component will move only in the limits of slot. The steps given next explain the procedure.

• Click on the **Slot** button from the **Mechanical Mates** rollout.
• Select the round face of slot and the follower, the mate will be applied.
• You can set the starting position of follower by using the options in the **Constraint** drop-down displayed below the **Slot** mate in **PropertyManager**. Figure-32 shows an example of slot mate.

Figure-32. Example of slot mate

Hinge

The **Hinge** button is used to make the two parts behave as hinged. The procedure to use this button is given next.

• Click on the **Hinge** button from the **Mechanical Mates** rollout. The selection boxes in the **Mate Selections** rollout are displayed as shown in Figure-33.

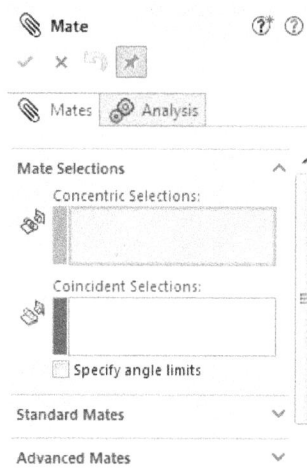

Figure-33. Hinge Mate Selection boxes

- Select the two round faces that you want to be concentric.
- Select the two flat faces that define the angular motion limit of the hinge.
- Select the **Specify angle limits** check box and specify the limits as explained in advanced mates. Figure-34 shows a hinge mate being applied to the assembly.

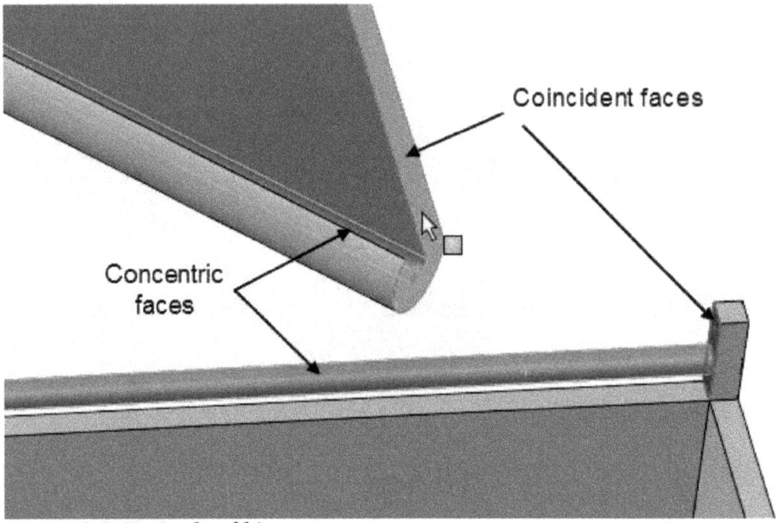

Figure-34. Example of hinge mate

Gear

The **Gear** button is used to create a joint between two gears. The procedure to use this button is given next.

- Click on the **Gear** button from the **Mechanical Mates** rollout.
- Select the flat faces of the gears. The **Gear Mate PropertyManager** will display as shown in Figure-35. Also, the gears will display as shown in Figure-36.

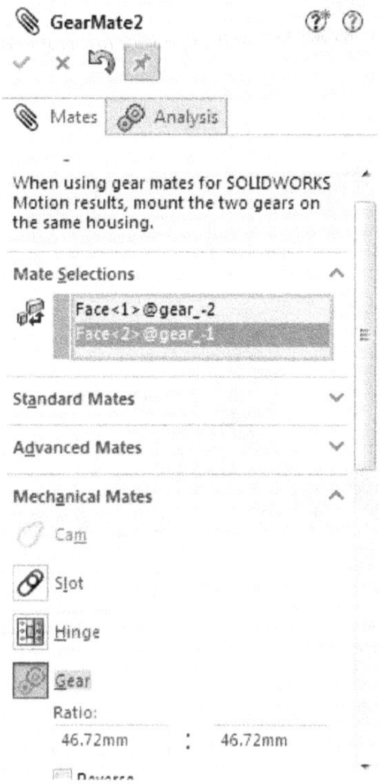

Figure-35. Gear Mate PropertyManager

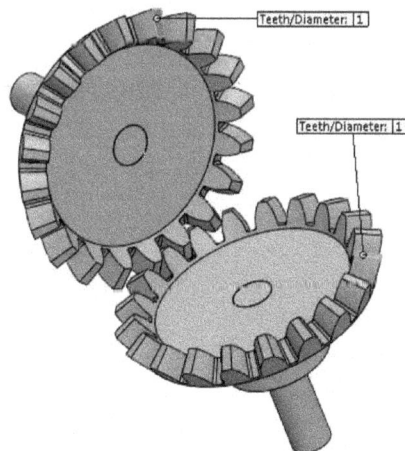

Figure-36. Example of gear mate

Rack Pinion

The **Rack Pinion** button is used to create joint between a rack and a pinion. The procedure to use this button is given next.

- Click on the **Rack Pinion** button from the **Mechanical Mates** rollout.
- Select edge of the rack and flat face of the pinion gear. The rack and pinion mate will be created; refer to Figure-37. Specify the desired ratio in the **PropertyManager**.

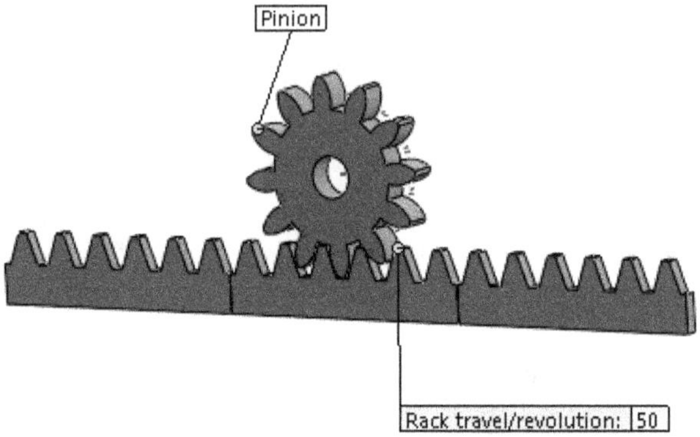

Figure-37. Rack pinion mate

Screw

The **Screw** button is used to create screw joint between a cylinder and a flat face. The procedure to use this button is given next.

- Click on the **Screw** button from the **Mechanical Mates** rollout.
- Select the cylindrical face of first component and flat face of other component. The preview will be displayed; refer to Figure-38.

- Specify the desired value in the box displayed below **Screw** button in the **PropertyManager**.

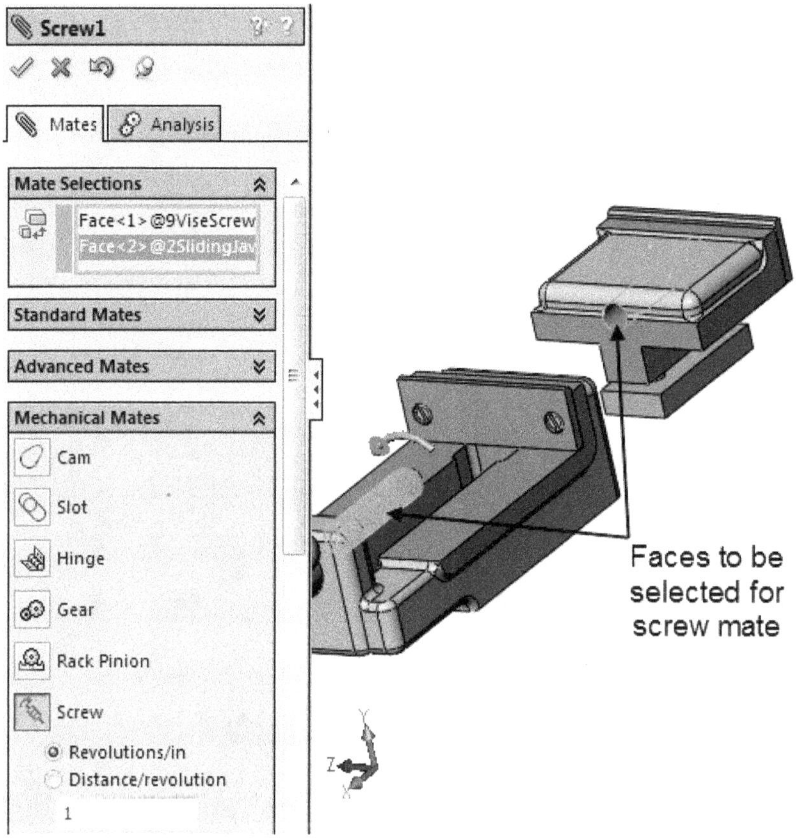

Figure-38. Screw Mate

Universal Joint

The **Universal Joint** button is used to create universal joint between two components. After clicking this button, select the two round faces. The universal joint will be created; refer to Figure-39.

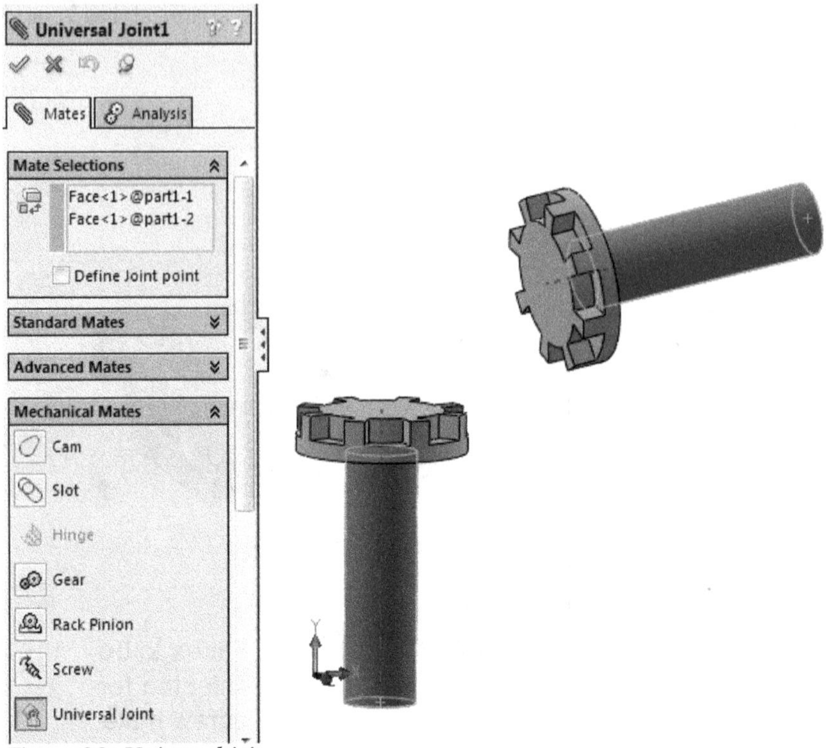

Figure-39. Universal joint

EXPLODED VIEW

After creating assembly, we are required to display the components of assembly expanded in the way they are assembled. Follow the steps given below to create the exploded view of assembly.

• Select all the components that you want to explode from the main assembly and select the **Exploded View** tool from the **Ribbon**. The **Explode PropertyManager** will display; refer to Figure-40.

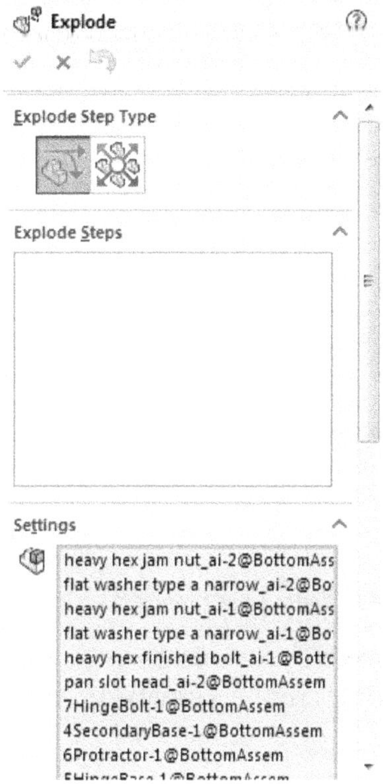

Figure-40. Explode PropertyManager

- Select the **Auto-space components on drag** check box from the **Options** rollout in the **Explode PropertyManager** and drag the components in the desired direction.
- Click on the **OK** button from the **PropertyManager**. The components will be exploded automatically.

Or

- Click on the **Exploded View** button and select the component you want to move. A triad will display on the component.
- Drag the component using the handles displayed to the desired distance or specify the desired distance and rotation in the edit boxes available in the **Settings** rollout of the **PropertyManager**.
- Repeat the above steps until you explode all the components.

To display the explode lines, click on the **Explode Line Sketch** button. The **Route Line PropertyManager** will display; refer to Figure-41.

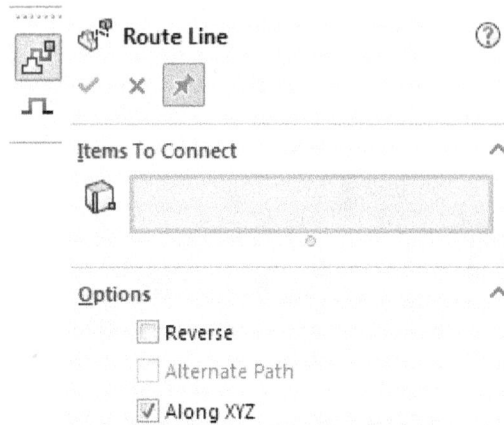

Figure-41. Route Line PropertyManager

- Click on the assembly reference of the component and then the corresponding assembly reference of the base component. The explode line will be created; refer to Figure-42.

Figure-42. Explode lines created

BILL OF MATERIALS

Bill of materials is used to list the total components of the assembly in the form of a table. To create bill of materials, follow the steps given below.

- Click on the **Bill of Materials** button from the **Ribbon**. The **Bill of Materials PropertyManager** will display; refer to Figure-43.

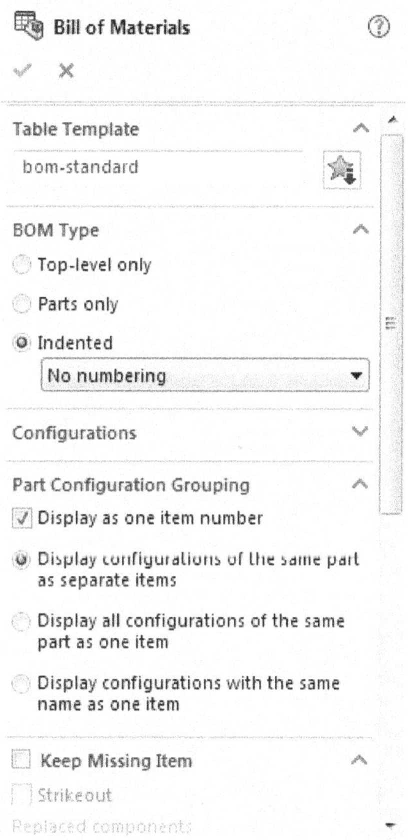

Figure-43. Bill of Materials PropertyManager

- Click on the ⬚ button to display templates. The **Open** dialog box will be displayed with default templates of bill of materials; refer to Figure-44.

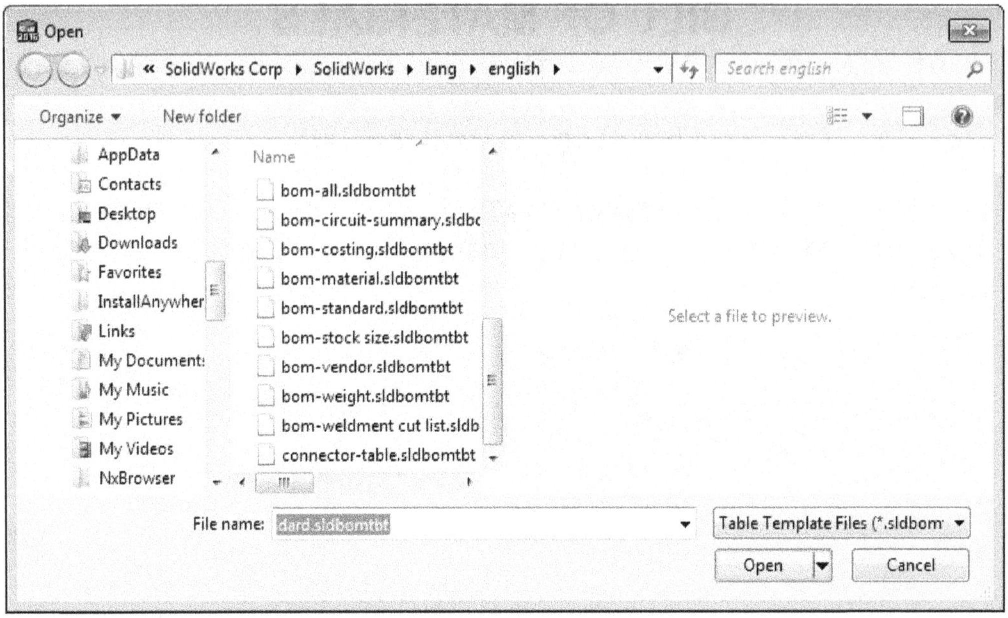

Figure-44. Default templates for BOM

- Select the desired template from the dialog box and click on the **Open** button. The template will be activated.
- Click on the **OK** button from the **PropertyManager**. You are asked to place the table of bill of materials.
- Click **OK** from the **Select Annotation View** dialog box.
- Click in the viewport to place the table; refer to Figure-45.

ITEM NO.	PART NUMBER	DESCRIPTION	WEIGHT	QTY.
1	Body			1
2	Washer			1
3	Valve spindle			1
4	Spring			1
5	Spring cap nut			1
6	Pre adjusting nut			1
7	Nozzle valve			1
8	Nozzle body			1
9	Cap nut			2
10	atomiser			1

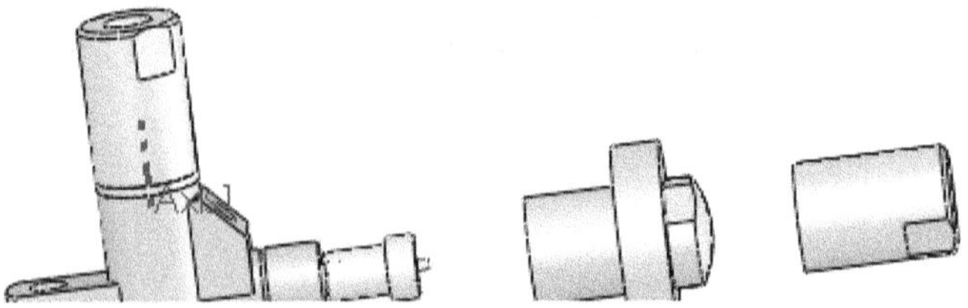

Figure-45. Bill of materials

MATE CONTROLLER

The **Mate Controller** tool is added in SolidWorks 2016. Using this tool, you can manage the distance and angle mates applied in the assembly at one place. The procedure to use this tool is given next.

- Click on the **Mate Controller** tool from the **Insert** menu; refer to Figure-46. The **Mate Controller PropertyManager** will be displayed; refer to Figure-47.

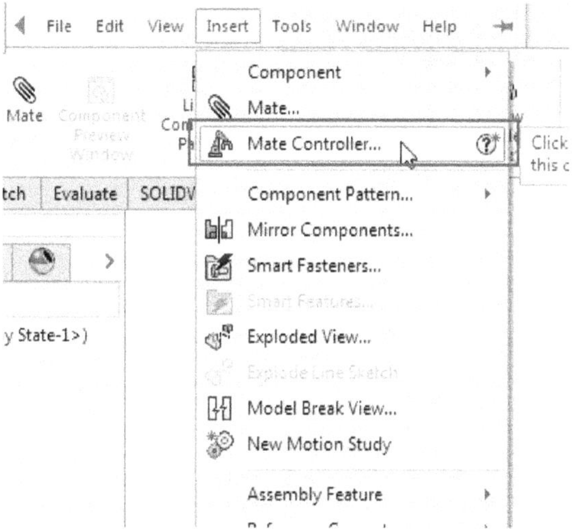

Figure-46. Mate Controller tool

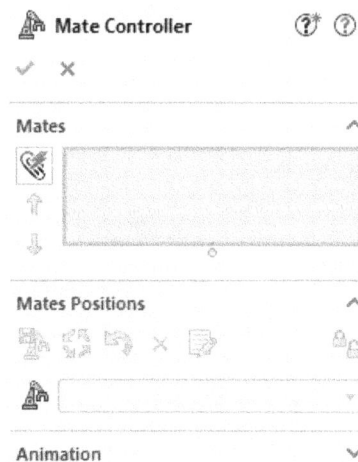

Figure-47. Mate Controller PropertyManager

- Click on the **Collect All Supported Mates** button from the **Mates** rollout of the **PropertyManager**. The mates that can be controlled with their angular or distance dimension will be displayed in the selection box; refer to Figure-48.

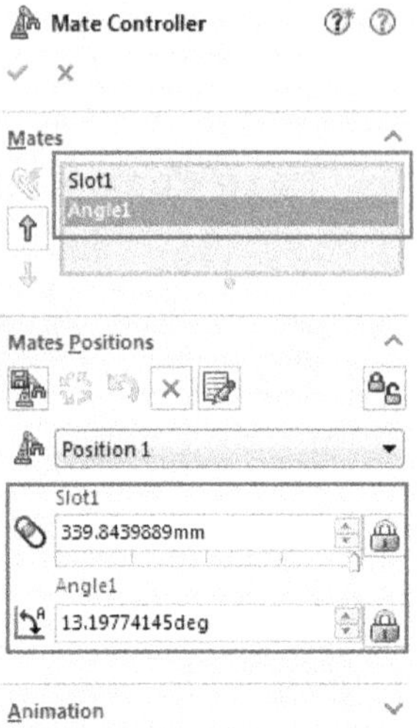

Figure-48. Mates selected automatically

- Set the desired dimension spinners available in the **Mates Positions** rollout.
- You can save the positions by using the **Add Position** button from the **Mates Positions** rollout. On doing so, the **Name Position** dialog box will be displayed; refer to Figure-49.

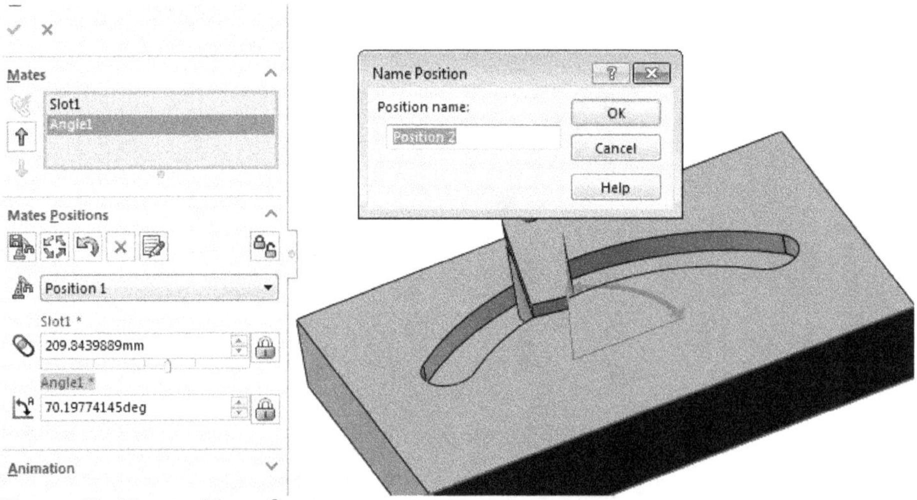

Figure-49. New position of mates

- Specify the desired name and click on the **OK** button from the dialog box.
- You can select the desired position from the drop-down in the **Mates Positions** rollout; refer to Figure-50.

Figure-50. Positions rollout

- To animate the position change, expand the **Animation** rollout in the **PropertyManager** and click on the **Calculation Animation** button from it; refer to Figure-51.

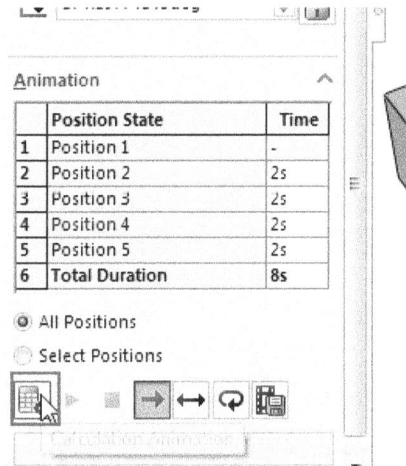

Figure-51. Animation rollout

- After calculation, you can use the **Play** button from the **Animation** rollout to play the animation.
- Click on the **OK** button from the **PropertyManager** to apply the mate control. The Mate Controller feature will be added in the **FeatureManager Design Tree**.

MOTION STUDY

Motion study is used to check the motion of components with respect to each other after applying the driving force the components. This feature of SolidWorks help to understand the mechanism of our assembly in real world conditions. To start the motion study, click on the **Motion Study 'x'** tab displayed at the bottom bar of the viewport. Note that here **'x'** is the sequence number. After clicking on this tab, the interface is displayed as shown in Figure-52.

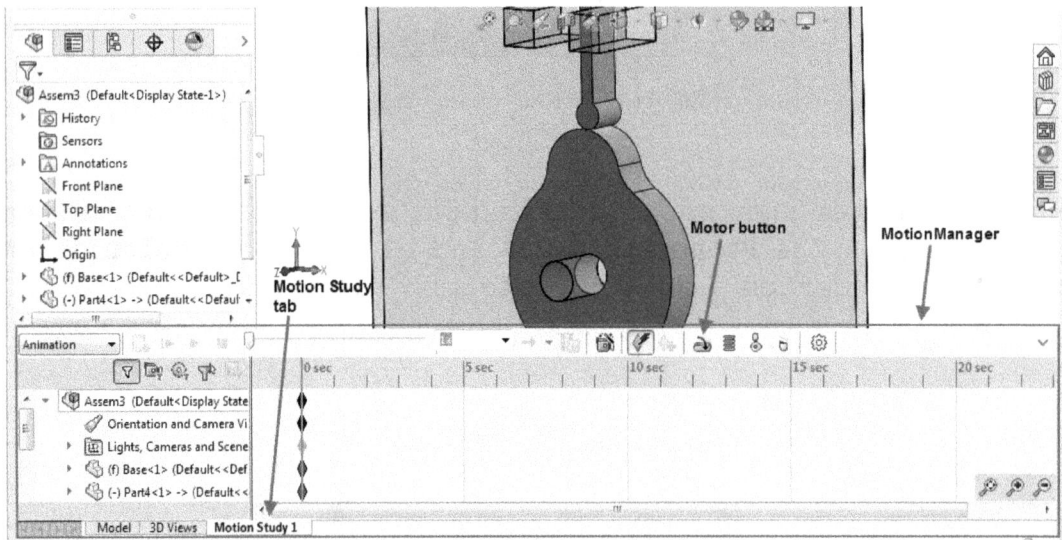

Figure-52. Motion study interface

Now, we need to apply some driving force to one component so that we can check the motion of other parts. In this case, we are using the cam-follower mechanism as displayed above. We need to add the rotary motion to the cam to check motion of follower. The steps to do so are given next.

• Click on the **Motor** button shown in the above figure. The **Motor PropertyManager** will display; refer to Figure-53.

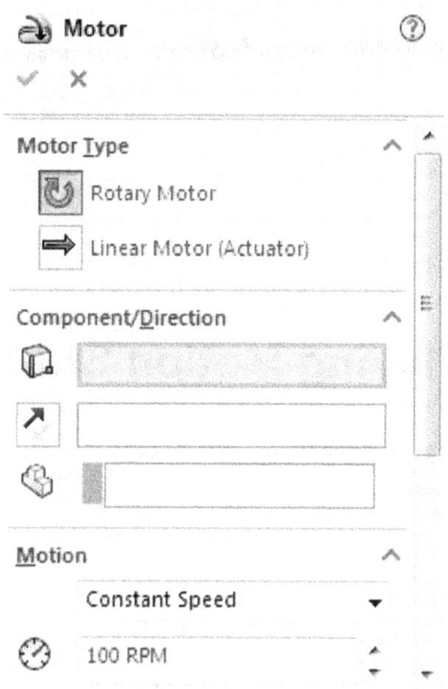

Figure-53. Motor PropertyManager

- Click on the circular face of the part to apply the rotary motion; refer to Figure-54.

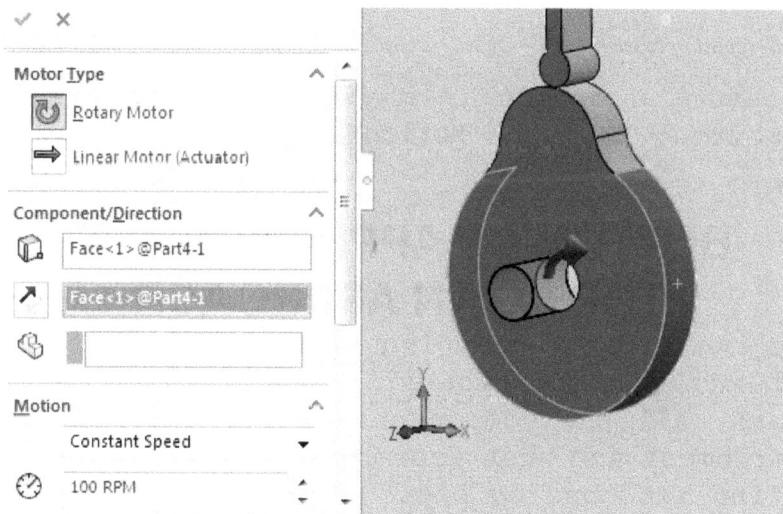

Figure-54. Circular face selected for rotary motion

- Specify the desired options for speed in the **Motion** rollout and click **OK** to apply the motion. The motor will be added in the **MotionManager**.

Note that in the same way you can apply linear motion to a flat face by selecting **Linear Motor (Actuator)** button from the **Motor PropertyManager** in place of **Rotary Motor** button.

If you want to start a new motion study, then click on the **New Motion Study** tool from the **Assembly** tab in the **Ribbon**. A new motion study tab will be added in the **MotionManager**.

Playing Motion Study

To animate the motion study, click on the **Calculate** (🖼) button from the **MotionManager** tool bar. The animation of mechanism will display in the viewport. To increase the time of animation, drag the key point of assembly to the desired point; refer to Figure-55. Click on the **Calculate** button again to run the animation.

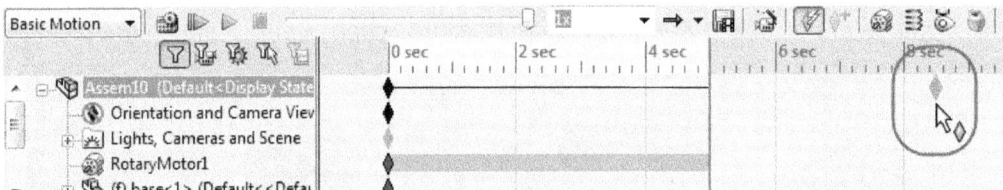

Figure-55. Keypoint of assembly

You can save the motion in movie format by selecting the **Save** (🖫)button from the **MotionManager** tool bar.

BOTTOM UP APPROACH AND TOP DOWN APPROACH

There are two ways of creating assembly in most of the CAD packages; Bottom Up approach and Top Down approach. In Bottom Up approach, all the components are created separately in Part environment and then assembled in Assembly environment by inserting them one by one.

In Top Down approach, we create all the parts in assembly environment and apply the mates on the spot. The procedure to create parts in assembly environment is given next.

Creating Parts in Assembly

- Click on the **New Part** tool from the **Insert Components** drop-down in the **Assembly** tab of **Ribbon**; refer to Figure-56. You are asked to select a face or plane on which the part is to be positioned.

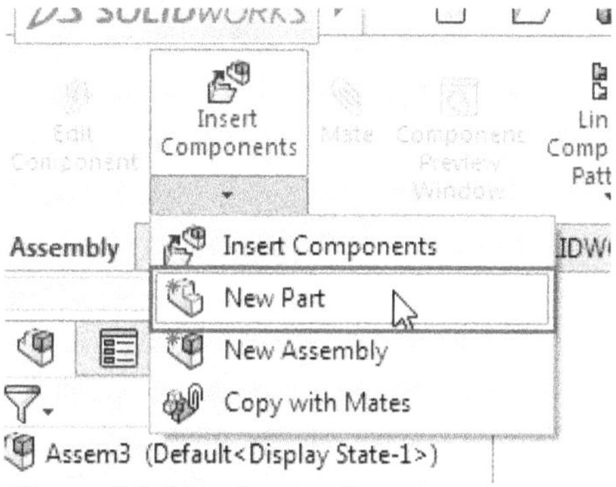

Figure-56. New Part tool

- Select the desired plane from the **FeatureManager Design Tree**. The Sketching environment will be displayed; refer to Figure-57.

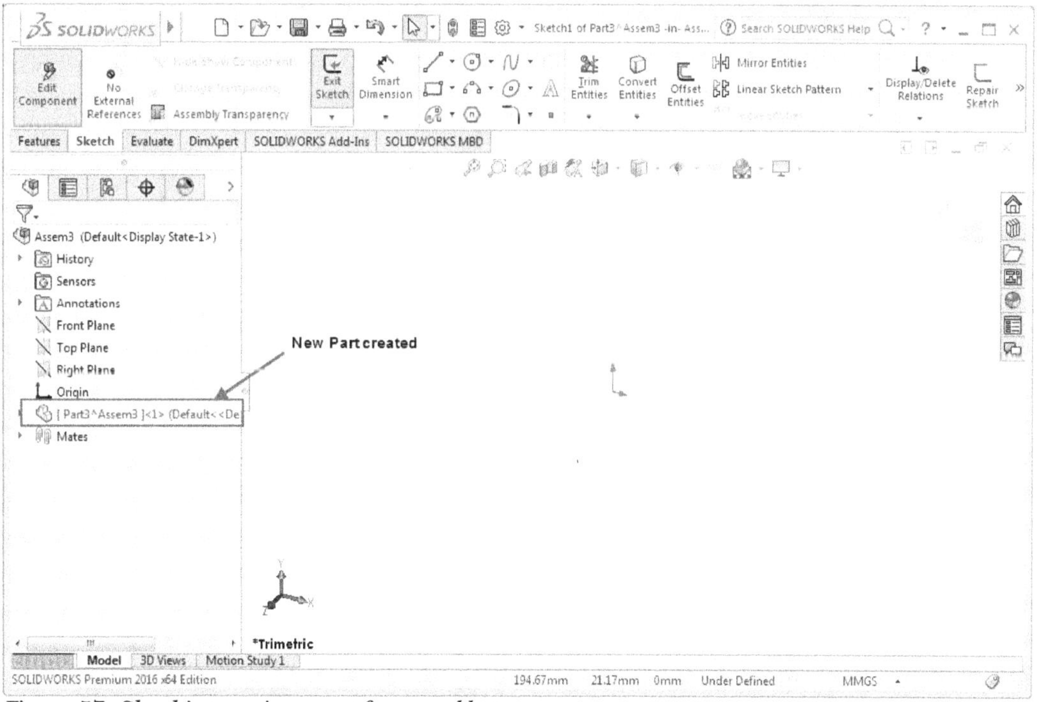

Figure-57. Sketching environment for assembly part

- Create desired sketch and perform operations using the tools in **Features** tab in the **Ribbon**; refer to Figure-58.

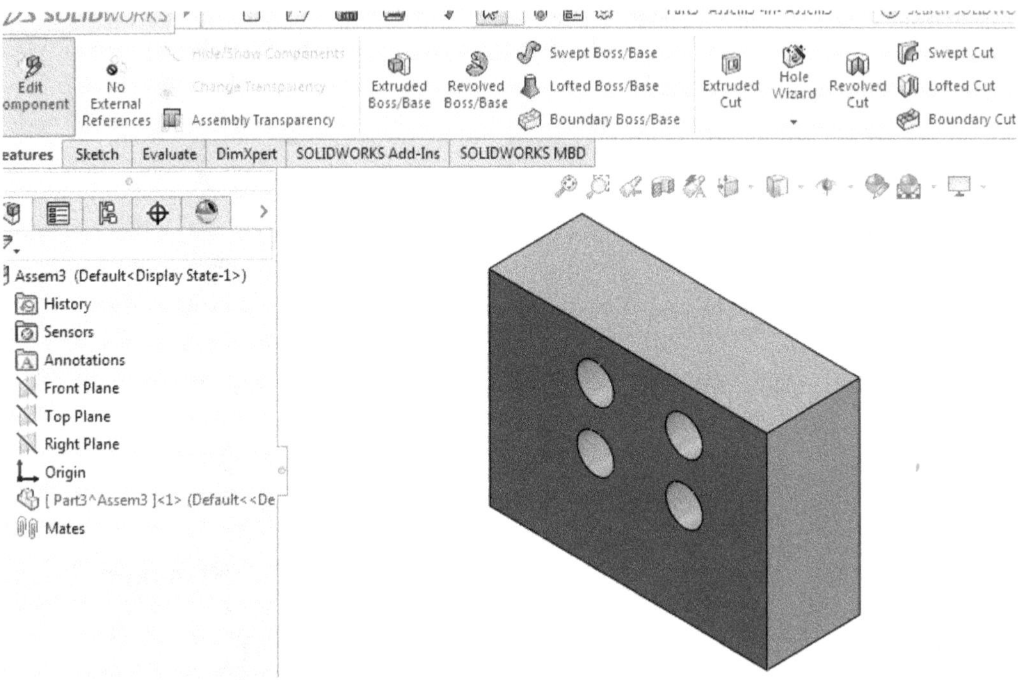

Figure-58. Part created

- Once you have created all the features of part, click on the **Exit editing component** button at the top-right in the viewport; refer to Figure-59.

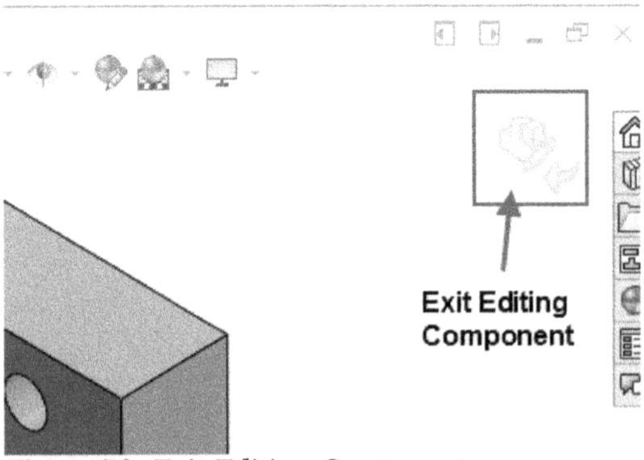

Figure-59. Exit Editing Component

- Similarly, you can create other components in assembly and apply the mates as discussed earlier.

Till this point, we have used various assembly tools. In the next chapter, we will apply these tools to perform practical and work out with practice questions.

Note that for the next chapter, you need to download some part. The link to these parts is provided on writing to us at cadcamcaeworks@gmail.com

SELF ASSESSMENT

Q1. Assembly is the combination of two or more components in any manner. (T/F)

Q2. Assembly Design is a separate environment in SolidWorks to perform operations related to assembly of components. (T/F)

Q3. The first component inserted in assembly is fixed by default. (T/F)

Q4. You can reinsert a component by holding the CTRL key from keyboard while drag the desired component. (T/F)

Q5. You can make the first component transparent while applying mates in SolidWorks 2016. (T/F)

Q6. Which of the following tool can be used to manage positions of various distance and angle constraints in assembly?

a. Edit Component
b. Mate
c. Mate Controller
d. Explode View

FOR STUDENT NOTES

ASSEMBLY PRACTICAL AND PRACTICE

Chapter 7

Topics Covered

The major topics covered in this chapter are:

- *Assembly Practical 1*
- *Assembly Practical 2*
- *Assembly and Motion Practical 3*
- *Practice Exercises.*

Note: Before starting this chapter, mail us at **cadcamcaeworks@gmail.com** to get the part files required to complete this chapter.

PRACTICAL 1

Assemble the parts of Fuel Injection Nozzle as shown in Figure-1. The exploded view if assembly is displayed as shown in Figure-2.

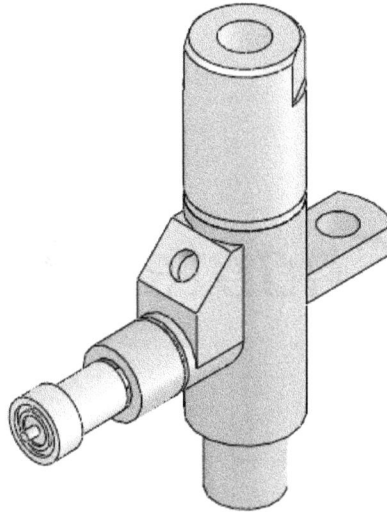

Figure-1. Assemble view of nozzle

Figure-2. Exploded view of assembly

All the part files can be downloaded from the provided link. In this practical, we will use parts of **Fuel injection nozzle** folder from downloaded files/folders.

The steps to assemble these parts are given next.

Inserting Body

- Start the assembly environment by selecting Assembly button from the **New SolidWorks Document** dialog box. You are asked to insert first part.
- Click on the Browse button from the **Begin Assembly PropertyManager** displayed in the left. The **Open** dialog box will display.
- Double click on the Body part in the **Fuel injection nozzle** folder. The part will attach to the cursor.
- Click in the viewport to place the part.

Inserting and constraining Washer

- Click on the **Insert Components** button from the **Ribbon**. And then click on the **Browse** button from the **PropertyManager**. The **Open** dialog box will display.
- Double click on the **Washer** part in the **Fuel injection nozzle** folder.
- The part will attach to the cursor.
- Click in the viewport to place the part.
- Click on the **Mate** button from the **Ribbon**. The **Mate PropertyManager** will display.
- Select the **Coincident** button from the **PropertyManager** and select the round edges as shown in Figure-3.

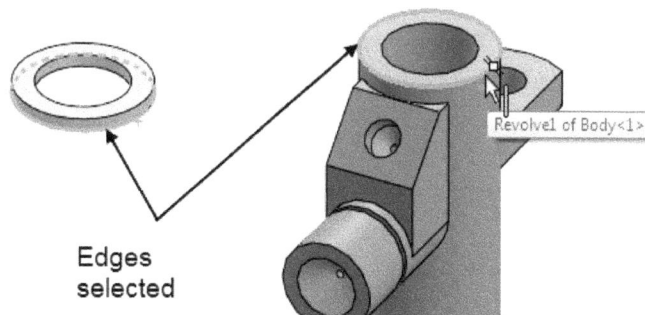

Figure-3. Edges to be selected

Inserting and constraining Spring

- Click on the **Insert Components** button from the **Ribbon**. And then click on the **Browse** button from the **PropertyManager**. The **Open** dialog box will display.
- Double click on the **Spring** part in the **Fuel injection nozzle** folder.
- The part will attach to the cursor.
- Click in the viewport to place the part.
- Click on the **Mate** button from the **Ribbon**. The **Mate PropertyManager** will display.
- Select the **Coincident** button from the **PropertyManager** and select the flat faces as shown in Figure-4.
- Click on the **Flip Mate Alignment** button(⬚) to align the spring properly.

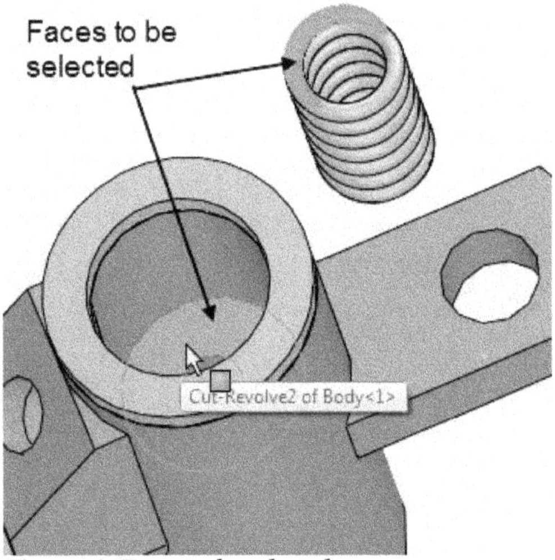

Faces to be selected

Figure-4. Faces to be selected

- Click on the **OK** button from the **PropertyManager** to apply the mate.
- Click on the **View Temporary Axes** button and **View Axes** button from the **Hide/Show Items** drop-down to display axes; refer to Figure-5.

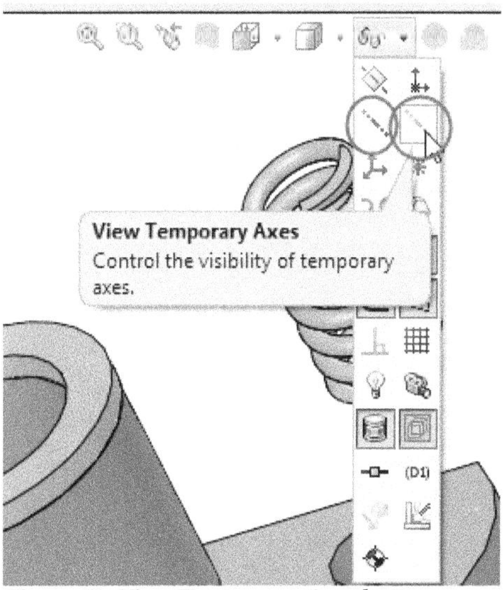

Figure-5. View Temporary Axes button

- Click on the **Coincident** button again from the **PropertyManager** and select the axes as shown in Figure-6.

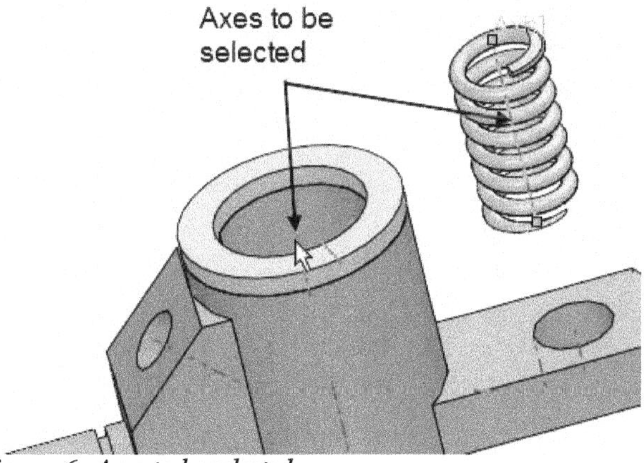

Figure-6. Axes to be selected

- Click **OK** button from the **PropertyManager**. The mate will be applied.

Inserting and constraining Pre adjusting Nut

- Click on the **Insert Components** button from the **Ribbon** and then click on the **Browse** button from the **PropertyManager**. The **Open** dialog box will display.

- Double-click on the **Spring** part in the **Fuel injection nozzle** folder.
- The part will attach to the cursor.
- Click in the viewport to place the part.
- Click on the **Mate** button from the **Ribbon**. The **Mate PropertyManager** will display.
- Select the **Coincident** button from the **PropertyManager** and select the flat faces as shown in Figure-7.

Faces to be selected

Revolve1 of Pre adjusting nut<1>

Figure-7. Faces to be coincident

- Click **OK** from the **PropertyManager** to apply the mate.
- Click on the **Coincident** button again from the **PropertyManager** and select the center axes as shown in Figure-8.

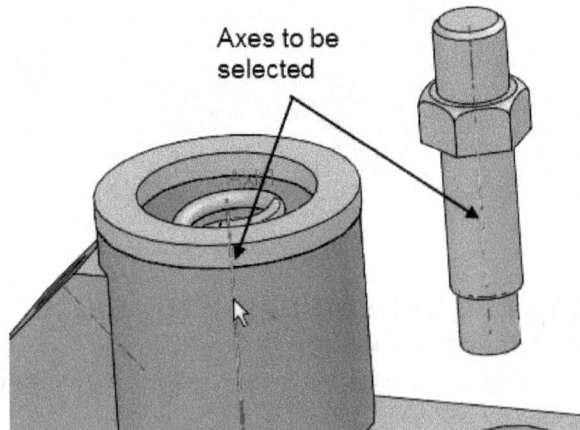

Axes to be selected

Figure-8. Axes to be coincident

- Click on the **OK** button from the **PropertyManager** to apply the mate.

Inserting and Constraining other components

In the same way, you can assemble the other components of the injection nozzle; refer to Figure-1 and Figure-2.

PRACTICAL 2

Assemble the parts of handle as shown in Figure-9. The exploded view if assembly is displayed as shown in Figure-10.

Figure-9. Handle assembled

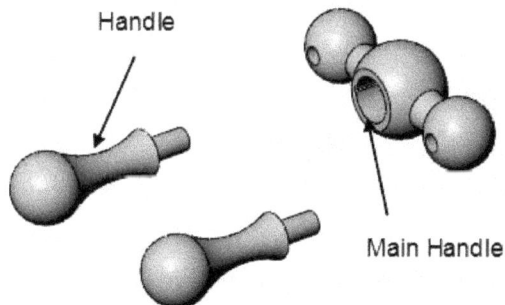

Figure-10. Exploded view of handle

All the part files can be downloaded from the provided link. In this practical, we will use parts of **Handle assembly** folder from downloaded files/folders.

The steps to assemble these parts are given next.

Inserting Main Handle

- Start the assembly environment by selecting **Assembly** button from the **New SolidWorks Document** dialog box. You are asked to insert first part.
- Click on the **Browse** button from the **Begin Assembly PropertyManager** displayed in the left. The **Open** dialog box will display.
- Double-click on the **Main Handle** part in the **Handle assembly** folder. The part will attach to the cursor.
- Click in the viewport to place the part.

Inserting and constraining Handle

- Click on the **Insert Components** button from the **Ribbon**. And then click on the **Browse** button from the **PropertyManager**. The **Open** dialog box will display.
- Double click on the **Handle** part in the **Handle assembly** folder.
- The part will attach to the cursor.
- Click in the viewport to place the part.
- Click on the **Mate** button from the **Ribbon**. The **Mate PropertyManager** will display.
- Select the **Concentric** button from the **PropertyManager** and select the round faces as shown in Figure-11.

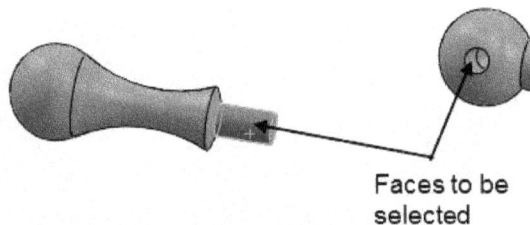

Faces to be
selected

Figure-11. Faces selected for concentric mate

- Click **OK** from the **PropertyManager** to apply the mate.
- Display the planes by selecting the **View Planes** button from the **Hide/Show Items** drop-down in the **Heads-up View** tool bar; refer to Figure-12.

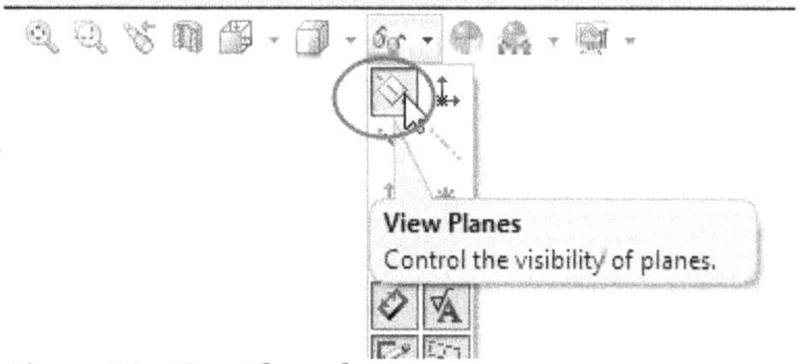

Figure-12. View Planes button

- Click on the **Coincident** button from the **Mate PropertyManager** and select the plane and edge as shown in Figure-13.

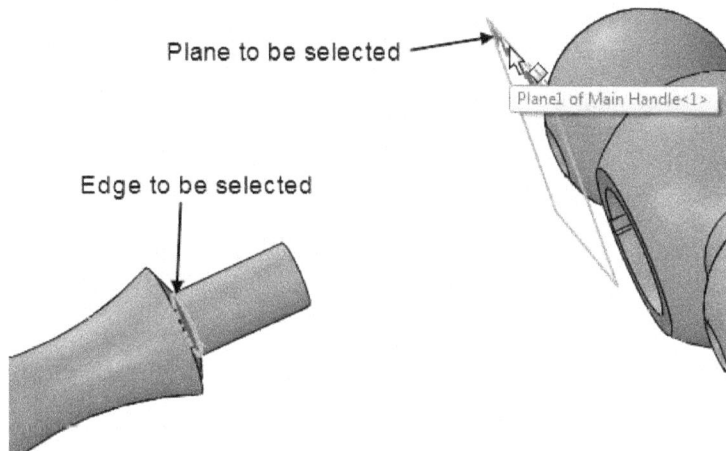

Plane to be selected

Plane1 of Main Handle<1>

Edge to be selected

Figure-13. Selection for coincident mate

- Click on the **OK** button from the **PropertyManager** to apply the mate.

The assembled handle will displayed as shown in Figure-14.

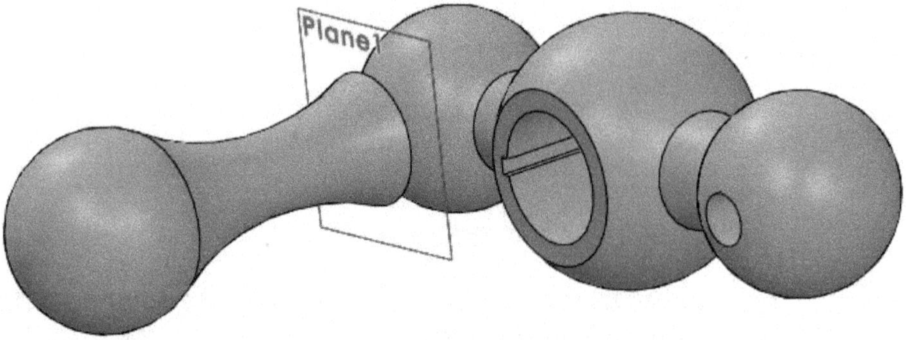

Figure-14. Assembled handle

Similarly, assemble the other handle so that the model displays as shown in Figure-9.

PRACTICAL 3

Assemble the parts of Epicyclic Gear Mechanism as shown in Figure-15. The exploded view if assembly is displayed as shown in Figure-16. Also, apply the simulation to mechanism as shown in video file provided in the resources.

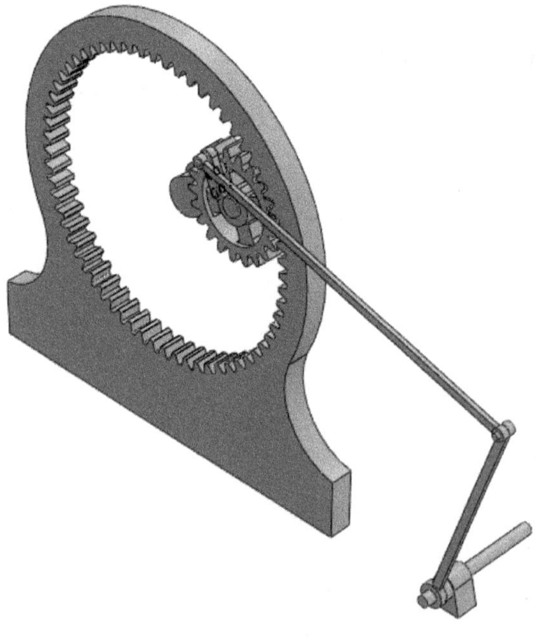

Figure-15. Epicyclic gear mechanism assembly

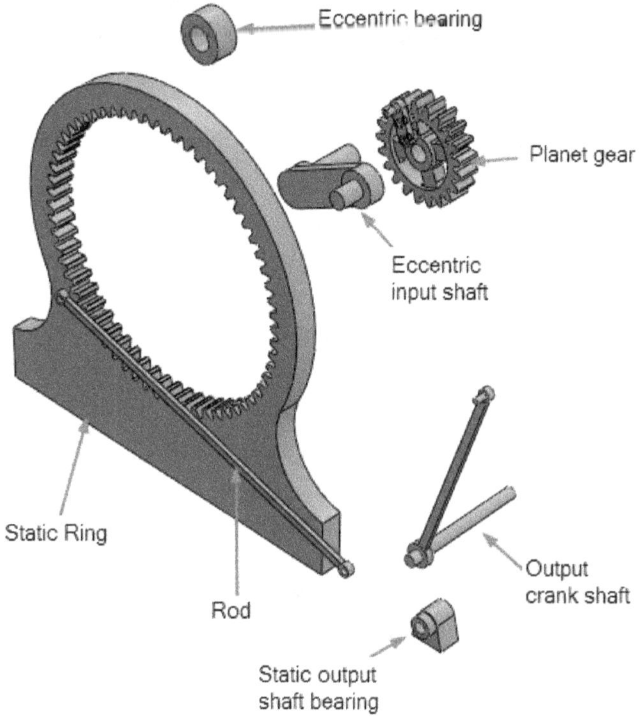

Figure-16. Exploded view of epicyclic gear mechanism

All the part files can be downloaded from the provided link. In this practical, we will use parts of **Epicyclic Gear Mechanism** folder from downloaded files/folders.

The steps to assemble these parts are given next.

Inserting Static Ring

- Start the assembly environment by selecting Assembly button from the **New SolidWorks Document** dialog box. You are asked to insert first part.
- Click on the **Browse** button from the **Begin Assembly PropertyManager** displayed in the left. The **Open** dialog box will display.
- Double-click on the Static Ring part in the **Epicyclic Gear Mechanism** folder. The part will attach to the cursor.
- Click in the viewport to place the part. The part will be fixes at the specified position.

Inserting and constraining Washer

• Click on the **Insert Components** button from the **Ribbon**. And then click on the **Browse** button from the **PropertyManager**. The **Open** dialog box will display.
• Double click on the **Eccentric Input shaft** part in the **Epicyclic Gear Mechanism** folder.
• The part will attach to the cursor.
• Click anywhere in the viewport to place the component.
• Select the round face of **Input shaft** and **Static ring**, and select the **Concentric** button from the **Mates** toolbar displayed; refer to Figure-17.

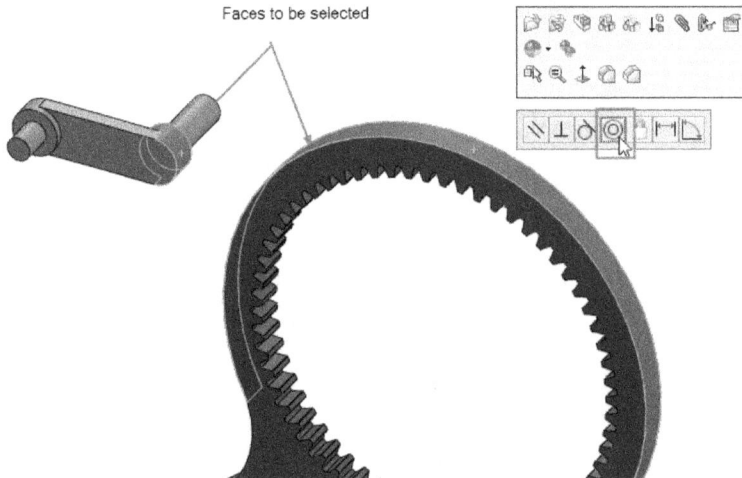

Figure-17. Applying concentric mate

• Select the flat face of Input shaft and Static ring as shown in and apply the **Coincident** mate from the **Mates** toolbar.

Inserting and constraining Eccentric static bearing

• Click on the **Insert Components** button from the **Ribbon**. And then click on the **Browse** button from the **PropertyManager**. The **Open** dialog box will display.
• Double click on the **Static Eccentric Bearing** part in the **Epicyclic Gear Mechanism** folder.
• The part will attach to the cursor.
• Click anywhere in the viewport to place the component.
• Select the round face of **Input shaft** and **Static Eccentric Bearing**, and select the **Concentric** button from the **Mates** toolbar displayed; refer to Figure-18.

Figure-18. Applying concentric mate on bearing

- Similarly, make the flat faces of the bearing and shaft coincident; refer to Figure-19.

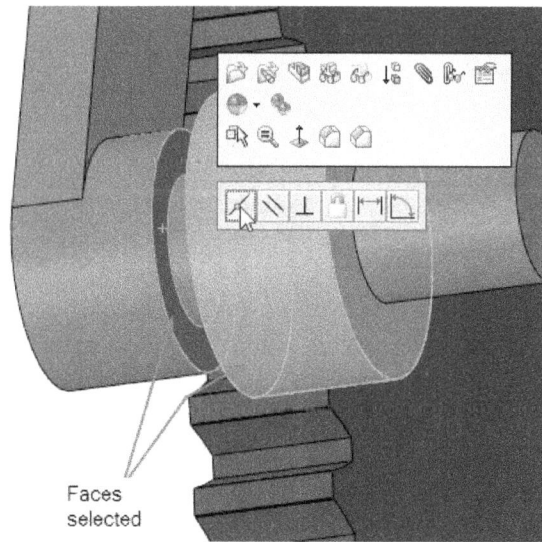

Figure-19. Making flat faces coincident

Inserting and constraining Planet gear

- Click on the **Insert Components** button from the **Ribbon**. And then click on the **Browse** button from the **PropertyManager**. The **Open** dialog box will display.
- Double click on the **Planet Gear** part in the **Epicyclic Gear Mechanism** folder.
- The part will attach to the cursor.
- Click anywhere in the viewport to place the component.
- Select the round face of **Input shaft** and **Plant Gear**, and select the **Concentric** button from the **Mates** toolbar displayed; refer to Figure-20.

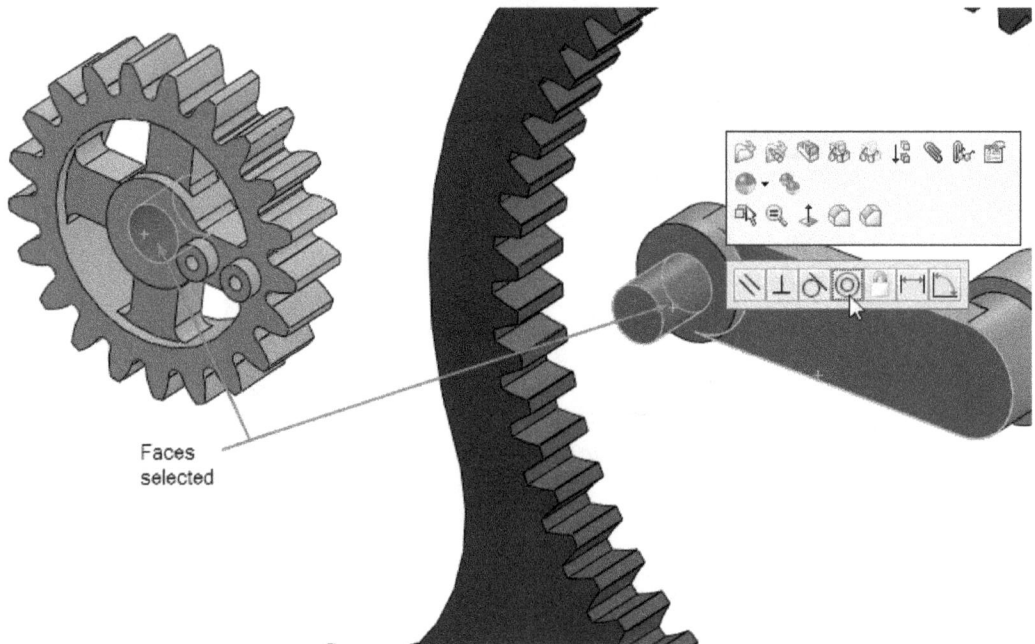

Faces selected

Figure-20. Applying concentric mate on planet gear

- Similarly, applying the coincident mate on the flat face of gear and Input shaft; refer to Figure-21.

Figure-21. Making flat faces of gear and shaft coincident

Inserting and constraining other components

In the same way, insert and constrain the other components refer to the figures given next for reference.

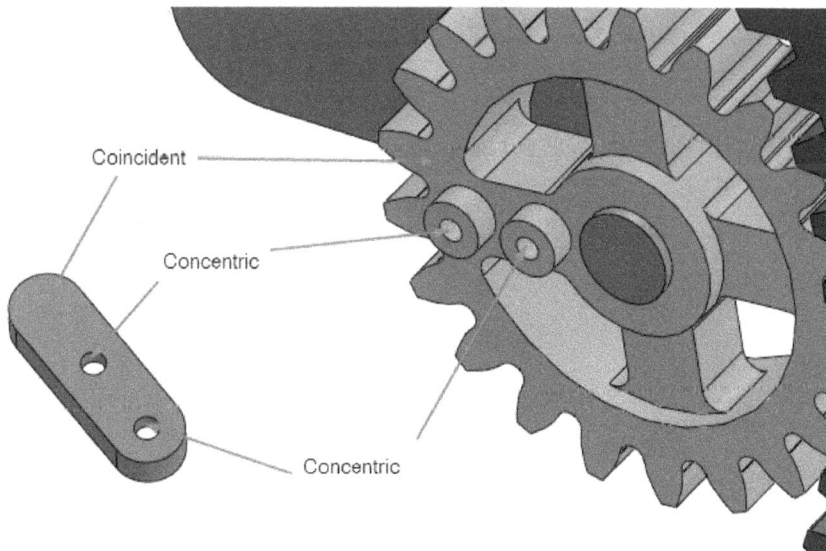

Figure-22. Placing arm pin

Figure-23. Placing the rod

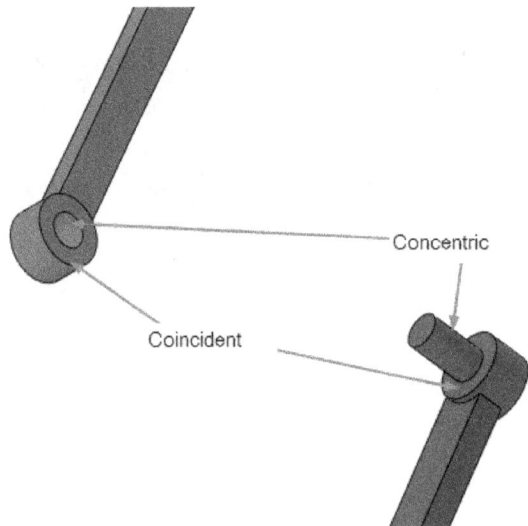

Figure-24. Placing output crank

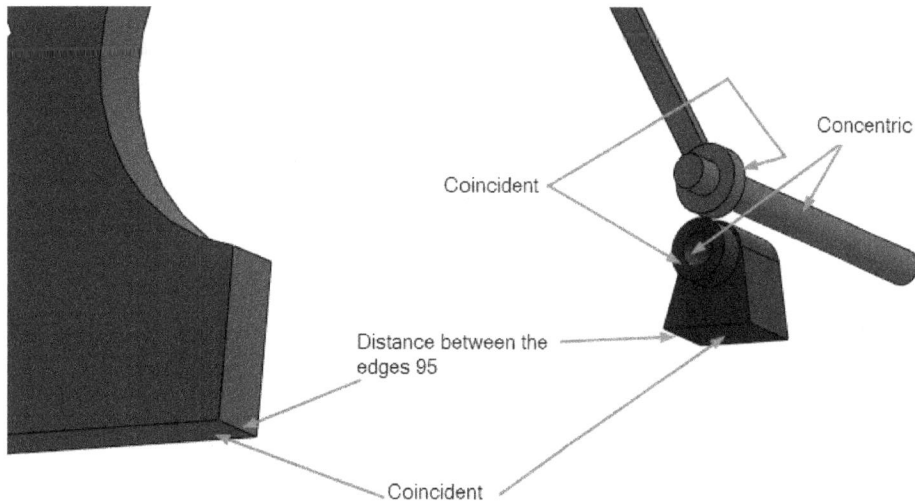

Figure-25. Placing the output shaft bearing

Applying relation between gears

• After placing all the components and applying the constraints, click on the **Mate** button from the **Ribbon**. The **Mate PropertyManager** will be displayed. Select the **Gearmate** button from the **Mechanical Mates** rollout in the **PropertyManager**; refer to Figure-26.

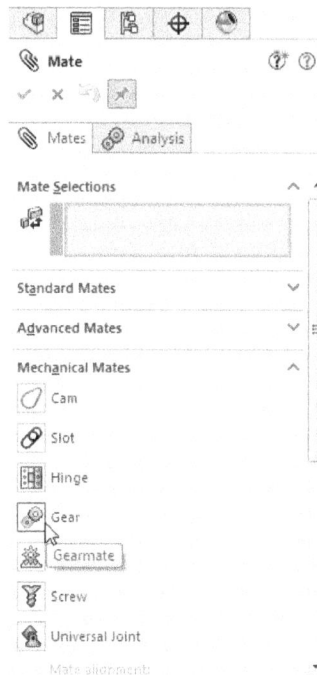

Figure-26. Gear mate

- Select the two edges of gear and static ring; refer to Figure-27.

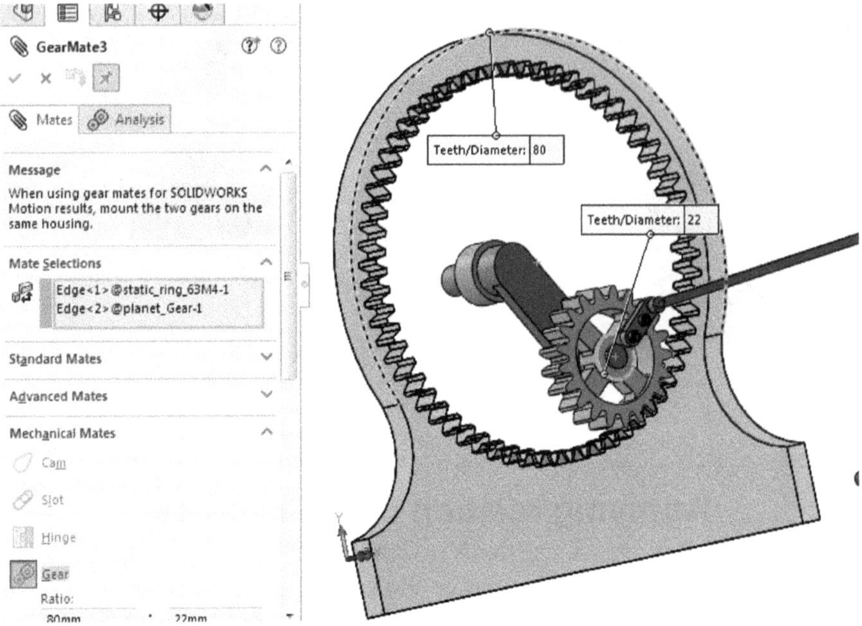

Figure-27. Edges selected for gear mate

- Specify the gear ratio as **80:22** in the **Mechanical Mates** rollout in the **PropertyManager** and click on the **OK** button from it twice.

Creating Motion Study

- Click on the **Motion Study 1** tab from the bottom bar. The Motion Study interface will be displayed; refer to Figure-28.

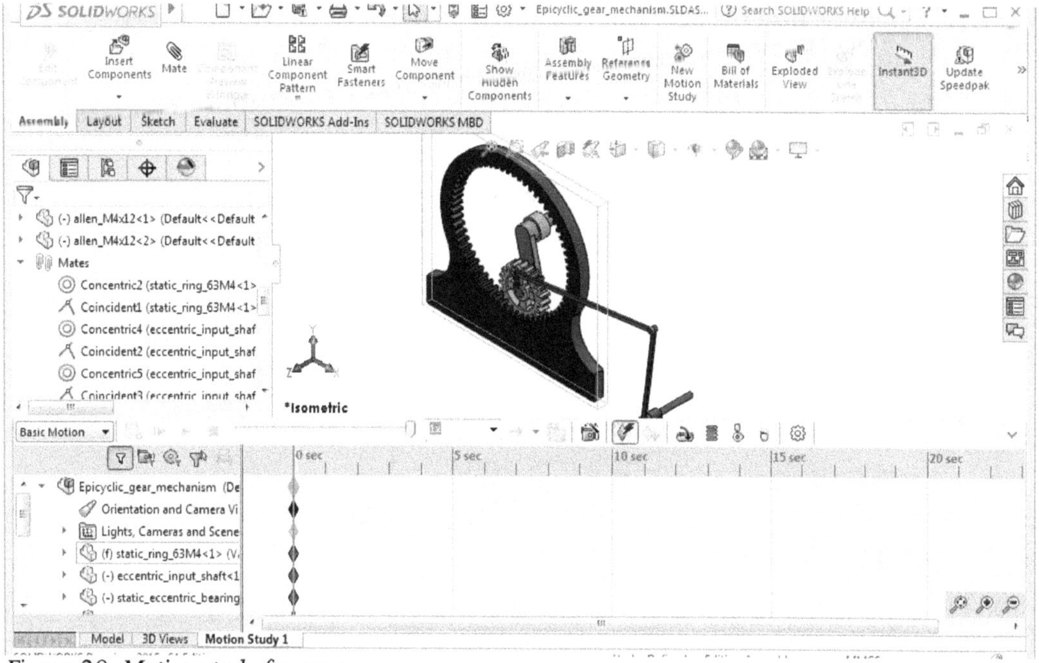

Figure-28. Motion study for gear

- Click on the **Motor** button from the Motion Study interface.
- Select the face shown in Figure-29. Specify the RPM as **10** and click on the **OK** button from the **Motor PropertyManager**.

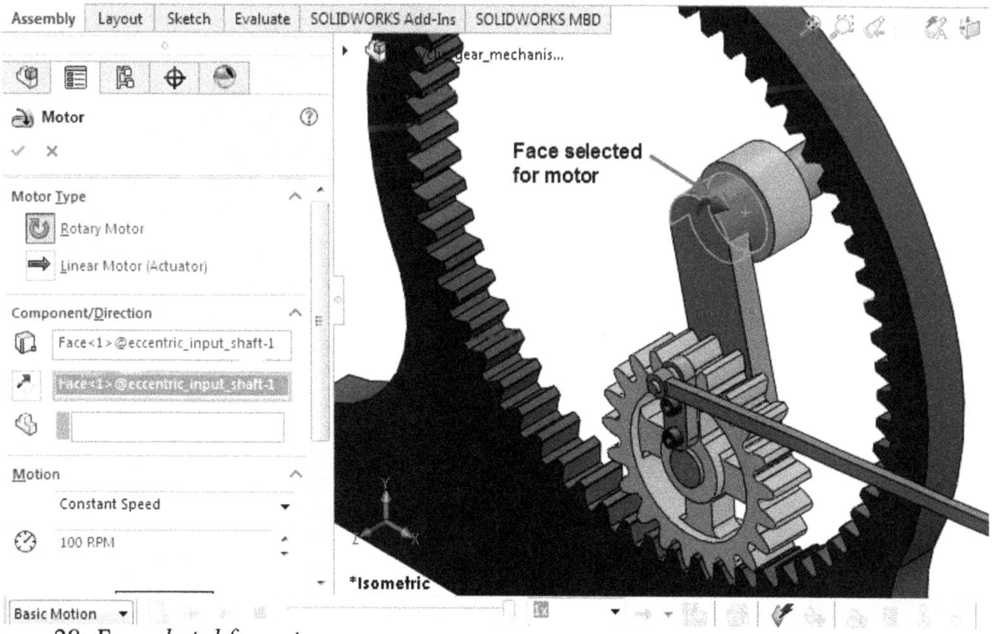

Figure-29. Face selected for motor

- Select the **Basic Motion** option from the drop-down; refer to Figure-30 and then click on the **Calculate** button to check the motion.

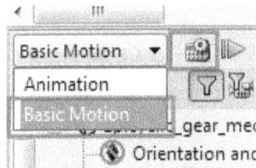

Figure-30. Basic Motion option

PRACTICE 1

Assemble the model as shown in Figure-31. The exploded view is given in Figure-32.

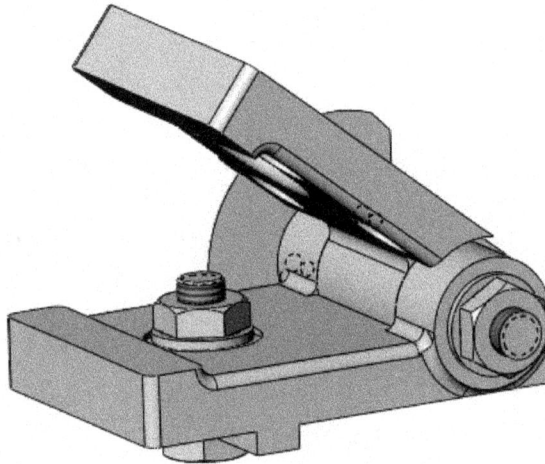

Figure-31. Bottom assembly

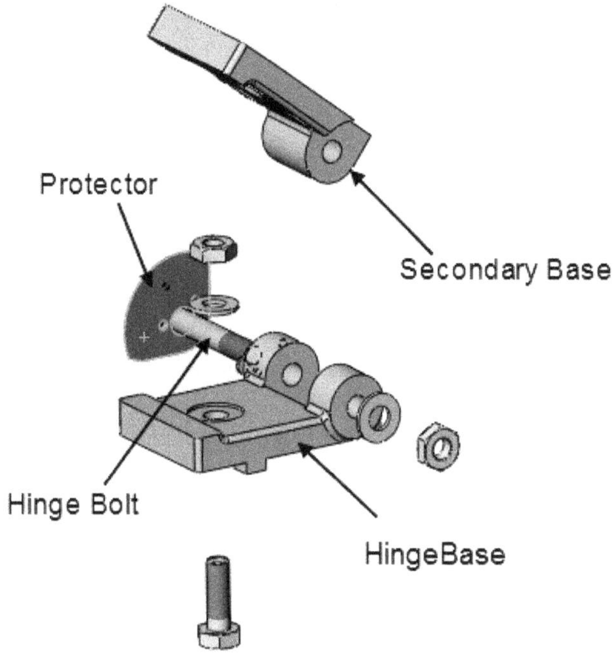

Protector

Secondary Base

Hinge Bolt

HingeBase

Figure-32. Exploded view of bottom assembly

Note that you need to import nuts and bolts from the toolbox provided by SolidWorks; refer to Figure-33. The toolbox is the standard library of components provided by SolidWorks. This library contains nuts, bolts, bearings, transmission parts, washer and a lot more with inch as well as mm specifications.

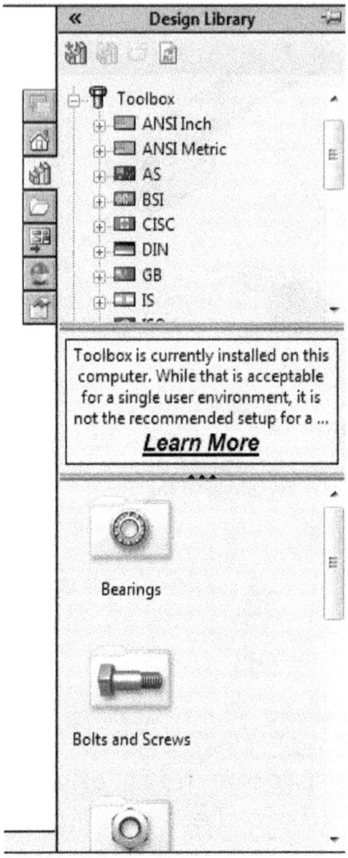

Figure-33. Toolbox

PRACTICE 2

Assemble the model as shown in Figure-34. The exploded view is given in Figure-35.

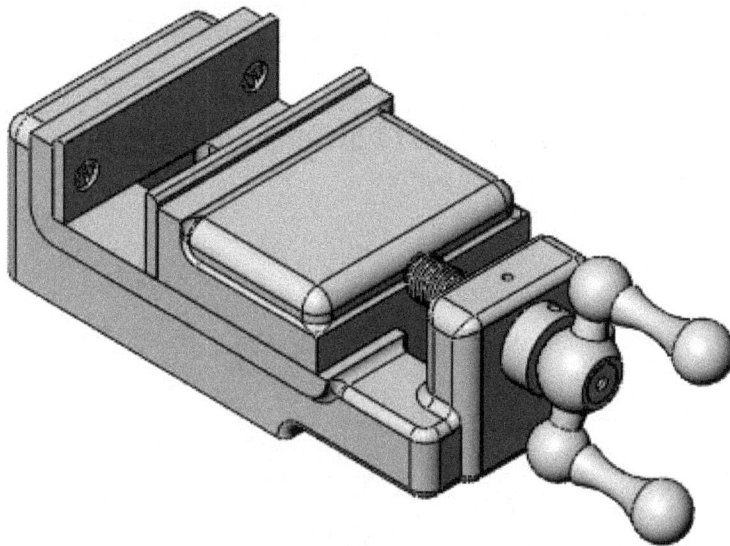

Figure-34. Top Assembly isometric view

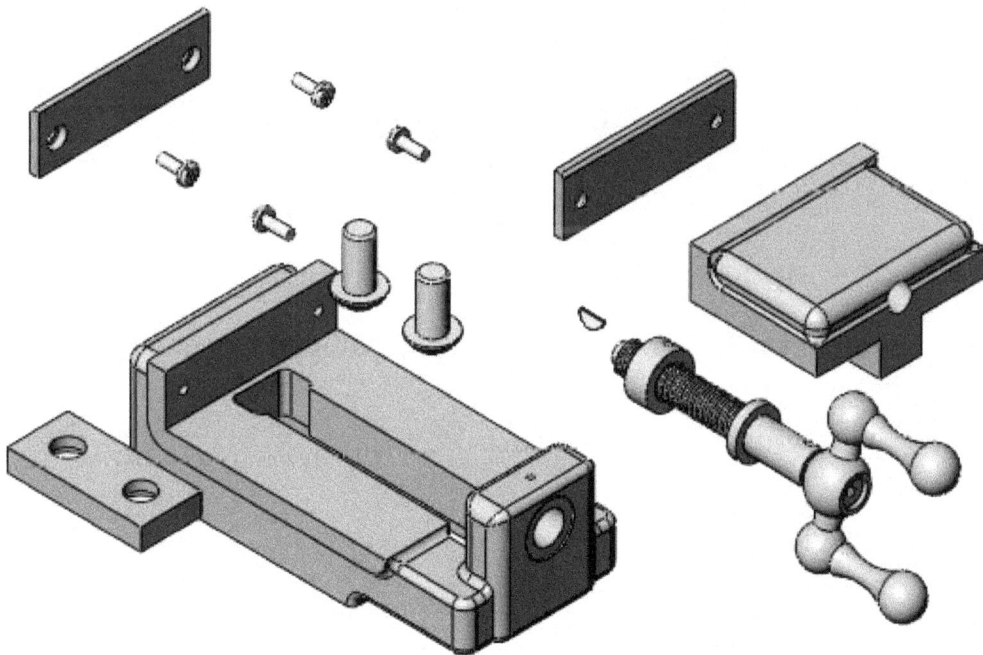

Figure-35. Exploded top assembly

PRACTICE 3

Assemble the Top Assembly and Bottom Assembly created in Practice 1 and Practice 2 to for the assembly as shown in Figure-36.

Note that in this practice, you will insert the complete assemblies in place of inserting single parts.

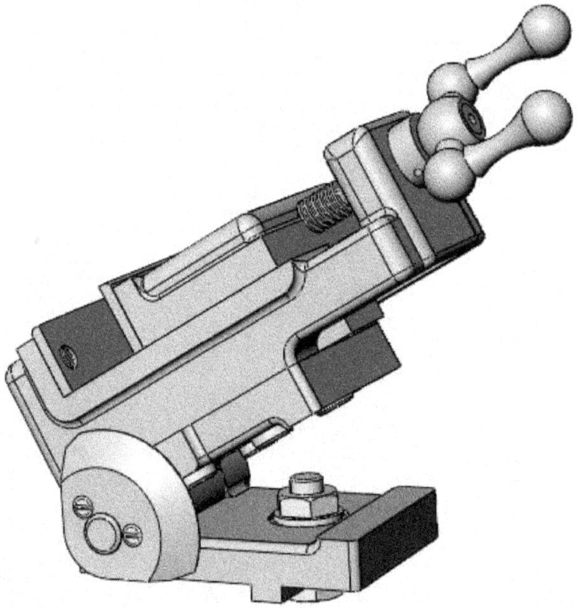

Figure-36. Assembly for practice 3

To get more parts for practice, write us at cadcamcaeworks@gmail.com

Surfacing and Practice

Chapter 8

Topics Covered

The major topics covered in this chapter are:

- *Surfacing Introduction.*
- *Surfacing tools similar to solid creation tools.*
- *Special tools for surfacing.*
- *Surface editing tools.*
- *Surface to solid conversion.*
- *Practical and Practice.*

SURFACING

Surfacing is a separate world in the field of CAD. The complicated shapes which are difficult for solid modeling are most of the time easy for surfacing. Basic tools of surfacing are very similar to the solid creation tools discussed earlier like, extrude, revolve, sweep and so on. To start surfacing in SolidWorks, click on the **Surfaces** tab in the **Ribbon** in Part environment. If this is not available by default, then right-click on any of the tab in the **Ribbon** and select the **Surfaces** option from the menu. The interface will display as shown in Figure-1. All the tools are used in SolidWorks for surfacing are explained as follow:

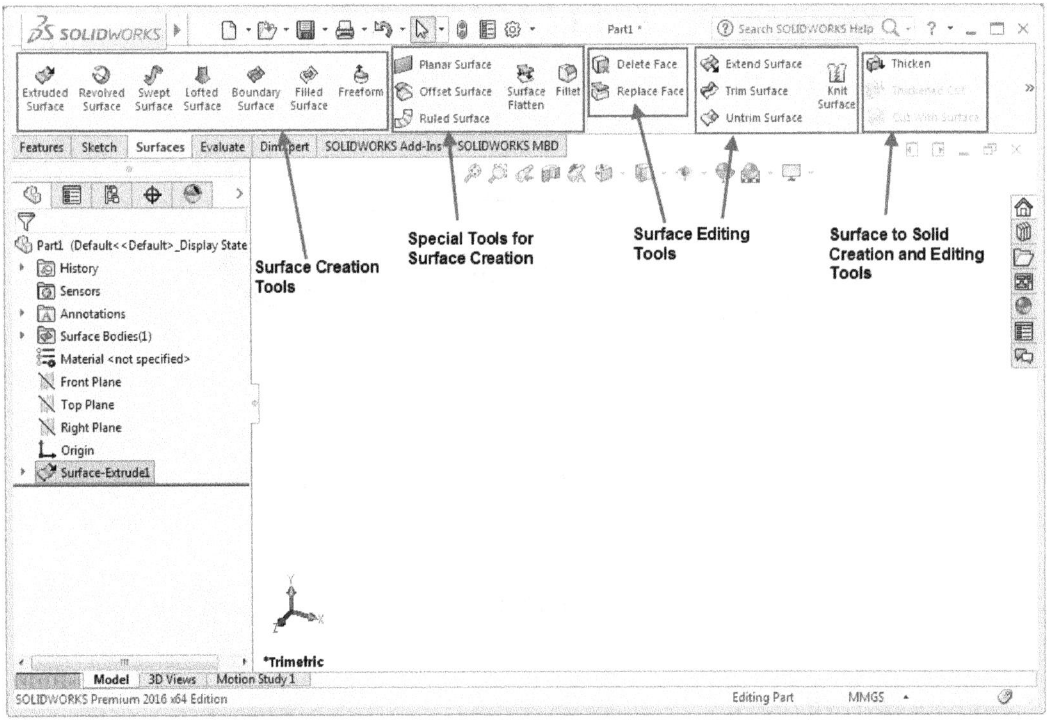

Figure-1. Surfacing interface

SURFACING TOOLS SIMILAR TO SOLID CREATION TOOLS

The list of these tools is given next.

- **Extruded Surface**   similar to   **Extruded Boss/Base**
- **Revolved Surface**   similar to   **Revolved Boss/Base**

- **Swept Surface**   similar to   **Swept Boss/Base**
- **Lofted Surface**   similar to   **Lofted Boss/Base**
- **Boundary Surface**   similar to   **Boundary Boss/Base**

These tools are one by one explained next.

Extruded Surface

The **Extrude Surface** tool is used to extrude a close or open sketch to the specified height to form a surface. The steps to use this tool are given next.

- Click on the **Extruded Surface** tool from the **Ribbon**. You are asked to select a plane to draw sketch.
- Draw open or close sketch of the surface on desired plane.
- Exit the sketch. Preview of the surface will display; refer to Figure-2.

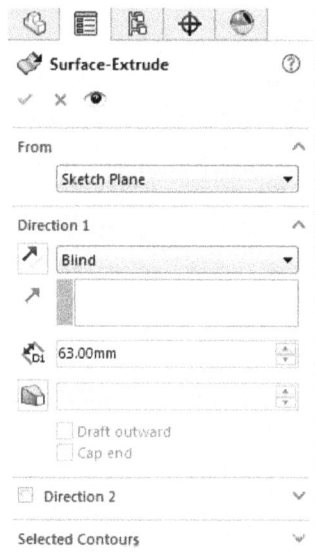

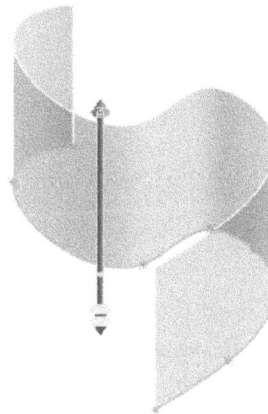

Figure-2. Extruded surface

- If you draw a close sketch, then you can select the **Cap end** check box to close the end edges.
- The other options are similar to **Extruded Boss/Base** tool.
- Click on the **OK** button from the **PropertyManager** to create the extruded surface.

Revolved Surface

The **Revolved Surface** tool is used to revolve a close or open sketch to the specified angle with respect to selected reference to form a surface. The steps to use this tool are given next.

- Click on the **Revolved Surface** tool from the **Ribbon**. You are asked to select a plane to draw sketch.
- Draw open or close sketch of the surface on desired plane. Make sure that you create a center line for revolving the sketch.
- Exit the sketch. Preview of the surface will display; refer to Figure-3. Options in the **PropertyManager** have be discussed already.

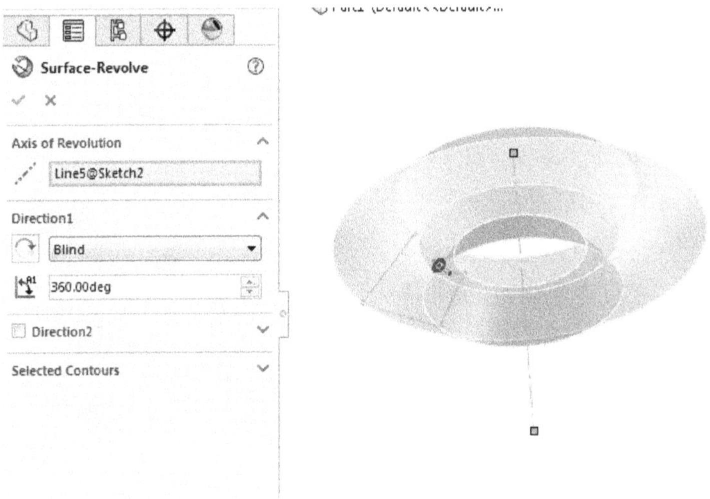

Figure-3. Revolved surface

- Click on the **OK** button from the **PropertyManager** to create the revolved surface.

Swept Surface

The **Swept Surface** tool is used to sweep a section along the selected path to form a surface. The steps to use this tool are given next.

- Click on the **Swept Surface** tool from the **Ribbon**. **Surface-Sweep PropertyManager** will display.

- Select the open or close section sketch.
- Select the open or close sketch for the path. Preview of the Swept surface will display; refer to Figure-4.

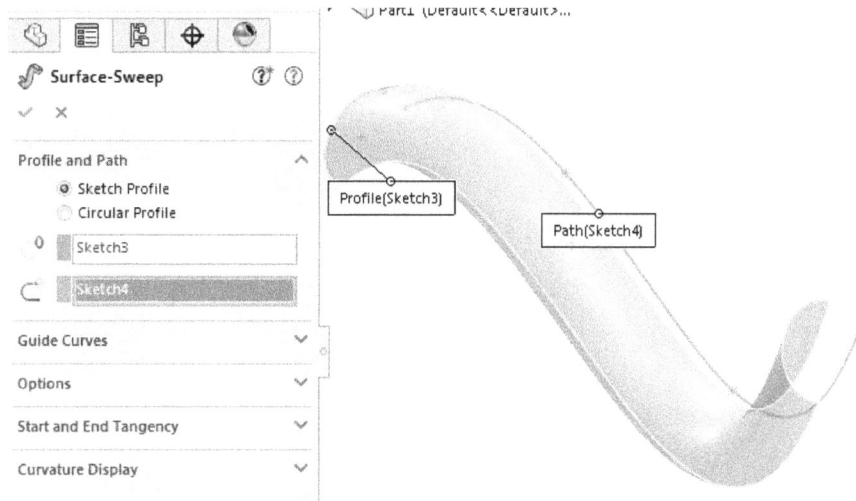

Figure-4. Swept surface

- The options in the **PropertyManager** are same as for **Swept Boss/Base** tool.
- Click on the **OK** button from the **PropertyManager** to create the surface.

Note that for creating **Swept Boss/Base** feature as well as for **Swept surface,** you can select edge of an existing model as a path if required.

Lofted Surface

The **Lofted Surface** tool is used to join two open/close sections to form surface. The steps to use this tool are given next.

- Click on the **Lofted Surface** tool from the **Ribbon. Surface-Loft PropertyManager** will display.

- Select the first open/close section sketch.
- Select the second open/close section sketch. Preview of the lofted surface will display; refer to Figure-5.

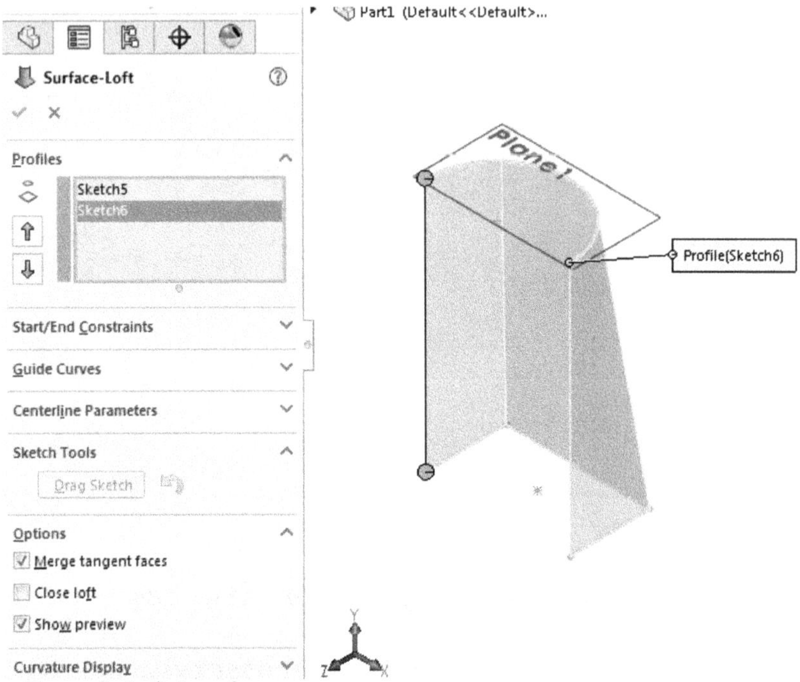

Figure-5. Lofted surface

- Click **OK** button to create the surface.

Note that if you are selecting first section as open sketch then the second section should also be open sketch. Same with the close sketches.

Boundary Surface

The **Boundary Surface** tool is used to join two open/close sections to form surface. The steps to use this tool are given next. This tool works in the same way as the lofted surface.

Note that using the Lofted Surface tool and Boundary Surface tool you can join two or more surfaces, edges of solids and face of solids. Refer to Figure-6 and Figure-7.

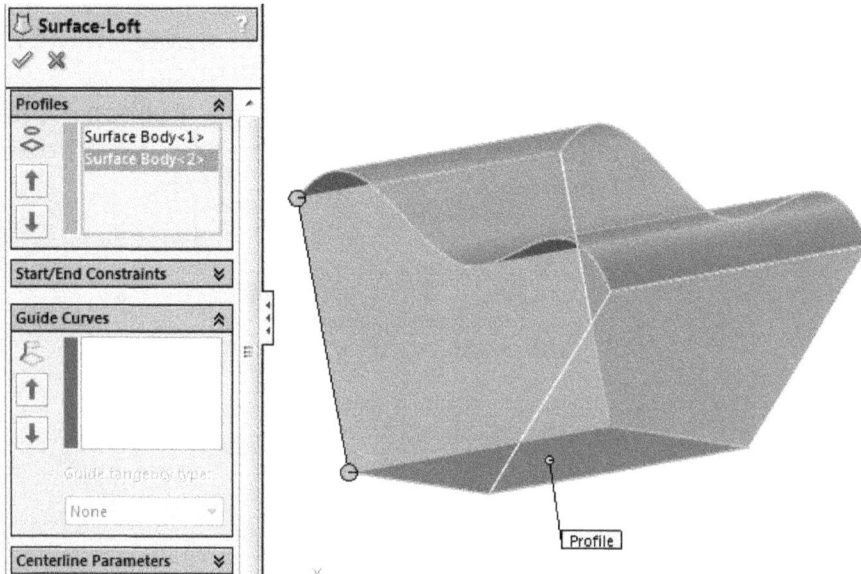

Figure-6. Surfaces joined by loft

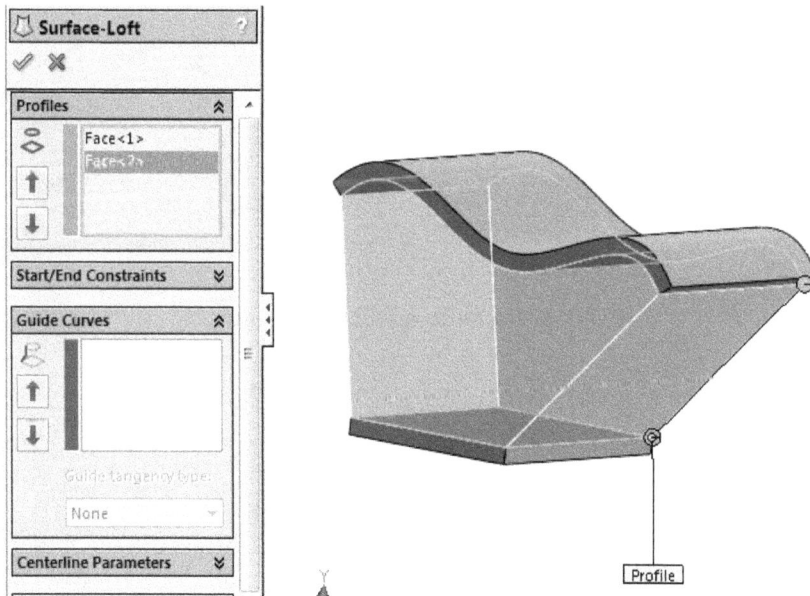

Figure-7. Faces of solids joined by loft

Filled Surface

The **Filled Surface** tool is use to fill gap by selecting the close boundary. This close boundary can be made by creating sketches or it can be formed by intersection of solids/surfaces. The steps to use this tool are given next.

- Click on the **Filled Surface** tool from the **Ribbon**. The **Fill Surface PropertyManager** will display; refer to Figure-8.

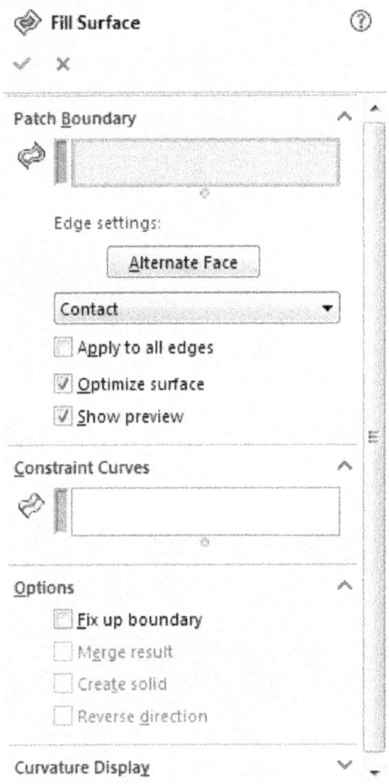

Figure-8. Fill Surface PropertyManager

- Select the boundary of area that you want to fill using the surface. Preview of the surface will display; refer to Figure-9.

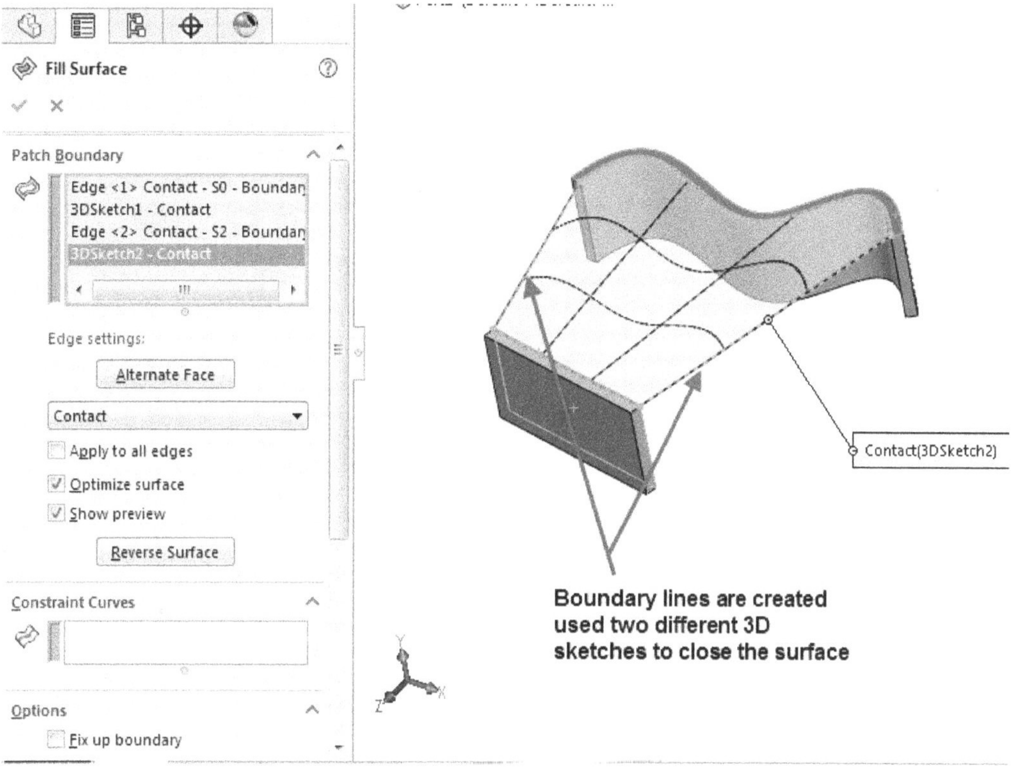

Figure-9. Filled surface

This tool is the most used one for closing surfaces. Using this tool, you can also close circular holes in surfaces/solids.

Freeform

The **Freeform** tool is used to freely deform solid faces/surfaces. The steps to use this tool are given next.

- Click on the **Freeform** tool from the **Ribbon**. The **Freeform PropertyManager** will display; refer to Figure-10.

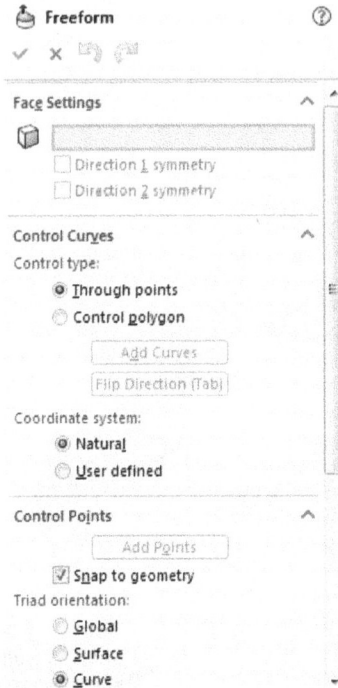

Figure-10. Freeform PropertyManager

- Select the surface/face you want to deform. Mesh on curves will display on the surface.
- Click on the **Add Curves** button and select at the desired position over the surface to set the highlighted line as control curve. You can select more than one curve for controlling the surface shape.
- Click again on the **Add Curves** button to exit the selection mode.
- Click on the **Add Points** button to add control points on the control curve.
- Click again on the **Add Points** button or Right-click to exit selection mode.

- Drag the point you have selected earlier to change the shape of the surface/face. Refer to Figure-11.

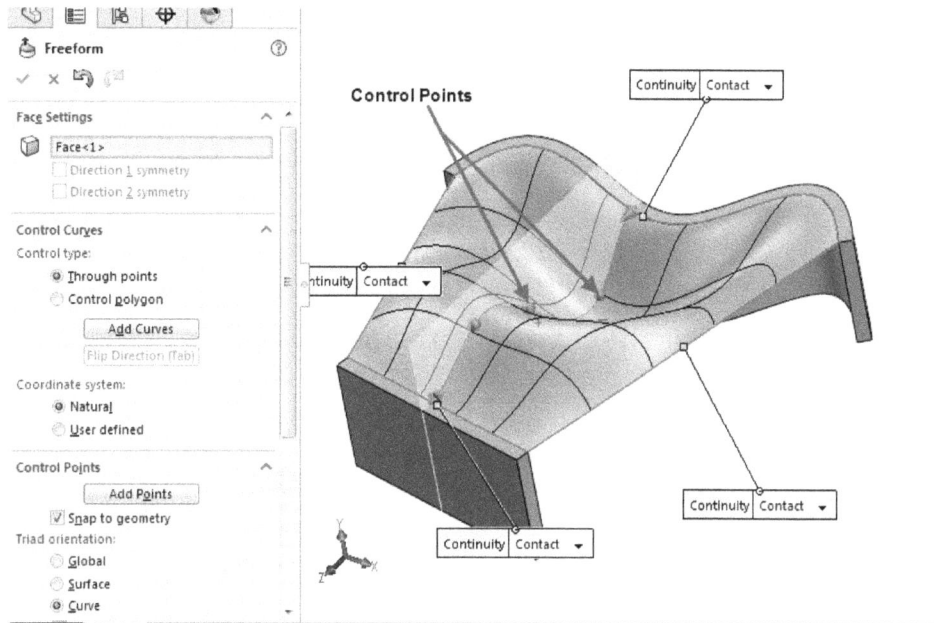

Figure-11. Freeforming surface

- Click on the **OK** button from the **PropertyManager** to apply the modification.

SPECIAL SURFACING TOOLS

Earlier, we have learned the tools that were similar to the Solid creation tools. Now, we will discuss about the special tools provided in SolidWorks to create surfaces. These tools are discussed next.

Planar Surface

The **Planar Surface** tool is used to create a surface joining two or more edges or a sketch that is in the same plane. The steps to use this tool are given next.

- Click on the **Planar Surface** tool from the **Ribbon**. The **Planar Surface PropertyManager** will display; refer to Figure-12.

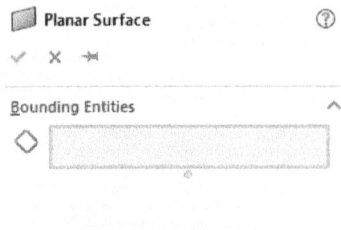

Figure-12. Planar Surface
PropertyManager

- Select the edges that form a planar surface. The preview will be displayed; refer to Figure-13.

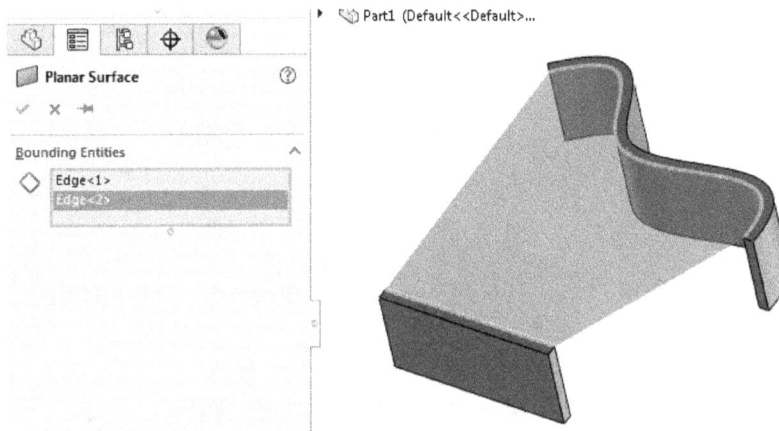

Figure-13. Preview of planar surface

- Click on the **OK** button from the **PropertyManager** to create the surface.

Offset Surface

The **Offset Surface** tool is used to create a surface at an offset distance from the selected face/surface. The steps to create offset surface are given next.

- Click on the **Offset Surface** tool from the **Ribbon**. The **Offset Surface PropertyManager** will display; refer to Figure-14.

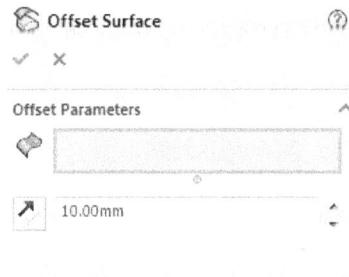

Figure-14. Offset Surface
PropertyManager

- Select a surface/face by which you want to create the offset surface. Preview of the surface will be displayed; refer to Figure-15.

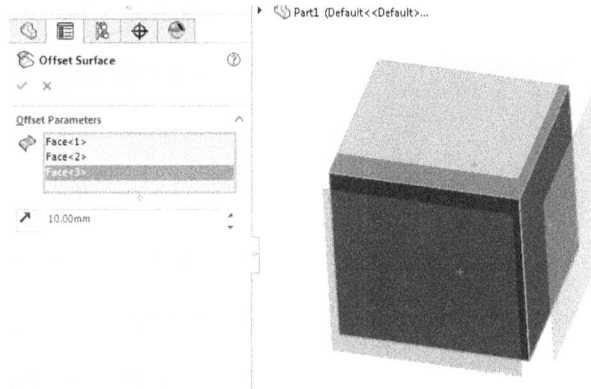

Figure-15. Preview of offset surface

- Specify the desired distance in the spinner and click on the **OK** button to create the offset surface.

Ruled Surface

The **Ruled Surface** tool is used to create a combination of surface adjoining to each other. This type of surface becomes very important while creating parting surface for molding/casting. The steps to use this tool are given next.

- Click on the **Ruled Surface** tool from the **Ribbon**. The **Ruled Surface PropertyManager** will display; refer to Figure-16.

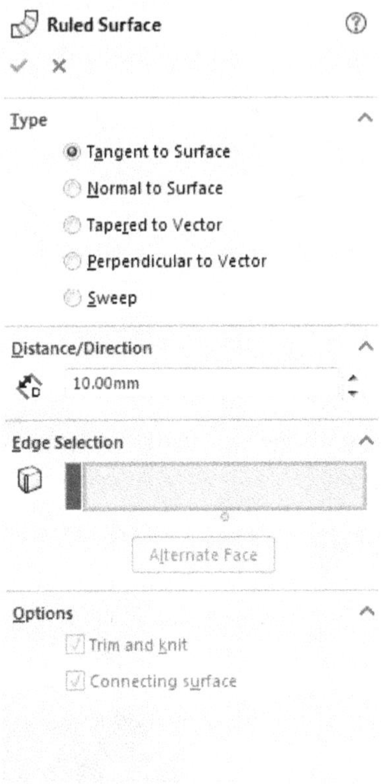

Ruled Surface ⑦

✓ ✕

Type ⌃

◉ Tangent to Surface

◯ Normal to Surface

◯ Tapered to Vector

◯ Perpendicular to Vector

◯ Sweep

Distance/Direction ⌃

10.00mm ⬍

Edge Selection ⌃

Alternate Face

Options ⌃

☑ Trim and knit

☑ Connecting surface

Figure-16. Ruled Surface PropertyManager

- Select the edges using which you want to create the ruled surface. Preview of surface will display; refer to Figure-17.
- Increase the length of surfaces by using the spinner in the **Distance/Direction** rollout.

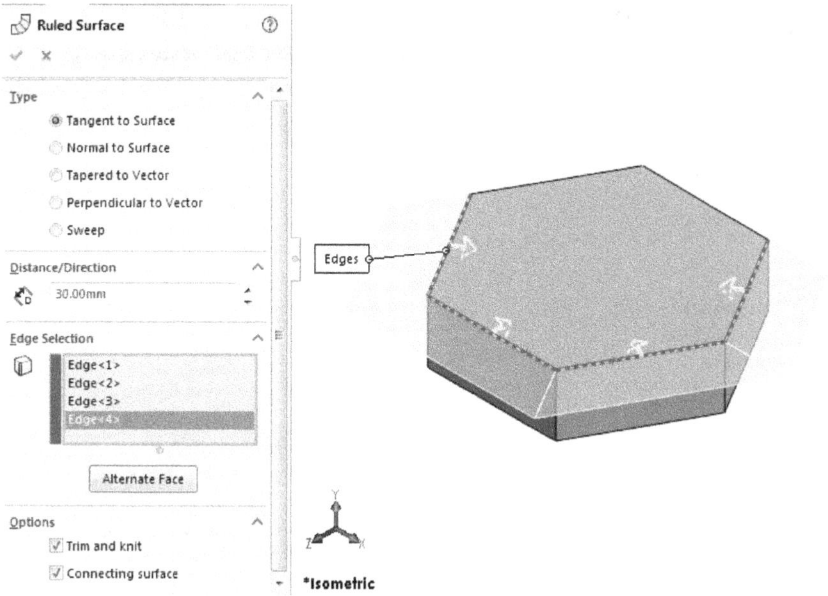

Figure-17. Preview of ruled surface

- If you select the **Normal to Surface** radio button, then the surfaces will be displayed as shown in Figure-18.

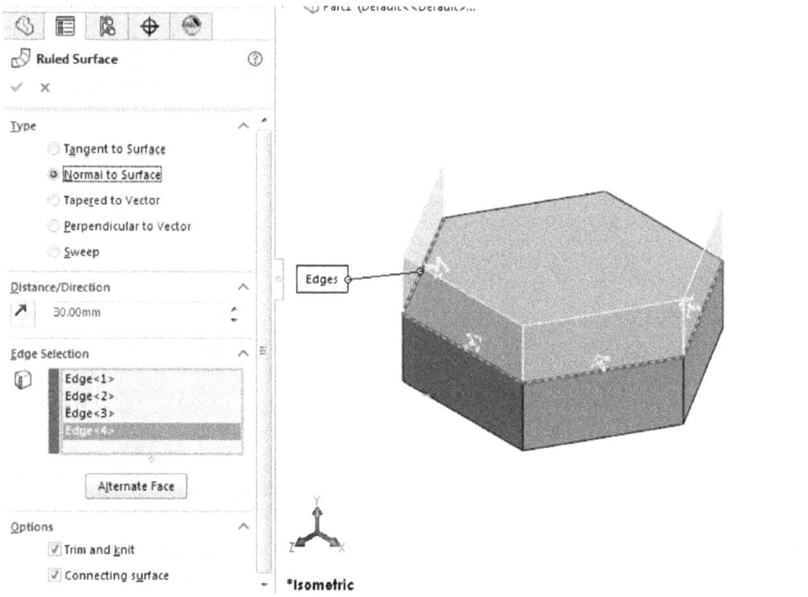

Figure-18. Preview of surface on selecting Normal to Surface radio button

- If you want to create surface tapered to the selected vector, then select the **Tapered to Vector** radio button and then select the edge to specify vector. Also, set the angle in the **Angle** spinner. Preview of the surface will be displayed; refer to Figure-19.

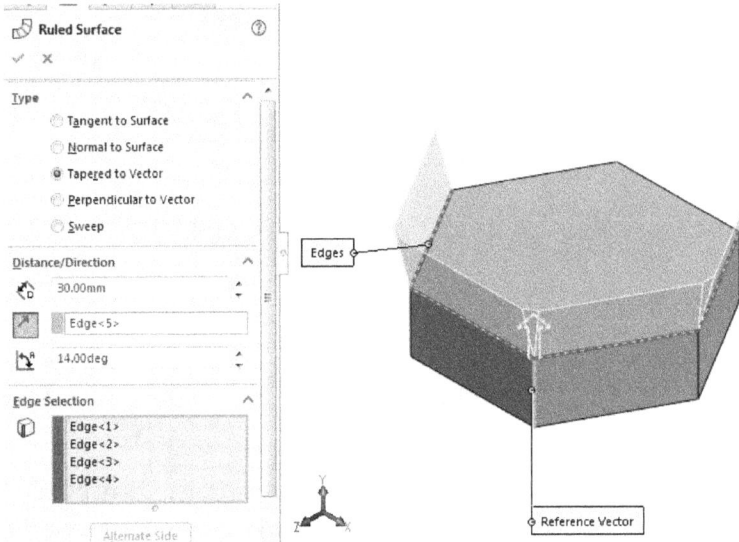

Figure-19. Preview of surface on selecting Tapered to Vector radio button

- Select the **Perpendicular to Vector** radio button to create the surfaces perpendicular to selected vector. Refer to Figure-20.

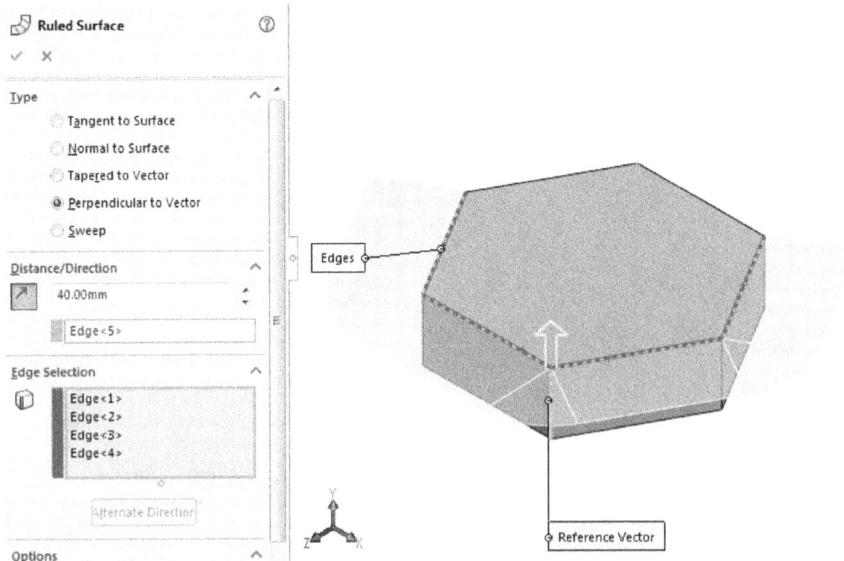

Figure-20. Preview of surface on selecting Perpendicular to Vector radio button

- Select the **Sweep** radio button to sweep the surface along the selected vector.

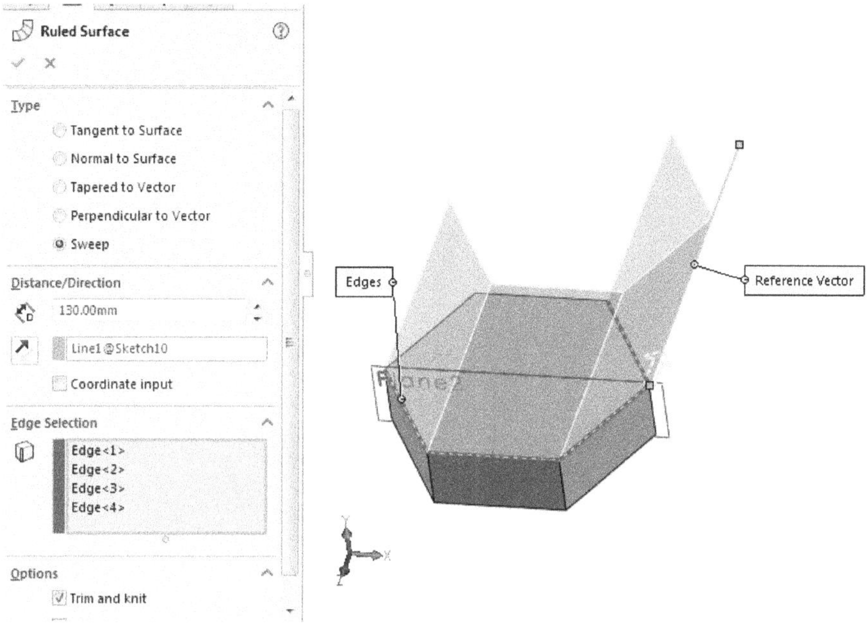

Figure-21. Preview of surface on selecting Sweep radio button

- After selecting the desired option, click on the **OK** button to create the surfaces.

Surface Flatten tool

The **Surface Flatten** tool was added in SolidWorks 2015 and is greatly enhanced in SolidWorks 2016. The **Surface Flatten** tool as the name suggests is used to make a flat surface of various interconnected surfaces. Note that before using this tool, there must be surfaces in the viewport. The procedure to use this tool is given next.

- Click on the **Surface Flatten** tool from the **Ribbon**. The **Flatten PropertyManager** will be displayed; refer to Figure-22.

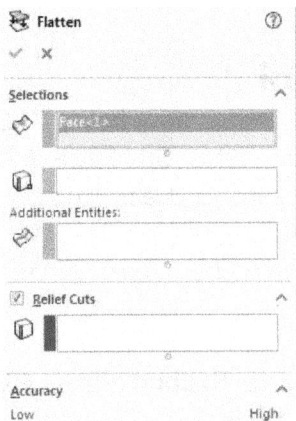

Figure-22. Flatten PropertyManager

- Select the surface/surfaces that you want to flatten; refer to Figure-23.

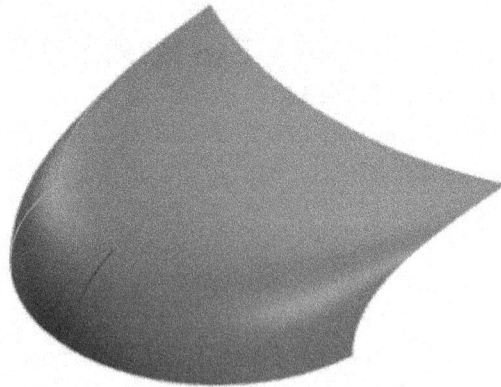

Figure-23. Surface selected

- Click in the **Vertex/Point** collector to select the fixed reference.
- Select the corner vertex of the surface. Preview of the surface will be displayed; refer to Figure-24.

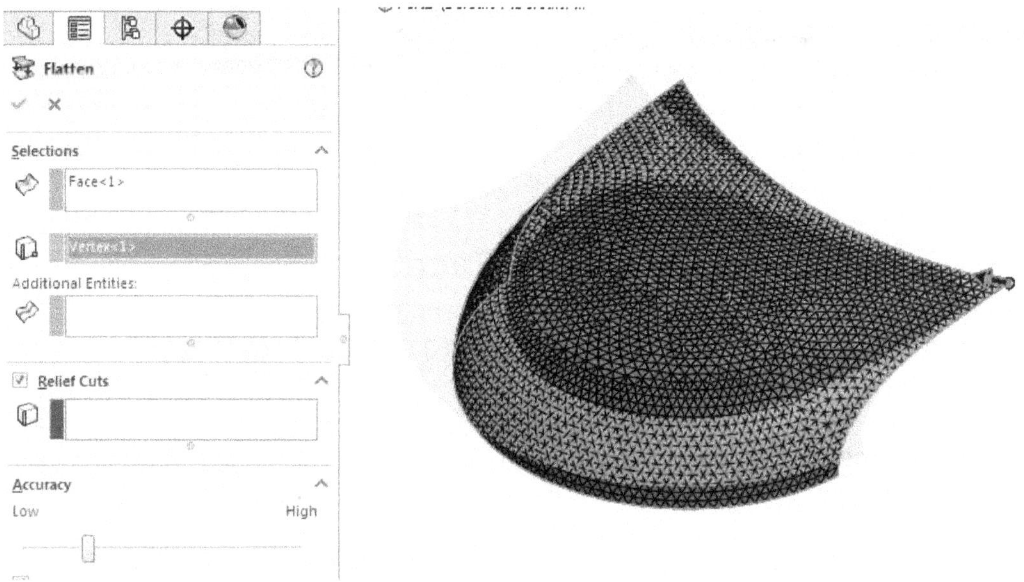

Figure-24. Preview of flattened surface

- If you want to apply relief in the surface, then select the **Relief Cuts** check box and click in the selection box in **Relief Cuts** rollout. You will be asked to select a curve.
- Click on the curve that you want to use for relief cut; refer to Figure-25.

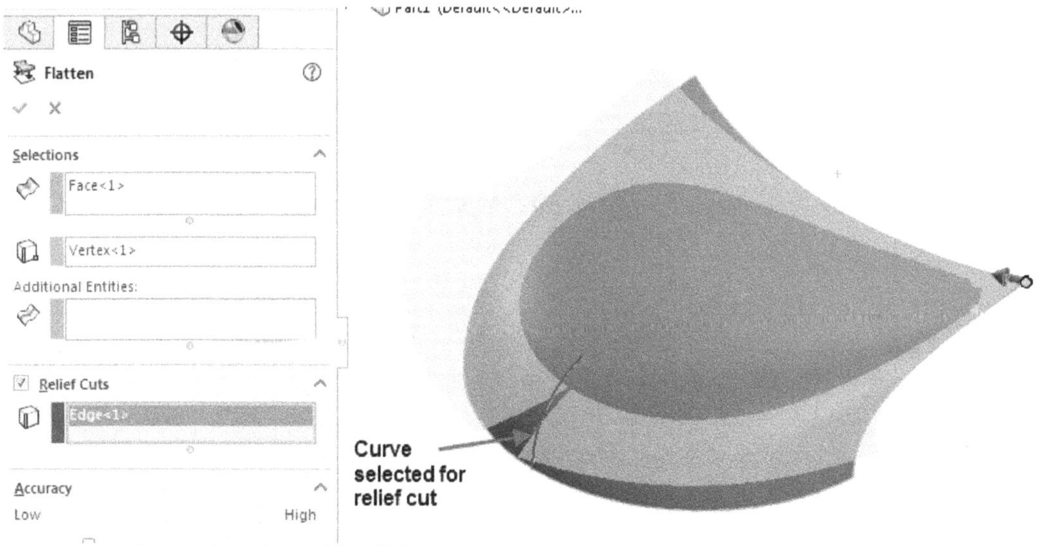

Figure-25. Flattened surface with relief cut

- Click on the **OK** button from the **PropertyManager** to create the flattened surface. Note that the parent surface will not be deleted by this operation.

SURFACE EDITING TOOLS

The tools in this category are used to edit the surfaces. For example, trimming the surface, deleting some portion of the surface, knitting two surfaces for applying fillet. These tools are explained next.

Delete Face

The **Delete Face** tool is used to remove a face/surface from the model. The procedure to use this tool are given next.

*   Click on the **Delete Face** tool from the **Ribbon**. The **Delete Face PropertyManager** will display as shown in Figure-26.
*   Select the face that you want to remove.

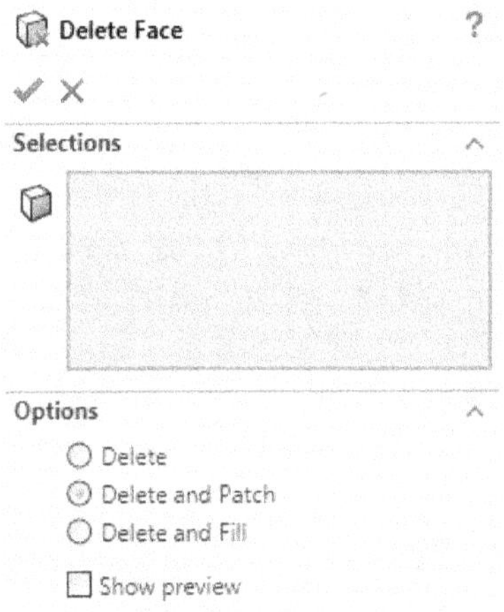

Figure-26. Delete Face PropertyManager

*   Select the desired radio button, if you want to patch or fill the gap created by deleting the surface.
*   Click on the **OK** button from the **PropertyManager**. The face will be deleted. Note that using this tool, you cannot delete a single face surface body.

Replace Face

The **Replace Face** tool is used to extend the selected face to the replacement face. The procedure to use this is given next.

- Click on the **Replace Face** tool. The **Replace Face PropertyManager** will display; refer to Figure-27.

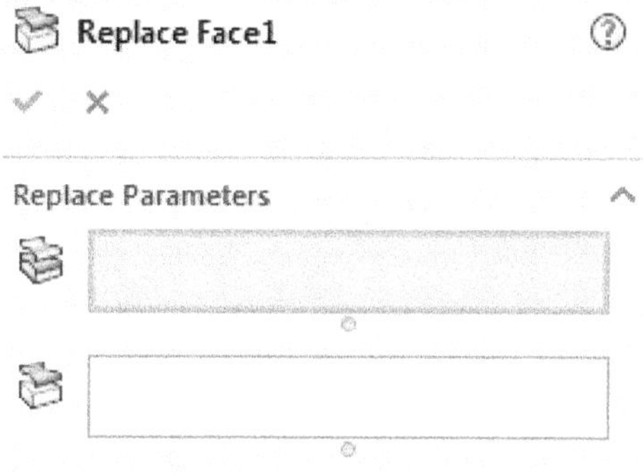

Figure-27. Replace Face PropertyManager

- Select the face/faces that you want to replace.
- Click in the next selection box in the **PropertyManager** and select the surface by which you want to replace the face.
- Click on the **OK** button to apply the replacement; refer to Figure-28.

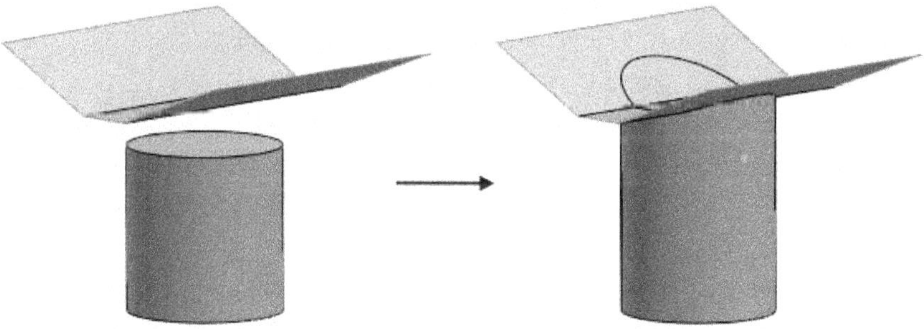

Figure-28. Output of replace face

Extend Surface

The **Extend Surface** tool is used to extend the selected surface by specified value. The procedure to use this tool is given next.

- Click on the **Extend Surface** tool. The **Extend Surface PropertyManager** will display; refer to Figure-29.

Figure-29. Extend Surface PropertyManager

- Select the edge of the surface that you want to extend. Preview of extension will display; refer to Figure-30.

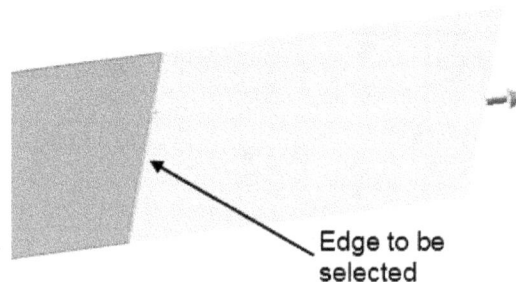

Figure-30. Preview of edge extension

- Specify the length of extension in the spinner. If you want to extend the surface to a selected point/surface then select the respective radio button and then the reference. The surface will extend to that reference.

Trim Surface

The **Trim Surface** tool is used to remove desired portion of the surface by using sketch/other surface. The procedure to use this tool is given next.

- Click on the **Trim Surface** tool. The **Trim Surface PropertyManager** will display; refer to Figure-31.
- Select the sketch or surface that you want to use as trimming tool. You are asked to select the portion that you want to keep; refer to Figure-32.

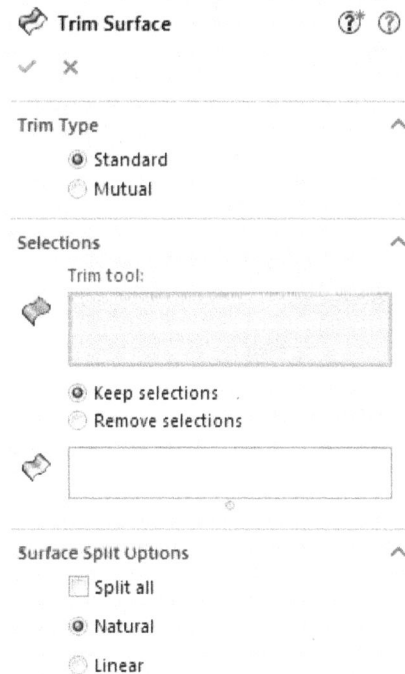

Figure-31. Trim Surface PropertyManager

Figure-32. Preview of trimmed surface

- Select the portion that you want to keep. The surface on the other side of trimming surface or sketch will be removed.

Untrim Surface

The **Untrim** tool is used to undo the trimmed surfaces. The following steps explain the procedure to untrim the surface.

- Select the **Untrim Surface** tool from the **Ribbon**. The **Untrim Surface PropertyManager** will display; refer to Figure-33.

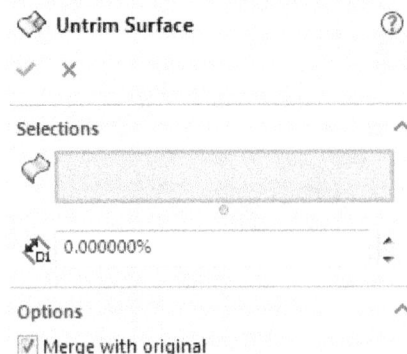

Figure-33. Untrim Surface PropertyManager

- Select the surface on which the trimming operations performed.
- Click on the **OK** button from the **PropertyManager**. The surface will be untrimmed and all the gaps will be patched.

Knit Surface

The **Knit Surface** tool is used to combine two or more surfaces at their common edges. If the surface form a close boundary then this tool turn the surfaces in to a solid. The steps to use this tool are given next.

- Click on the **Knit Surface** tool from the **Ribbon**. The **Knit Surface PropertyManager** will display; refer to Figure-34.

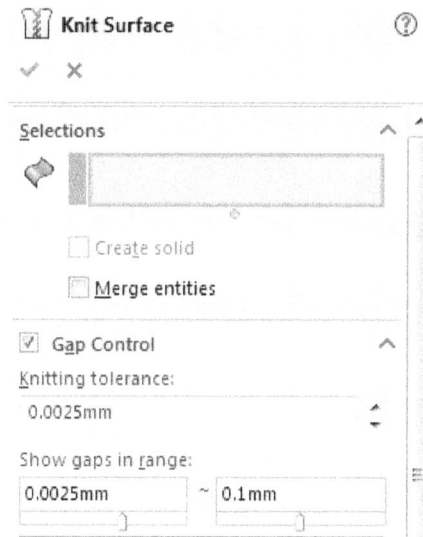

Figure-34. Knit Surface PropertyManager

- Select the surfaces that you want to knit with each other.
- Click **OK** button. The surfaces will be combined into a single entity. All the knit surfaces are added in the **Surface Bodies** folder; refer to Figure-35.

Note that the knitted surfaces are used as parting surfacing in molding and casting. Also, if the surfaces form a closed boundary then select the **Create solid** check box to create solid from surface.

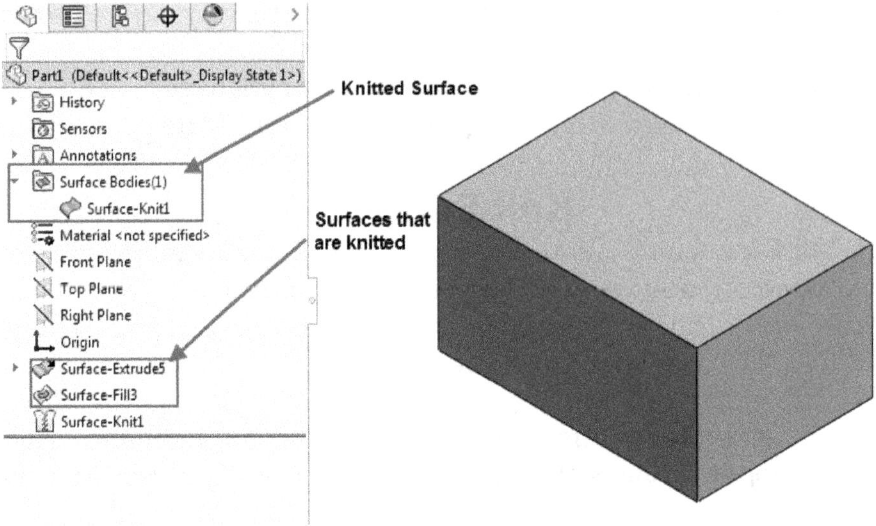

Figure-35. Knitted Surfaces

Thicken

The **Thicken** tool is used to add thickness to the surface. The procedure to use this tool is given next.

- Click on the **Thicken** tool from the **Ribbon**. The **Thicken PropertyManager** will display; refer to Figure-36.

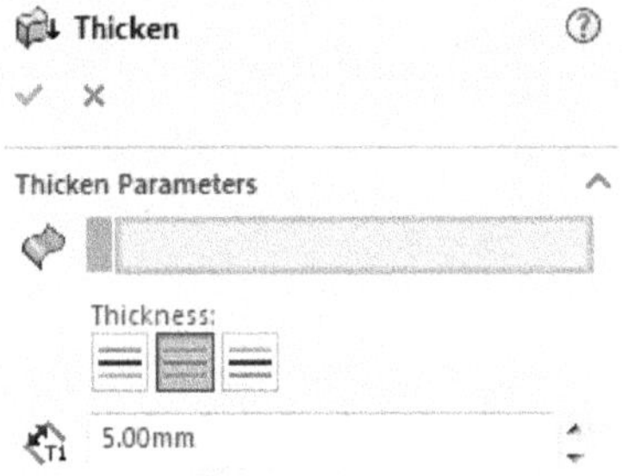

Figure-36. Thicken PropertyManager

- Select the surface to which you want to add thickness. The preview of thickened surface will display; refer to Figure-37.

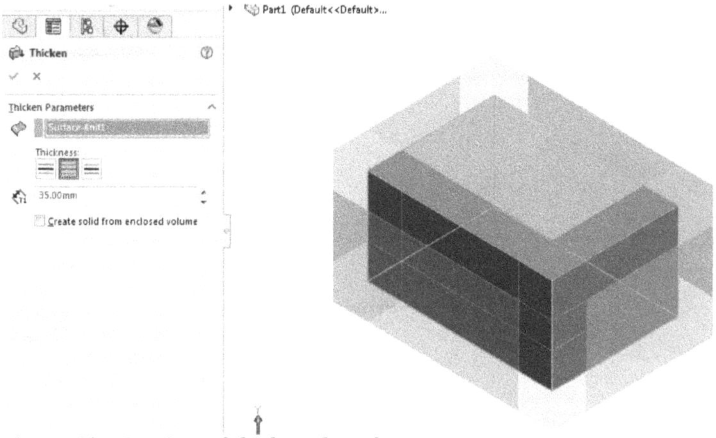

Figure-37. Preview of thickened surface

- Specify the desired thickness in the spinner and change the side of thickness application by selecting the desired button.

Thickened Cut

The **Thickened Cut** tool is used to remove material by thickening the surface. This tool works in the same way as **Thicken** tool but it removes material in place of adding.

Cut with Surface

The **Cut with Surface** tool is used to cut solids by using the surface. The steps to use this tool are given next.

- Click on the **Cut with Surface** tool from the **Ribbon**. The **SurfaceCut PropertyManager** will display; refer to Figure-38.

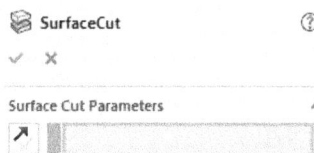

Figure-38. SurfaceCut PropertyManager

- Select the surface which you want to use for cutting the solids.
- The solids in the direction of arrow will be removed. Click on the arrow to flip the direction.

- Click on the **OK** button to cut the solids; refer to Figure-39.

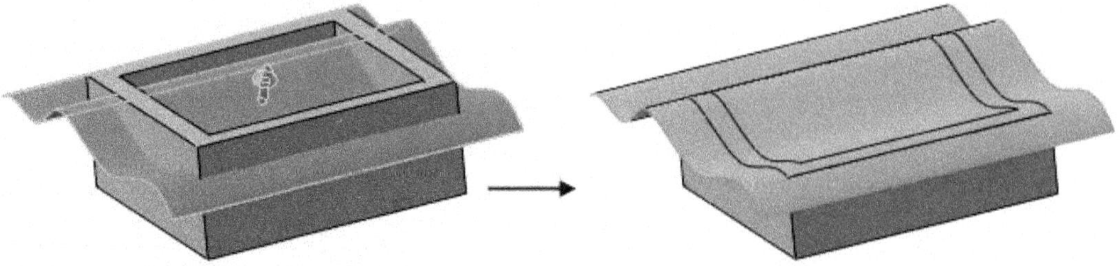

Figure-39. Solid cut by surface

We have covered all the important tools that are used for surfacing in SolidWorks. Now, we will practice on some models to apply these tools.

PRACTICAL 1

Create the model of helmet glass as shown in Figure-40. The dimensions of the model are given in Figure-41.

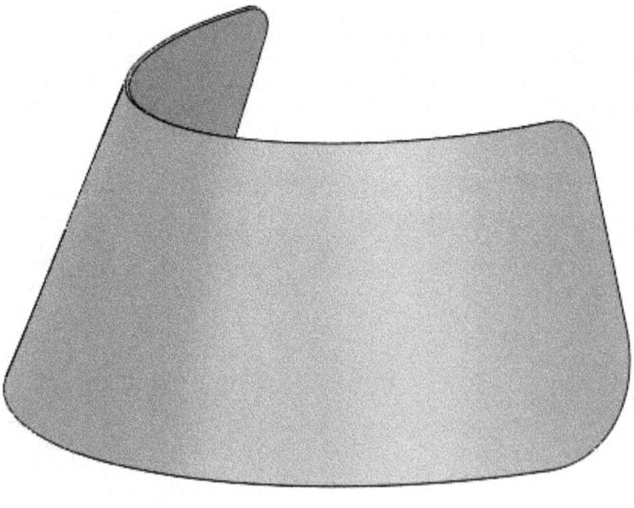

Figure-40. Practical1 model

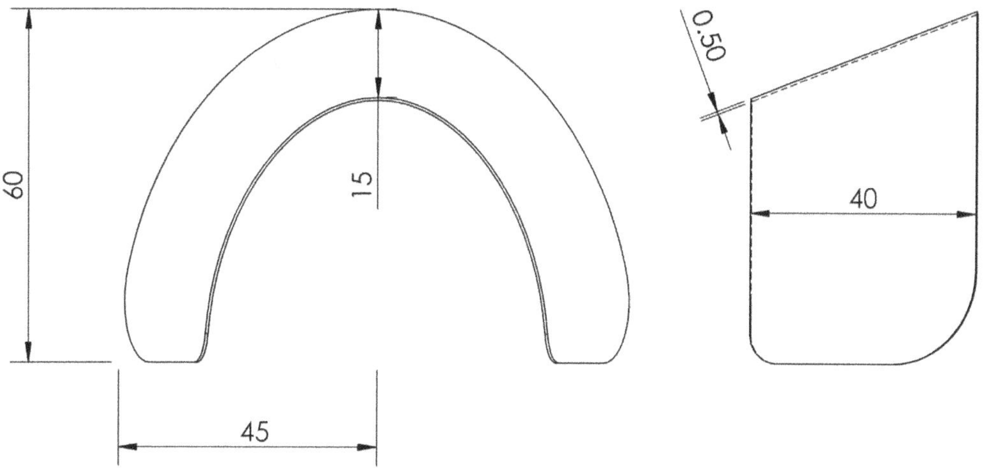

Figure-41. Practical1 drawing

The model displayed is having very low thickness and its having complex 3D shape. So, it is a good idea to use surfacing in this case.

We can create this model by lofted surface easily. For that we need to have two sketches.

Creating first sketch

- Click on the **Sketch** tab and select the **Sketch** button. You are asked to select sketching plane.
- Click on the **Top** plane. The sketching environment will display.
- Click on the **Ellipse** tool from the **Ribbon** and draw an ellipse as shown in Figure-42.
- Draw center line passing through coordinate system and trim the bottom portion of the ellipse; refer to Figure-43.

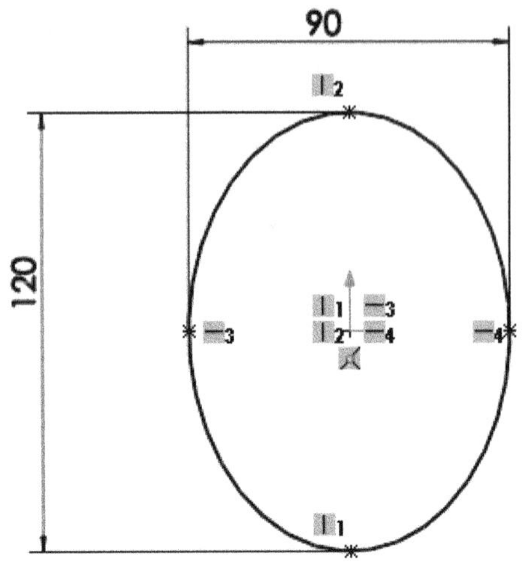

Figure-42. Ellipse to be drawn

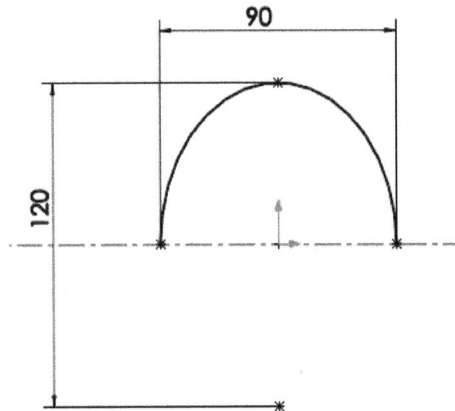

Figure-43. Trimmed ellipse

- Click on the **Exit Sketch** button to exit the sketch.

Creating second sketch

- Click on the **Plane** tool from **Reference Geometry** drop-down in **Surfaces** tab of the **Ribbon** and create a plane at an offset distance of **40** above the **Top** plane; refer to Figure-44.

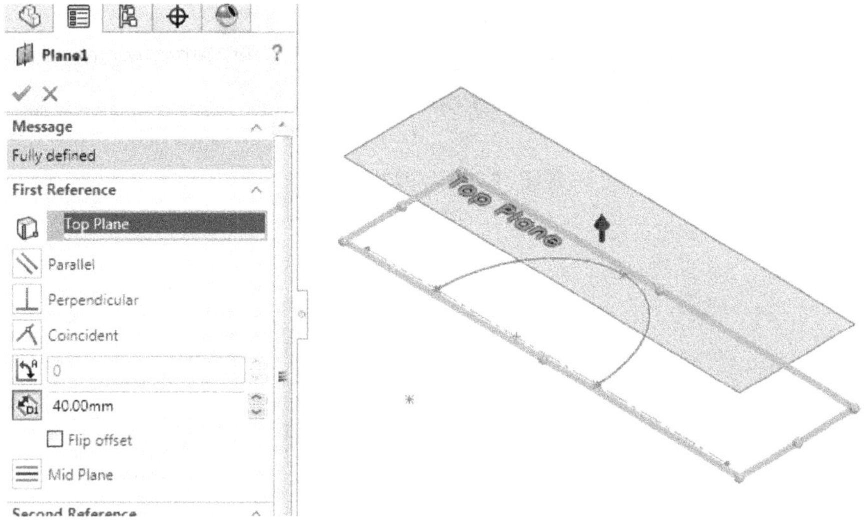

Figure-44. Offset plane to be created

- Click on the **Sketch** button from the **Sketch** tab and select the **Top** plane as sketching plane. Press **CTRL+8** if sketching plane is not parallel.
- Click on the **Offset Entities** tool from the **Ribbon** and selected the earlier created sketch.
- Click on the **Reverse** check box and specify the value as **15** in the spinner; refer to Figure-45.

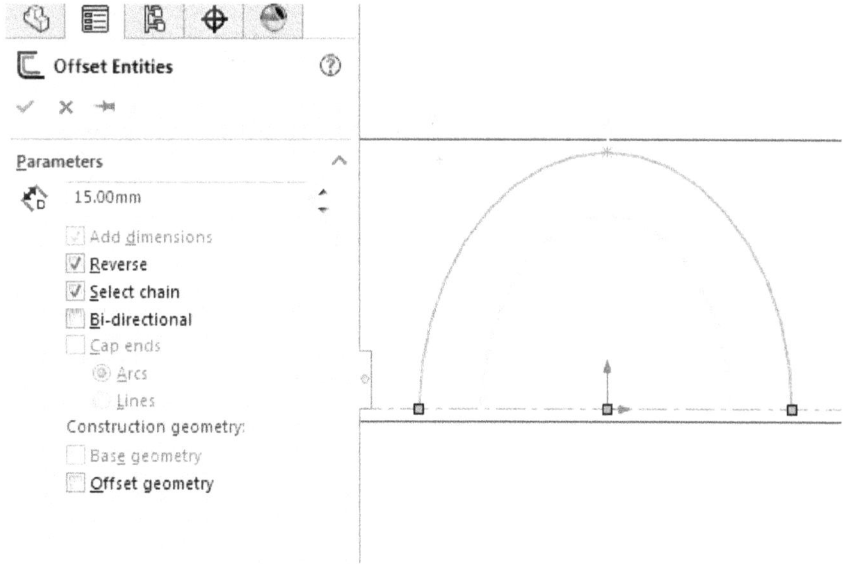

Figure-45. Preview of offset entities

- Click on the **OK** button to create the offset curve.
- Click on the **Exit Sketch** button.

Creating lofted surface

- Click on the **Lofted Surface** tool from **Surfaces** tab of the **Ribbon**. The **Surface-Loft PropertyManager** will display.
- Select the two sketches one by one. Preview of the surface will display; refer to Figure-46.

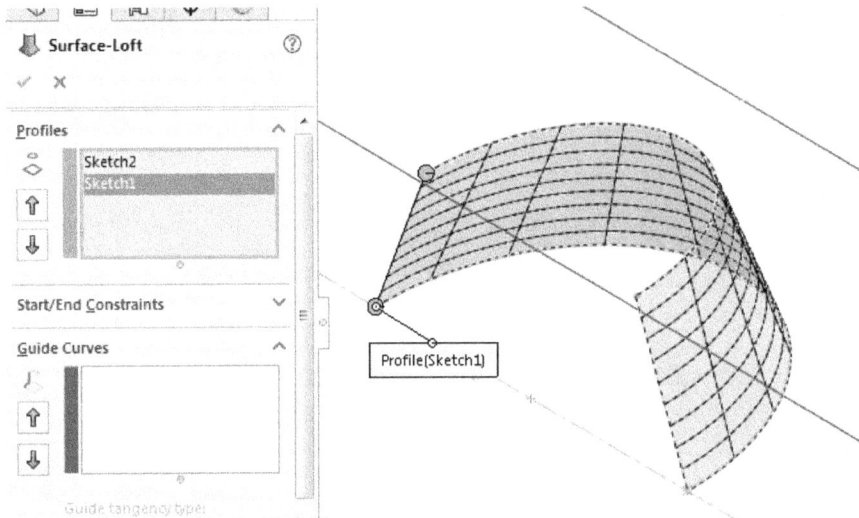

Figure-46. Preview of lofted surface

- Click on the **OK** button to create the surface.
- To hide the plane, select it and right-click on it. Select the **Hide** button from the shortcut menu box.

Thickening surface and applying fillet

- Click on the **Thicken** tool from the Ribbon and select the surface. Preview of thickened surface will display.
- Enter the thickness value as **0.5**.
- Click **OK** to create the solid.
- Click on the **Fillet** tool and apply suitable fillets at the small edges on the corners. Refer to Figure-40.

PRACTICAL 2

Create the model of flower vase as shown in Figure-47. The dimensions of the model are given in Figure-48.

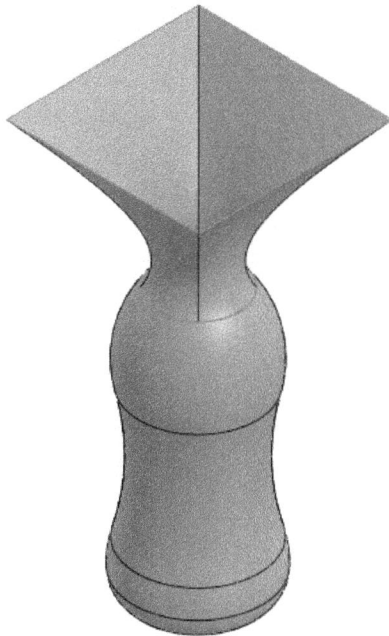

Figure-47. Flower vase model

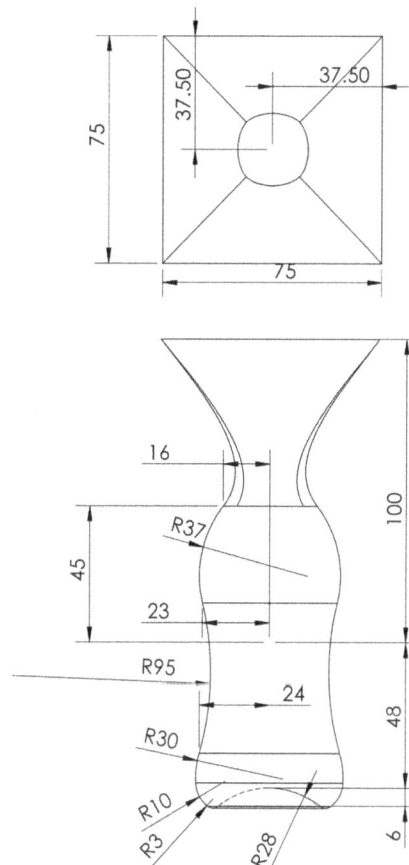

Figure-48. Practical 2 Drawing

The model displayed is having no thickness and its having complex 3D shape. So, we will be using surfacing to create this model.

We can create this model in two steps. In step 1, we will create a revolved surface and in step 2 we will connect the revolved surface to a rectangle by using **Lofted Surface** tool. The procedure to create this surface model is given next.

Creating Revolved Surface

- Click on the **Sketch** tab and select the **Sketch** button. You are asked to select sketching plane.
- Click on the **FRONT** plane. The sketching environment will display.
- Using the **3 Point Arc** tool create the sketch as shown in Figure-49. Make sure you create a centerline as shown in the figure.

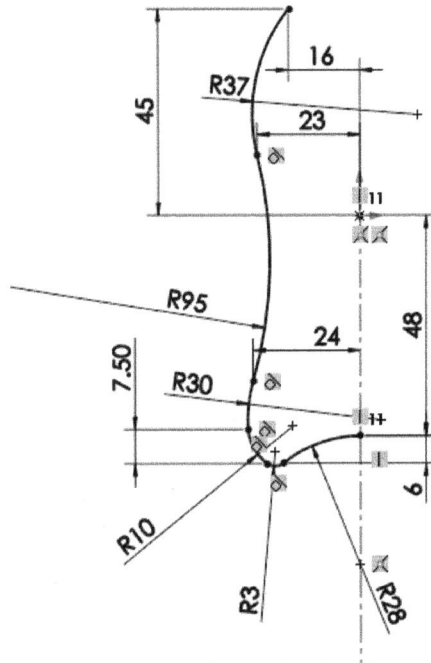

Figure-49. Sketch for revolved surface

- Click on the **Exit Sketch** button and Select the **Revolved Surface** tool from the **Surfaces** tab in the **Ribbon**.
- Select the sketch created recently. Preview of the revolved surface will be displayed; refer to Figure-50.

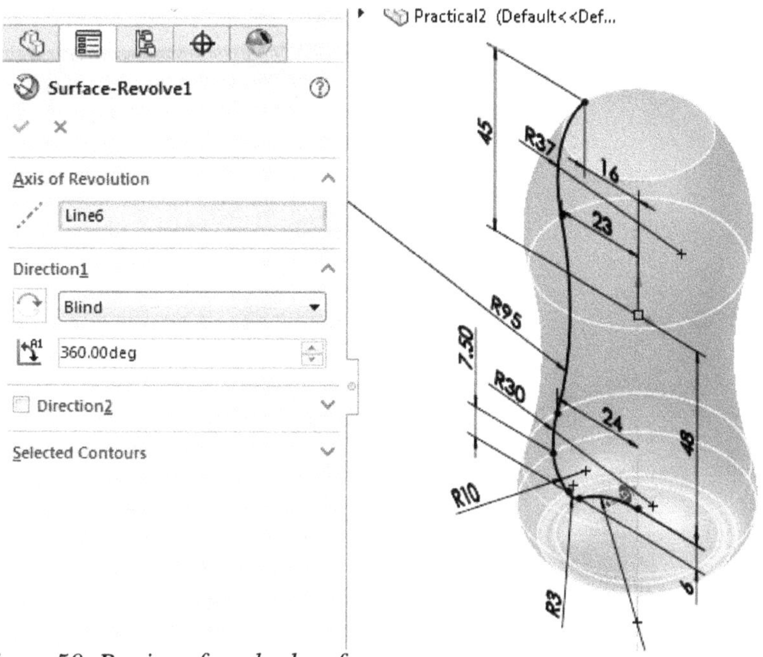

Figure-50. Preview of revolved surface

- Click on the **OK** button from the **PropertyManager** displayed to create the surface.

Creating Lofted Surface

- Click on the **Plane** tool from the **Reference Geometry** drop-down in the **Ribbon** and create an offset plane at distance of **100** from the TOP plane; refer to Figure-51.

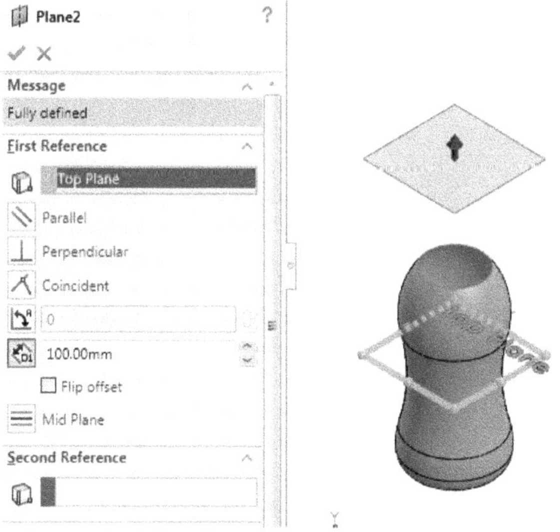

Figure-51. Offset plane created

- Create a rectangle on the selected plane; refer to Figure-52.

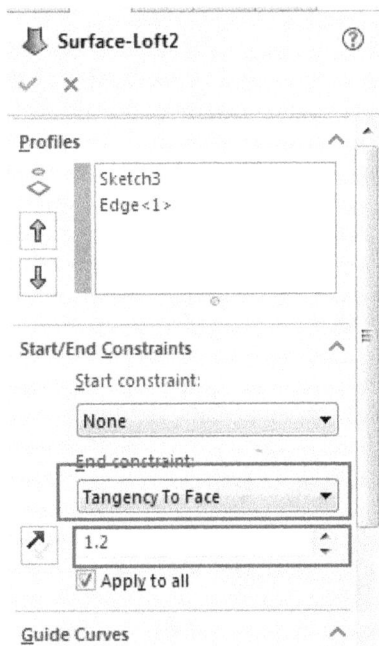

Figure-52. Rectangle to be created

- Click on the **Lofted Surface** tool from the **Ribbon**. The **Surface-Loft PropertyManager** will be displayed.
- Select the recently created sketch and edge of the surface. Preview of the lofted surface will be displayed; refer to Figure-53.

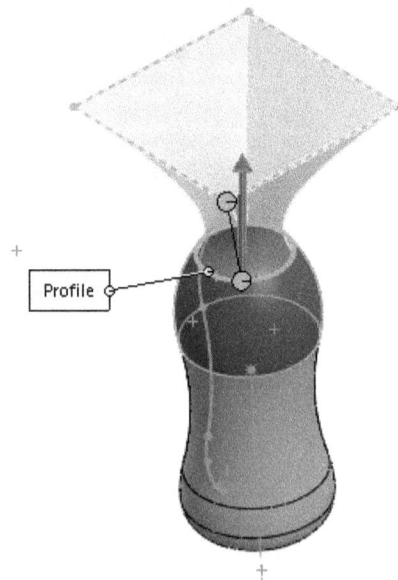

Figure-53. Preview of the lofted surface

- Make sure you set the same values as highlighted in red boxes in the Figure-53. Click on the **OK** button from the **PropertyManager** to create the feature.

PRACTICE 1

Create the surface model of tank as shown in Figure-54. The dimensions of the model are given in Figure-55.

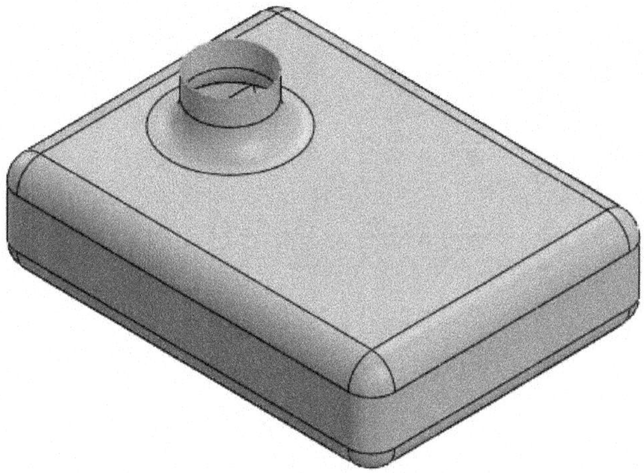

Figure-54. Practice1 model

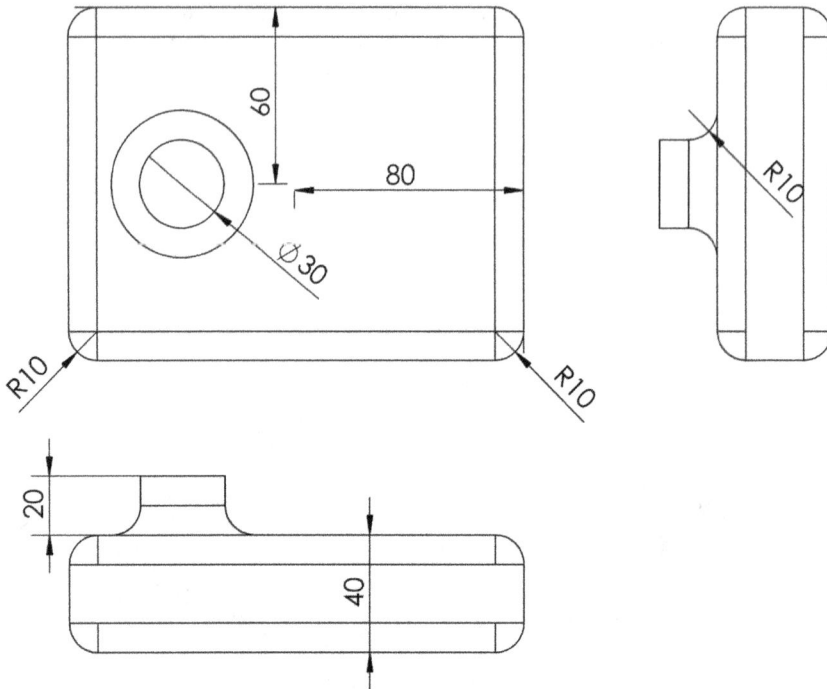

Figure-55. Practice1 drawing

PRACTICE 2

Create the surface model of car bumper as shown in Figure-56. The dimensions of the model are given in Figure-57. **Assume the missing dimensions.**

Figure-56. Practice 2 model

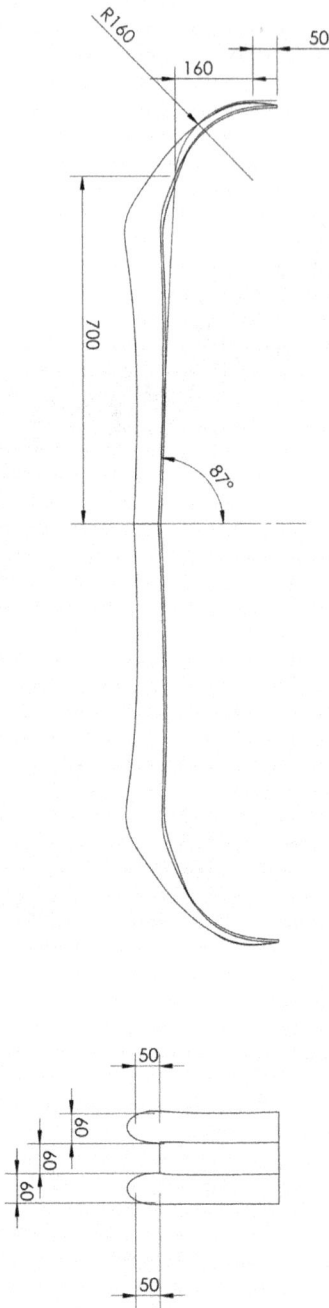

Figure-57. car bumper

SELF ASSESSMENT

Q1. We can create an extruded surface by using an open section. (T/F)

Q2. We do not need any centerline for creating revolved surface. (T/F)

Q3. We can create a swept surface by using open section and closed path. (T/F)

Q4. The **Freeform** tool can not deform face os a solid body. (T/F)

Q5. The **Ruled Surface** tool is used to create a combination of surfaces adjoining to each other. (T/F)

Q6. Which of the following tool can be used to create solid from the surface?

a. **Knit Surface**
b. **Untrim Surface**
c. **Trim Surface**
d. **Extend Surface**

Q7. Which of the following tool is used to remove desired portion of the surface by using sketch/other surface.

a. **Thickened Cut**
b. **Trim Surface**
c. **Cut with Surface**
d. **Delete Face**

Q8. The _____ tool is used to extend the selected face to the replacement face.

Q9. The _____ tool is used to undo the trimmed surfaces.

Q10. The _____ tool is used to combine two or more surfaces at their common edges.

FOR STUDENT NOTES

FOR STUDENT NOTES

FOR STUDENT NOTES

Drawing and Practice

Chapter 9

Topics Covered

The major topics covered in this chapter are:

- *Drawing Introduction.*
- *Drawing Sheet Selection.*
- *Adding Views to sheet.*
- *Annotating Views.*
- *Exploded View and Bill of Material.*
- *Balloons and Title Block*
- *Practice.*

INTRODUCTION

Drawing is the engineering representation of a model on the paper. For manufacturing a model in real world, we need some means by which we can tell the manufacturer what to manufacture. For this purpose, we create drawings from the models. These drawings have information like dimensions, material, tolerances, objective, precautions and so on. In SolidWorks, we create drawings by using the Drawing environment. To activate this environment, click on the **New** button from the **Menu Bar**. The **New SolidWorks Document** dialog box will display. Double-click on the **Drawing** button; refer to Figure-1. The Drawing environment will open.

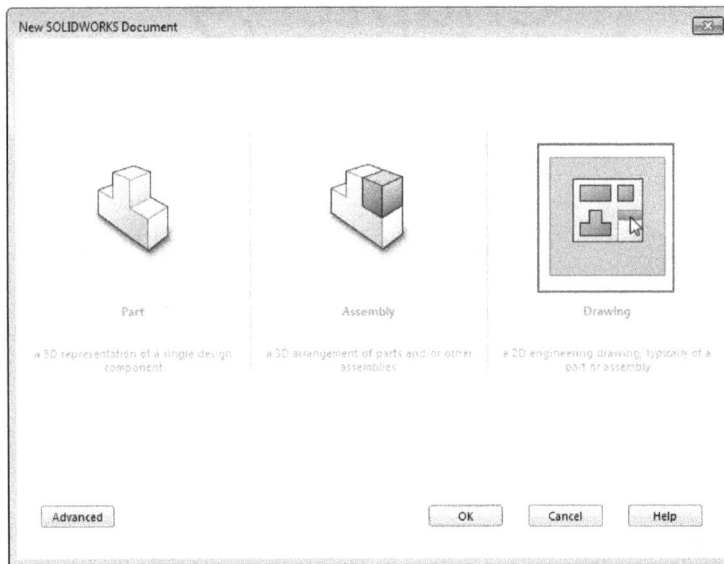

Figure-1. Drawing button to be selected

DRAWING SHEET SELECTION

On starting the Drawing environment, the interface will display as shown in Figure-2. Also, **Sheet Format/Size** dialog box will display. Using this dialog box, you can set the size of sheet for placing the drawing views.

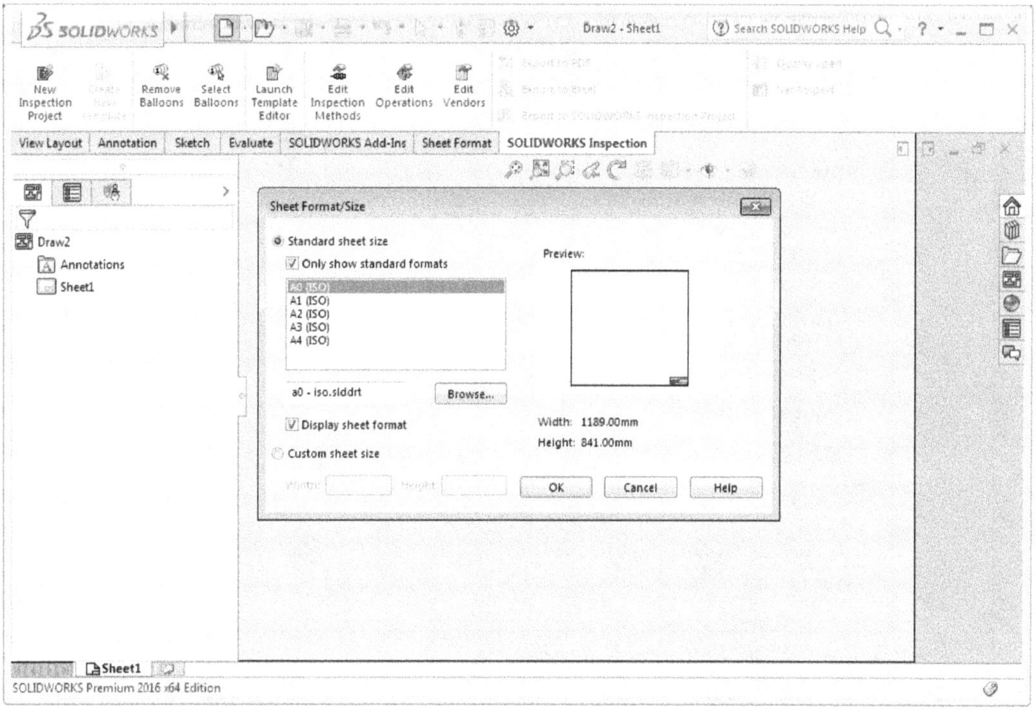

Figure-2. Drawing interface

Select the desired size from the list of sheet sizes and click on the **OK** button from the list. If you want to change the sheet size, follow the steps given below.

- Right-click on the **Sheet** tab from the bar displayed below drawing area. Shortcut menu will display; refer to Figure-3.

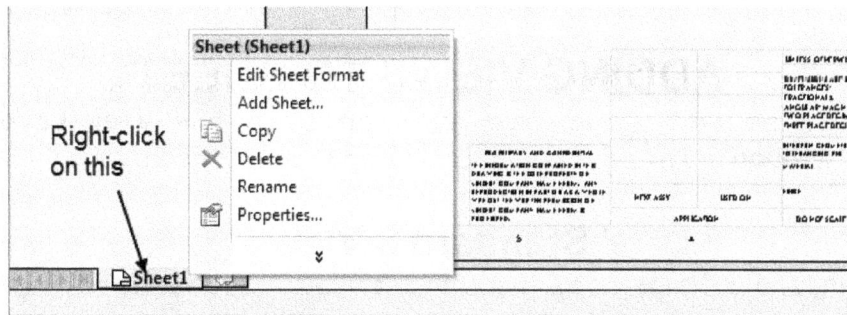

Figure-3. Shortcut menu displayed

- Select the **Properties** button from the menu. The **Sheet Properties** dialog box will display; refer to Figure-4.

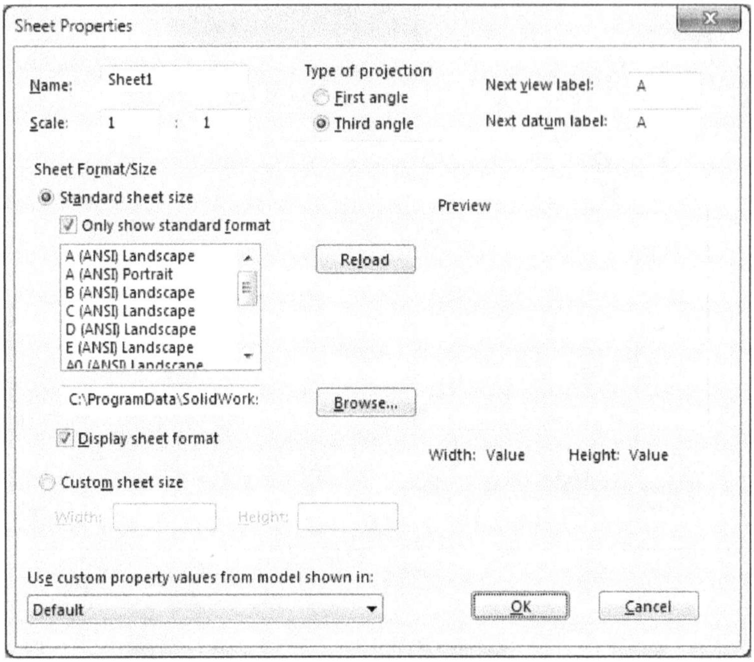

Figure-4. Sheet Properties dialog box

- Select the desired sheet size from the list in this dialog box. You can also set the scale for view and projection type (First angle or Third angle) by using the options in this dialog box.
- You can set you custom dimensions by selecting the **Custom sheet size** radio button. Click on the **OK** button to accept the sheet size.

After setting the sheet size, the next step is to add various views of the model in the sheet.

ADDING VIEWS TO SHEET

The tools to add views are available in the **View Layout** tab of the **Ribbon.** These tools are explained next.

Standard 3 View

The **Standard 3 View** tool is used to add 3 standard view (Top view, Front view, and Right view) to the sheet. To add these views, follow the steps given next.

- Click on the **Standard 3 View** button from the **Ribbon**. The **Standard 3 View PropertyManager** will display; refer to Figure-5.

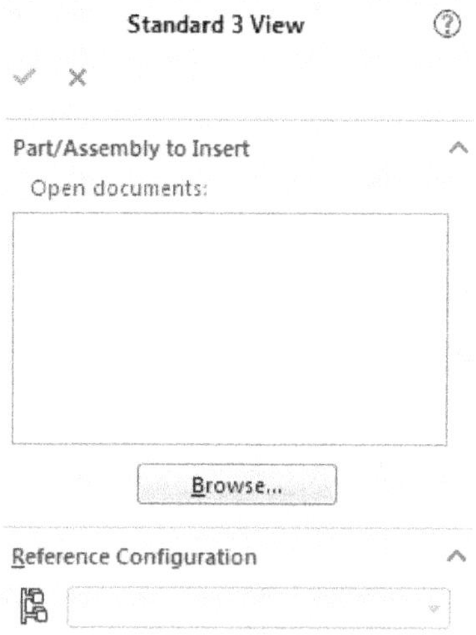

Figure-5. Standard 3 View PropertyManager

- Click on the **Browse** button from the **PropertyManager**. You are asked to select part/assembly model.
- Double-click on the file of model for which you want to create the views. Views of the model will be placed automatically. Figure-6 shows the views placed in Third angle projection.
- You can drag the view by selecting any curve of model to place at your desired location. The other views will be shifted accordingly; refer to Figure-7.

Figure-6. 3 Views placed automatically

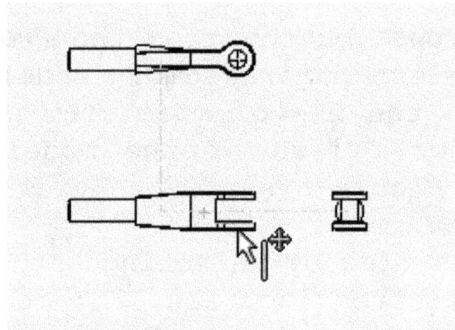

Figure-7. Moving views

To change the properties of any of the view placed, select it from the drawing. The **Drawing View PropertyManager** will display. If you select the main view, then the **PropertyManager** will display as shown in Figure-8. If you select any projected view then the **PropertyManager** will display as shown in Figure-9.

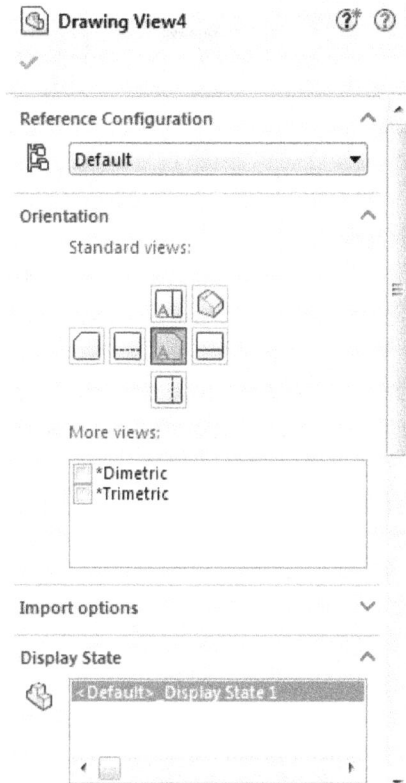

Figure-8. Drawing View Property-
Manager on selecting main view

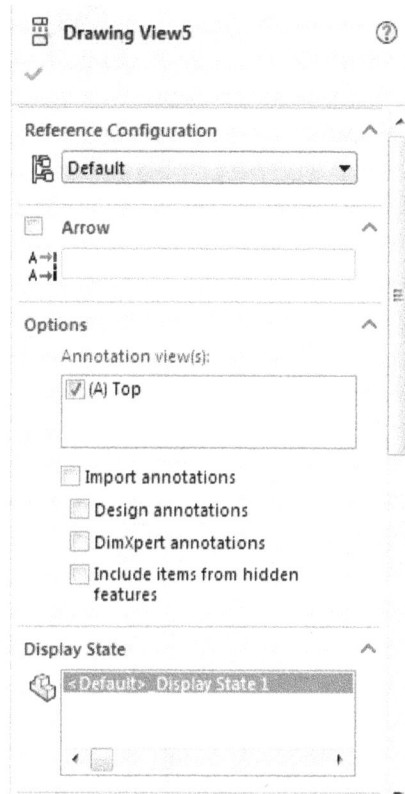

Figure-9. Drawing View Property-
Manager on selecting projected view

The options in the **PropertyManager** are explained next.

Orientation rollout

Select the desired button to change the view orientation of
the model. You can display dimetric or trimetric orientation
by selecting the respective check box.

Import Options rollout

Options in this rollout are used to import information from
the Part environment. If you want to use dimensions or
other annotations created in Part environment, then select
the **Import annotations** check box. Similarly, you can select
other check boxes to include design annotations, DimXpert
annotations and hidden features from Part environment.

Display Style rollout

Options in this rollout are used to change the display of model view. If you want to display hidden lines then select the **Hidden Lines Visible** button from the rollout. Similarly, you can select the other buttons as per your requirement.

Scale rollout

Options in this rollout are used to change the size of view by using the scale value. To set the desired scale, click on the **Use custom scale** radio button from the rollout and enter the desired ratio in the edit box below it or select the desired ratio from the drop-down below the radio buttons.

Similarly, you can set the thread display quality and dimension type from the respective rollouts.

If you have selected a projected view for changing properties, then you can add the projection arrows by selecting the **Arrow** check box from the **PropertyManager**; refer to Figure-10. Enter the name of projection arrows in the edit box displayed in the rollout.

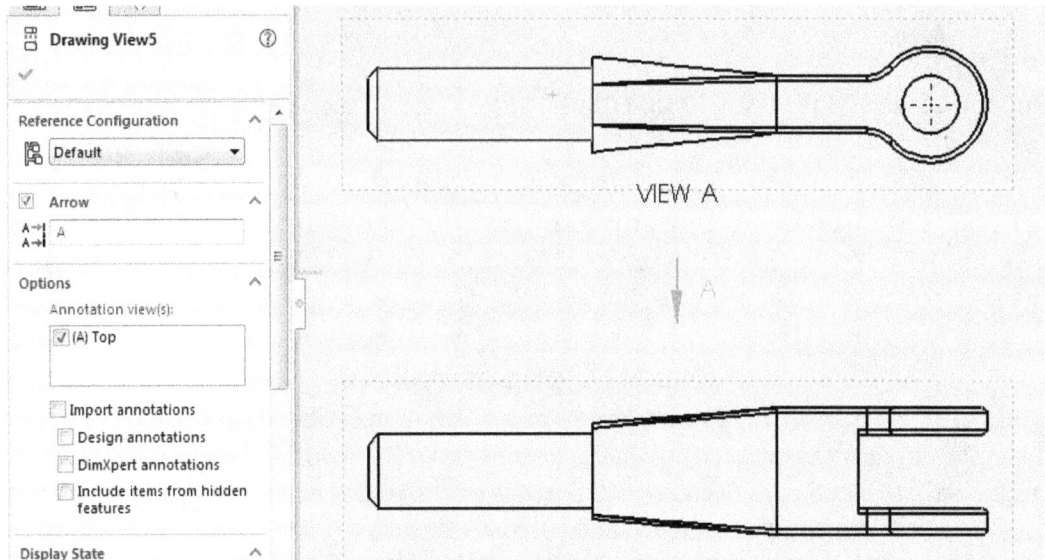

Figure-10. Arrow for projection

Now, we will learn to place the main view in the sheet. If you do not want to use the Standard 3 Views then the next tool is the first step for drawing. **Note that if you want to delete any view/views from the sheet, then select it and press DEL key from keyboard.**

Model View

The **Model View** tool is used to add main view in the drawing. All the other views in the drawing are projection of it and are related to it. The main view can be of any orientation like, front, left, top and so on. To use this tool, follow the steps given next.

- Click on the **Model View** button from the **Ribbon**. The **Model View PropertyManager** will display.
- Click on the **Browse** button from the **PropertyManager**. The **Open** dialog box will display.
- Double click on the part/assembly model file to generate the view. The first view of model will get attached to the cursor; refer to Figure-11.
- Select the desired orientation from the **PropertyManager** and click in the sheet to place the view.

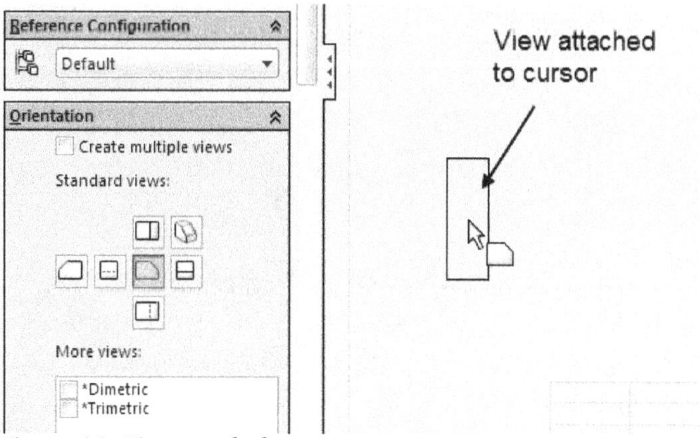

Figure-11. View attached to cursor

- Projection mode of the tool will start and you will be prompted to click in the sheet to place the projection view.
- Click to place the projection view.
- Press **ESC** key to exit the tool.

Projected View

The **Projected View** tool is used to create projected view of the selected view in the sheet. The procedure to use this tool is given next.

- Click on the **Projected View** tool from the **Ribbon**. The **Projected View PropertyManager** will be displayed.
- Click in the sheet to place the projection of view available in the sheet. Note that to use this tool you must have a view in the sheet.
- If you have more than one view in the sheet then select the view whose projection is to be generated and then click to place the projection; refer to Figure-12.

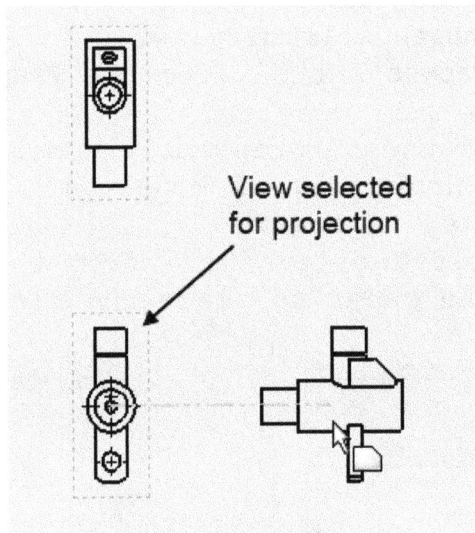

Figure-12. View selected for projection

Auxiliary View

The **Auxiliary View** tool is used to create projected view from a view by making selected edge of view parallel to screen. The procedure to use this tool is given next.

- Click on the **Auxiliary View** button from the **Ribbon**. The **Auxiliary View PropertyManager** will display and you will be asked to select an axis, edge or sketch line to generate the view.
- Select the axis/edge/sketch line. The preview of view will be displayed; refer to Figure-13.

Figure-13. Auxiliary view

Section View

The **Section View** tool is used to create section view by cutting the model using section lines. The procedure to use this tool is given next.

- Click on the **Section View** button from the **Ribbon**. The **Section View Assist PropertyManager** will display; refer to Figure-14.

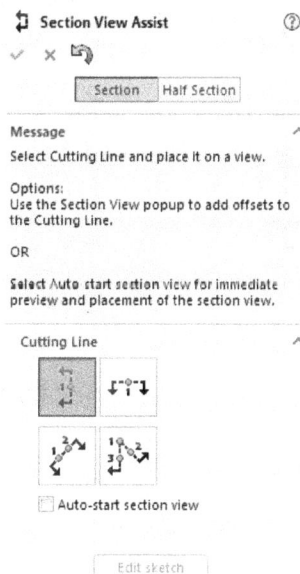

Figure-14. Section View Assist PropertyManager

- Select the desired button from the **Cutting Line** area of the **PropertyManager**.
- Click in the view to specify the start point of line. The toolbar will display; refer to Figure-15.

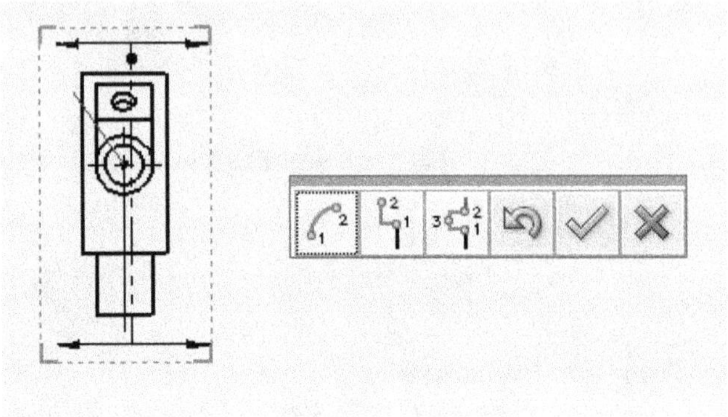

Figure-15. Section view toolbar

- Select the desired button from toolbar to change the style of cutting line and specify the points.
- Click on the **OK** button from the toolbar, the preview will display; refer to Figure-16.

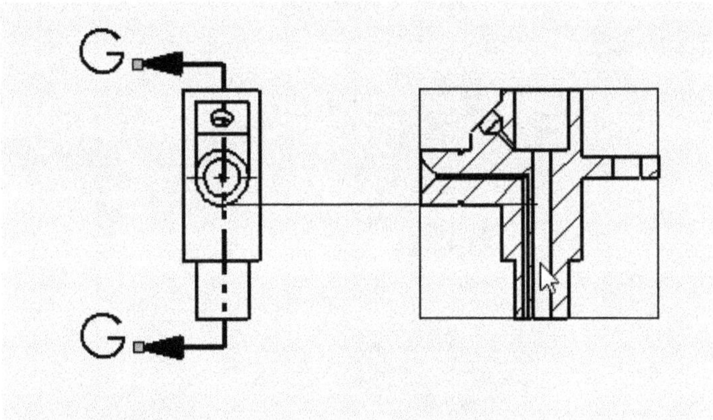

Figure-16. Preview of section view

- Click to place the view. The **Section View PropertyManager** will display; refer to Figure-17.

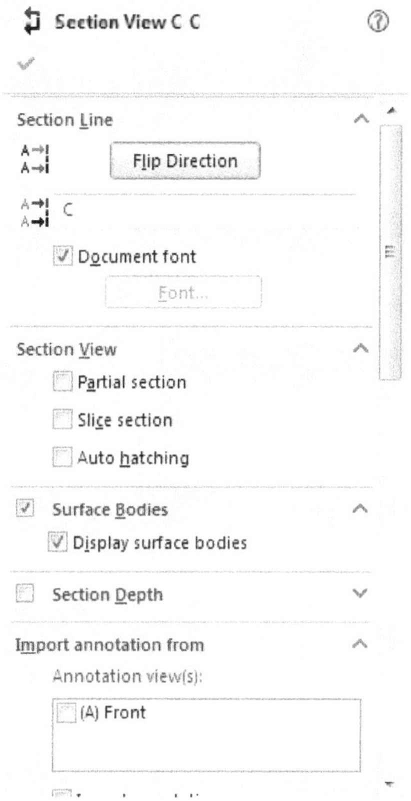

Figure-17. Section View PropertyManager

- Set the desired option in the **PropertyManager** and then click on **OK** button from the **PropertyManager** to create the section view.

Detail View

The **Detail View** tool is used to create detailed view of a specific portion of the model. The steps to create detail view are given next.

- Click on the **Detail View** button from the **Ribbon**. The **Detail View PropertyManager** will display. Also, you are asked to specify center point of circle for creating detail view circle.
- Click in the view to specify center of the detail circle.
- Click to specify the radius of the circle. Preview of detail view will be attached to the cursor.

- Change the scale of the detail view by using the options in the **PropertyManager**; refer to Figure-18.

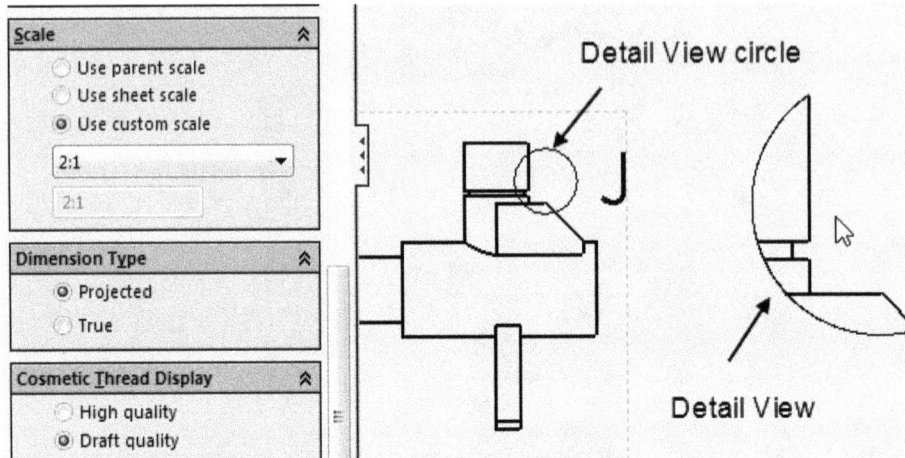

Figure-18. Detail view

- Click to place the detail view. Press **ESC** to exit the tool.

Broken-out Section

The **Broken-out Section** tool is used to create section in the selected view to display inner detail of the model. Follow the steps to use this tool.

- Click on the **Broken-out Section** tool from the **Ribbon**. You are asked to specify the start point of the section spline.
- Click in the view to start spline and create a close spline; refer to Figure-19.

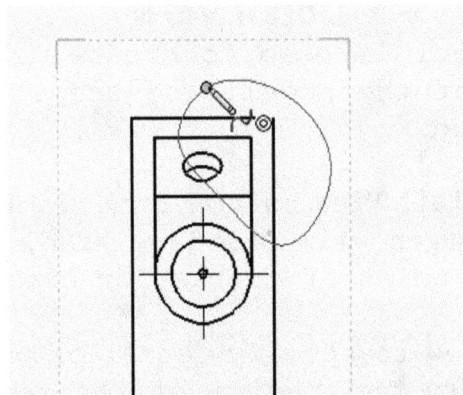

Figure-19. Spline for broken-out section

- Click on the **Preview** check box and specify the depth in the spinner in the **PropertyManager**. The preview of broken-out section will display; refer to Figure-20.

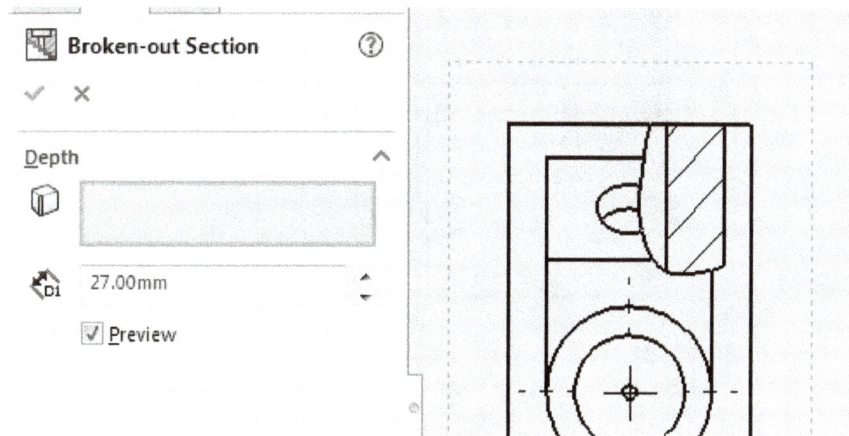

Figure-20. Preview of broken-out section

- Click on the **OK** button to create the section.

Break

The **Break** tool is used to represent very long objects in the drawing by breaking them at specific span. The procedure to use this tool is given next.

- Click on **Break** tool from the **Ribbon**.
- Click on the view that you want to break. The break line will attach to the cursor; refer to Figure-21.
- Set the desired option from the **PropertyManager** and click to specify starting point of break span.
- Click to specify the end of break span. The selected span will be removed and the broken view will display with specified gap; refer to Figure-22.

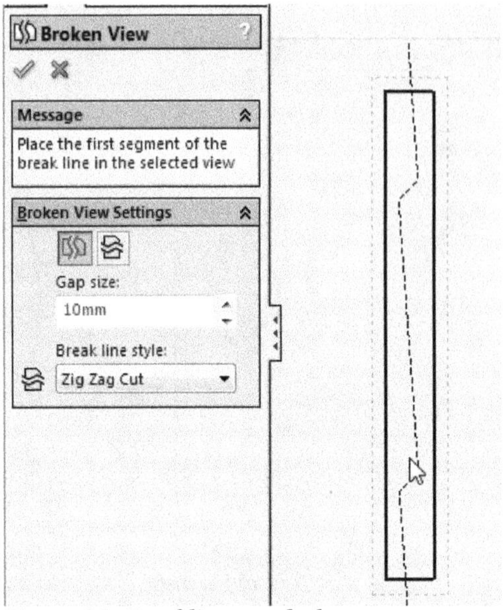

Figure-21. Breakline attached to cursor

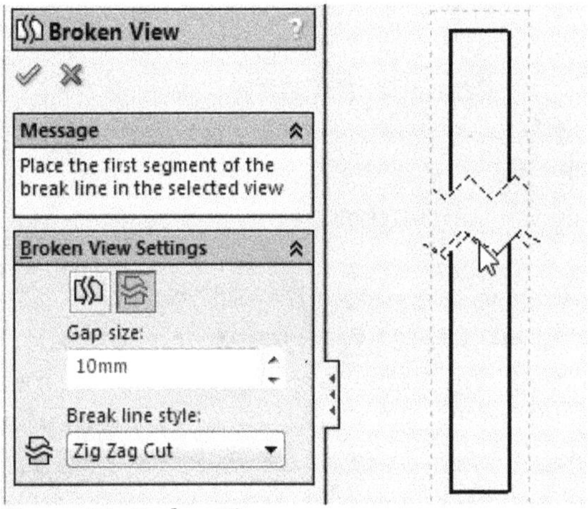

Figure-22. Broken View

- Click on the **OK** button to create the view.

ADDING ANNOTATIONS TO VIEW

Till this point, we have learned to place views but without dimensions and annotations these views are of no use for manufacturers. So, we will now add annotations to views. The tools to apply annotations are available in the **Annotation** tab of the **Ribbon**; refer to Figure-23.

Figure-23. Annotation tab of Ribbon

Smart Dimension

The **Smart Dimension** tool is used to dimension each entity of the model on paper. This tool works in the same way as it does for sketches.

Model Items

The **Model Items** tool is used to import all the dimensions/annotations applied to the model in Part modeling environment. To import the dimensions/annotations, follow the steps given next.

- Click on the **Model Items** tool from the **Ribbon**. The **Model Items PropertyManager** will display; refer to Figure-24.
- Select the desired buttons from the **PropertyManager** to import the respective annotations in the view.

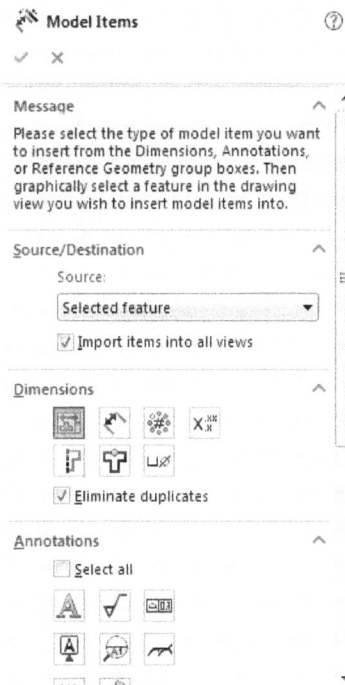

Figure-24. Model Items PropertyManager

- Click in the **Source** drop-down in the **Source/Destination** rollout to specify the entities that you want to annotate in the view.
- Click on the **OK** button to generate the annotations from the model. The automatic annotations will be generated. Drag the annotations to place them properly.

Select one of the dimension, the **Dimension PropertyManager** will be displayed. The options in this **PropertyManager** are same as discussed in **Advanced Dimensioning Chapter**.

Note

The **Note** tool is used to specify extra information in the drawing that are not mentioned in the dimensions. For example, if you want to say "All dimensions are in mm" then this is the tool to do so. The steps to use this tool are given next.

- Click on the **Note** tool from the **Ribbon**. The **Note PropertyManager** will display and note box will get attached to the cursor; refer to Figure-25.

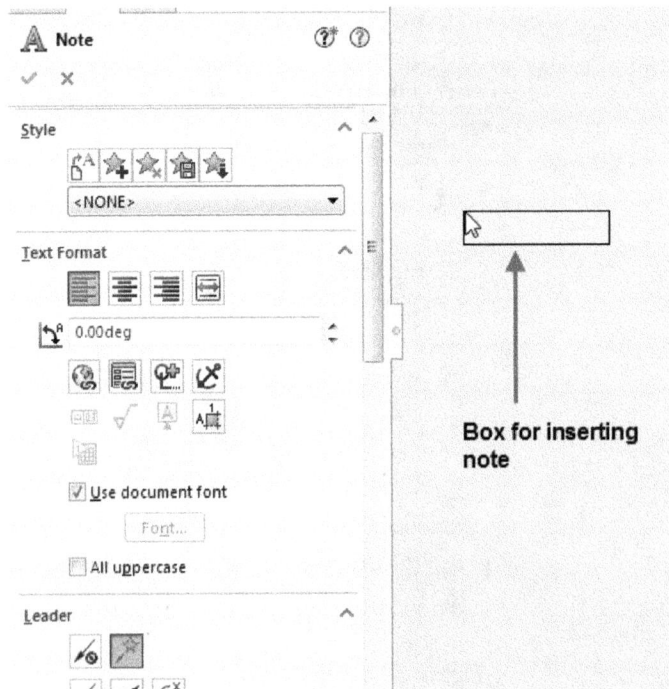

Figure-25. Note PropertyManager

- If you move the cursor over any entity in the view, then the leader will be added before the note box; refer to Figure-26.

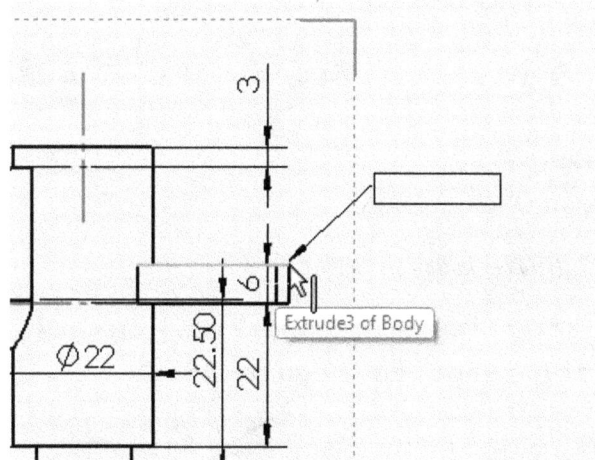

Figure-26. Note box with leader

- Click to place the box. If leader is attached then click again to place the note box.
- On clicking, the editing mode of note will activate. Enter the desired text in the box. Apply desired formatting by using the options displayed.
- Click on the **OK** button to create the note.

Flag Notes

The Flag Notes are used to cross-reference listed notes to specific area or feature in the drawing. In other words, flag notes are numbered list of items referenced to different area of drawing. The procedure to create flag notes is given next.

- Click on the **Note** tool from the **Ribbon** and click in the drawing area to specify insertion point.
- Click on the Number button from the toolbar displayed; refer to Figure-27.

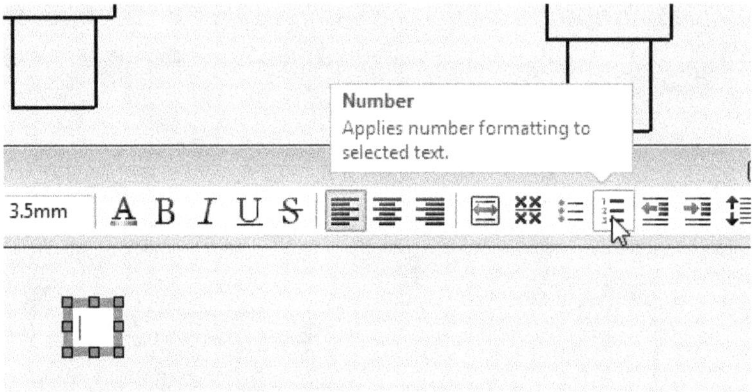

Figure-27. Number button in toolbar

- Type the desired note in the text box in the form of numbered list; refer to Figure-28.

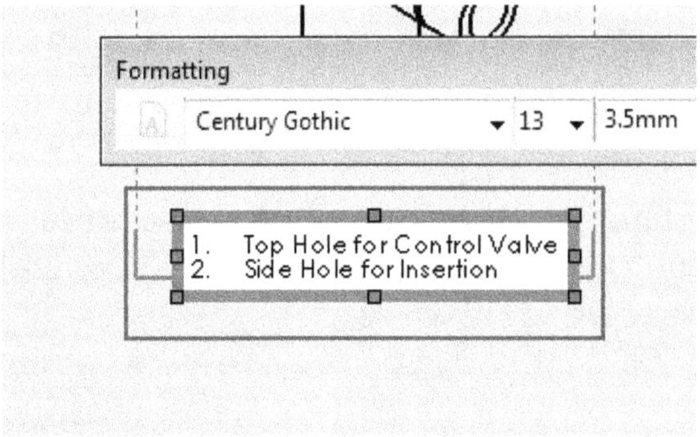

Figure-28. List of notes created

- Click on the number in the list which is to be added in the Flag Note Bank and select the **Add to Flag Note Bank** check box from the **PropertyManager** displayed; refer to Figure-29.

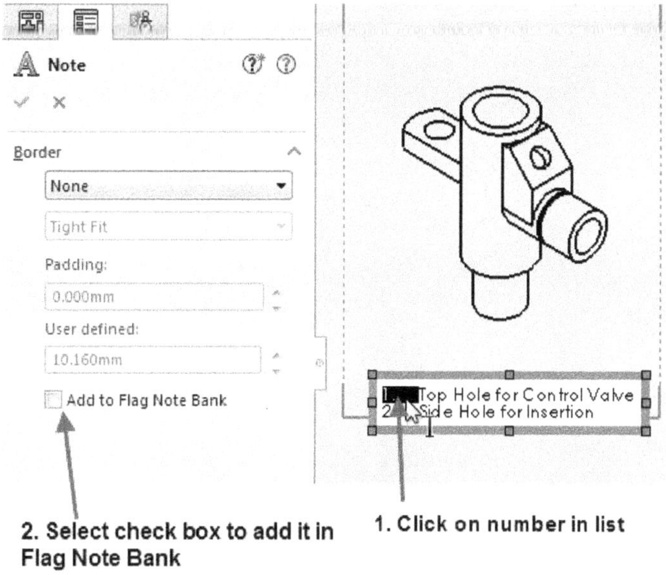

2. Select check box to add it in Flag Note Bank

1. Click on number in list

Figure-29. Adding notes flag note bank

- Similarly, you can add other numbered notes in the Flag Note Bank.
- Click on the **OK** button from the **Note PropertyManager** after adding all the notes in the Flag Note Bank.

Now, we will add balloons in the drawing as per the notes. The procedure to add balloons is given next.

- Click on the **Balloon** tool from the **Annotation** tab in the **Ribbon**. The **Balloon PropertyManager** will be displayed; refer to Figure-30.

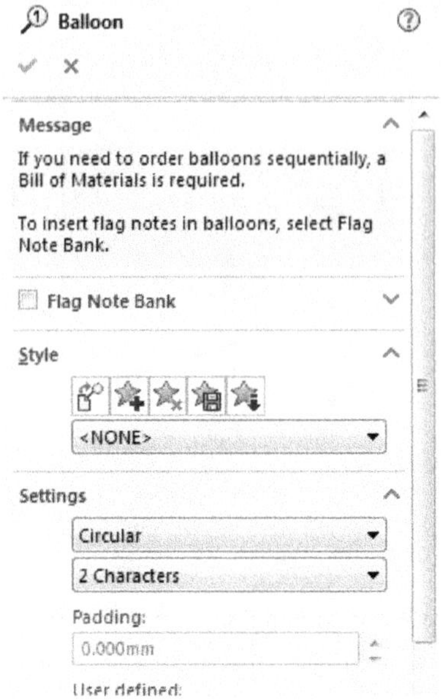

Figure-30. Balloon PropertyManager

- Select the **Flag Note Bank** check box. The notes earlier saved will be displayed in the list; refer to Figure-31.

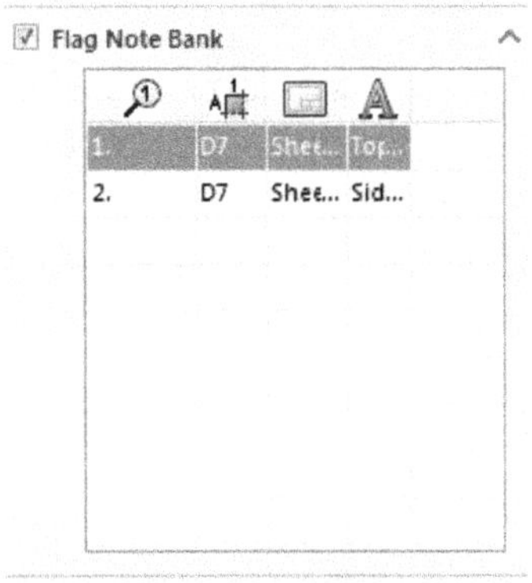

Figure-31. Flag Note Bank

- Select the desired note from the list. A balloon will get attached to the cursor.
- Click on the desired reference in the drawing and place the balloon; refer to Figure-32.

If you need to order balloons sequentially, a Bill of Materials is required.

To insert flag notes in balloons, select Flag Note Bank.

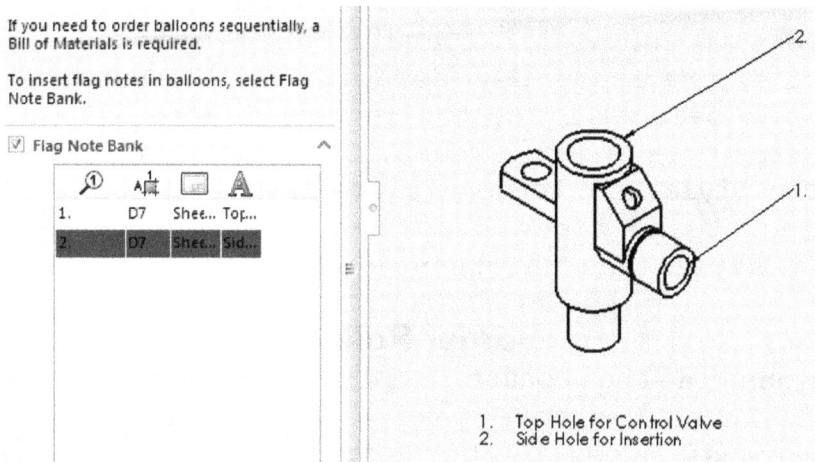

Figure-32. Flag notes created

- Click on the **OK** button from the **Balloon PropertyManager** to exit.

Note that using the options in the **Note PropertyManager**, you can generate all type of geometric annotations like, Datum reference, surface finish symbol, Geometric Tolerance box and so on. The options in the **Note PropertyManager** are discussed next.

Style Rollout

You can save a note as favorite using the options in the **Style** rollout. The options available in this rollout are the same as discussed earlier.

Text Format Rollout

The **Text Format** rollout is used to set the format of the text such as font, size, justification, and rotation of the text. You can also add symbols and hyperlinks to the text using the options available in this rollout.

Leader Rollout

The options in the **Leader** rollout are used to define the style of arrows and leaders that are displayed in the notes.

Leader Style Rollout

The options in this rollout are used to define the style and thickness of the leader. By default, the **Use document display** check box is selected. So, the leader will be displayed with the default style and thickness. On clearing this check box, the **Leader Style** and **Leader Thickness** drop-down lists will be enabled. Using these drop-down lists, you can specify different styles and thickness for the leader.

Border Rollout

The options in the **Border** rollout are used to define the border in which the note text will be displayed. You can assign various types of borders from the **Style** drop-down list. The **Size** drop-down list available in this rollout is used to define the size of the border in which the text will be placed.

Parameters Rollout

The **Parameters** rollout is used to specify the X and Y coordinate values of the note center.

Wordwrap Rollout

Select the Wordwrap check box to expand the **Wordwrap** rollout. Specify the desired wordwrap width in the edit box.

Layer Rollout

This rollout is used to assign existing layer or create new layer to the notes.

Datum Feature

The **Datum Feature** tool is used to add reference for measuring all the geometric tolerances. For example, if you want to check perpendicularity of a face then you need to give a reference with respect to which the perpendicularity will be measured. The steps to create datum features are given next.

- Click on the **Datum Feature** button from the **Ribbon**. The datum symbol will attach to the cursor and the **Datum Features PropertyManager** will display; refer to Figure-33.

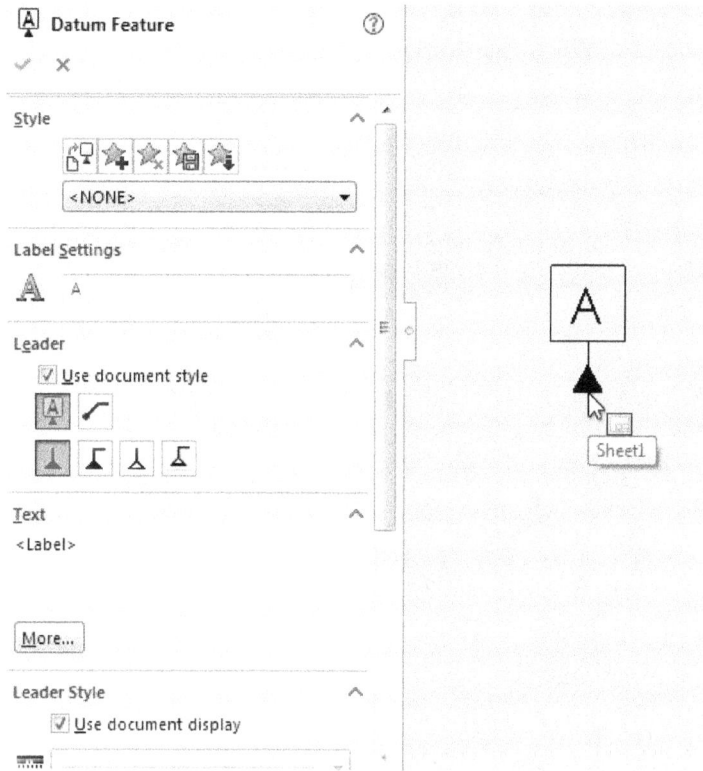

Figure-33. Datum Feature PropertyManager

- Click on the reference to place the leader's start point. Move the cursor to specify length of the leader.
- Click to place the label of the datum feature. Next, datum feature box will attach to the cursor. Press **ESC** to exit the tool. Refer to Figure-34 for datum feature placement.

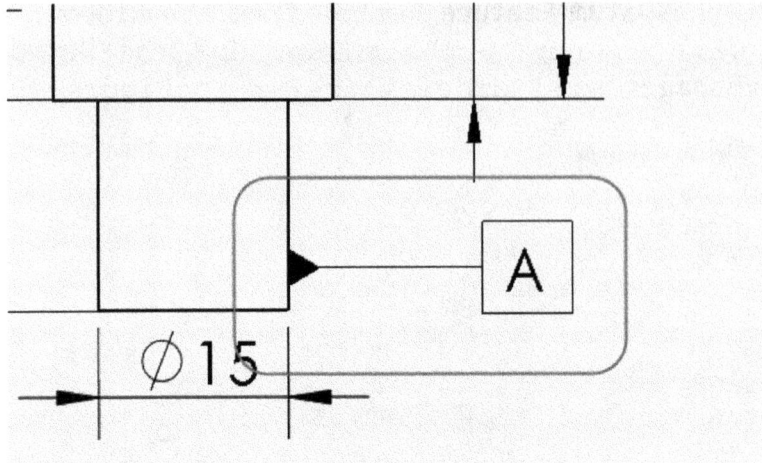

Figure-34. Datum feature placement

Figure-35 shows the break-up of a datum feature symbol.

Figure-35. Datum feature symbol

Datum Target

Datum targets are circular frames divided in two parts by a horizontal line. The lower half represents the datum feature, and the upper half is for additional information, such as dimensions of the datum target area; refer to Figure-36.

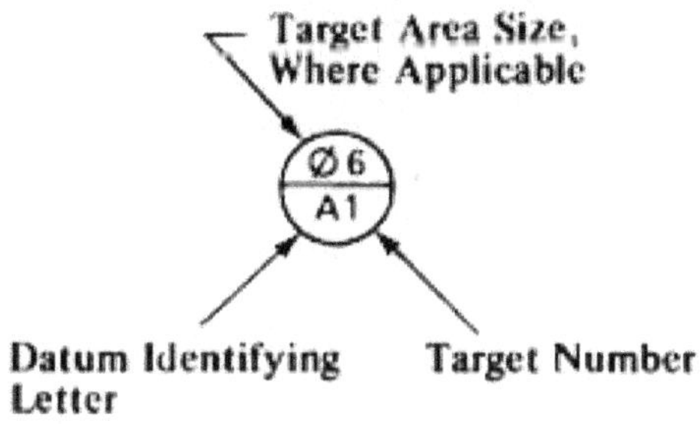

Figure-36. Datum target symbol

The **Datum Target** tool is used to add datum target to the drawing. The steps to do so are given next.

- Click on the **Datum Target** button from the **Ribbon**. The **Datum Target PropertyManager** will display; refer to Figure-37.

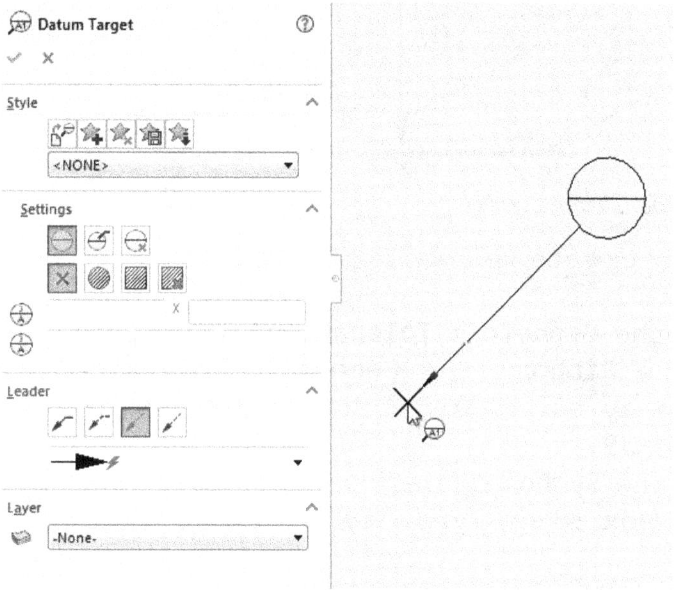

Figure-37. Datum Target PropertyManager

- Specify the desired value and click in the drawing view to place the leader start point.
- Stretch the symbol and click to place at the desired position.

Geometric Tolerance

Geometrical tolerance is defined as the maximum permissible overall variation of form or position of a feature. Geometrical tolerances are used,
(i) to specify the required accuracy in controlling the form of a feature.
(ii) to ensure correct functional positioning of the feature.

Figure-38 shows a geometric symbol with meaning of each box.

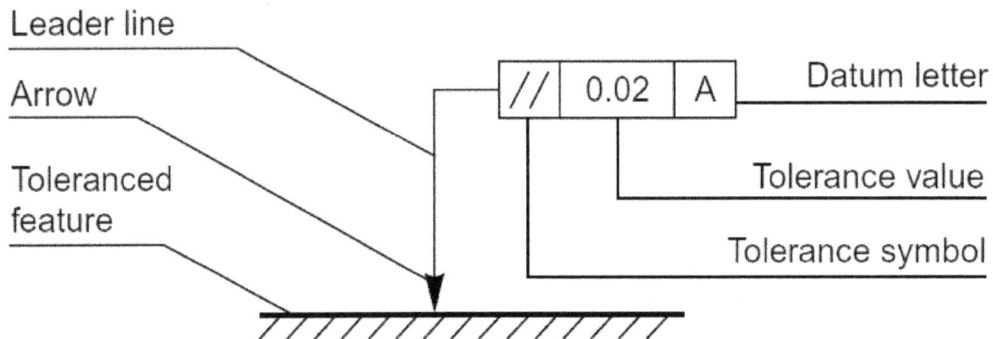

Figure-38. Geometric Tolerance symbol

The steps to create geometric tolerance box are given next.

- Click on the **Geometric Tolerance** button from the **Annotations** tab in the **Ribbon**. The **Geometric Tolerance PropertyManager** will display along with the **Properties** dialog box; refer to Figure-39.
- Click on the **Symbol** drop-down and select the desired button from the box displayed; refer to Figure-40.

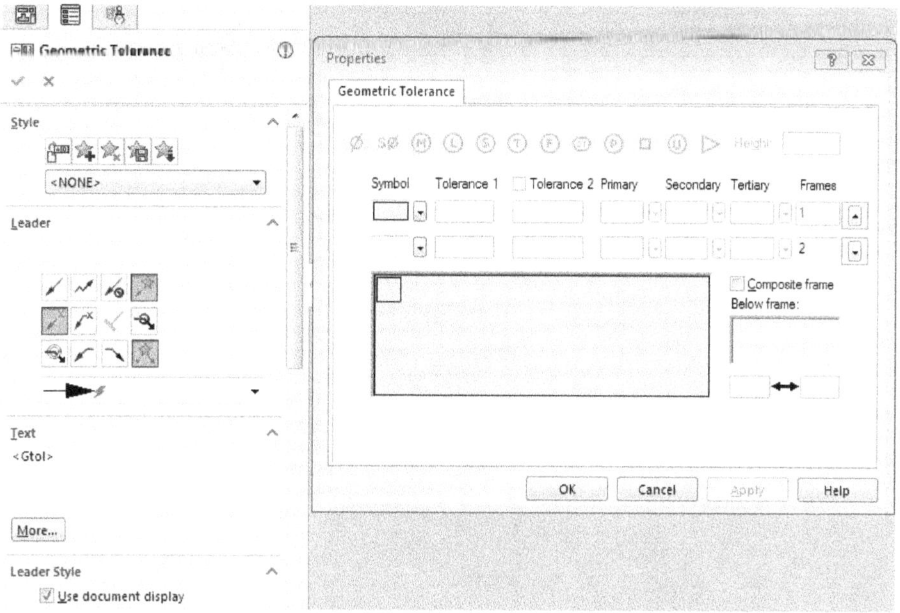

Figure-39. Geometric Tolerance PropertyManager and Properties dialog box

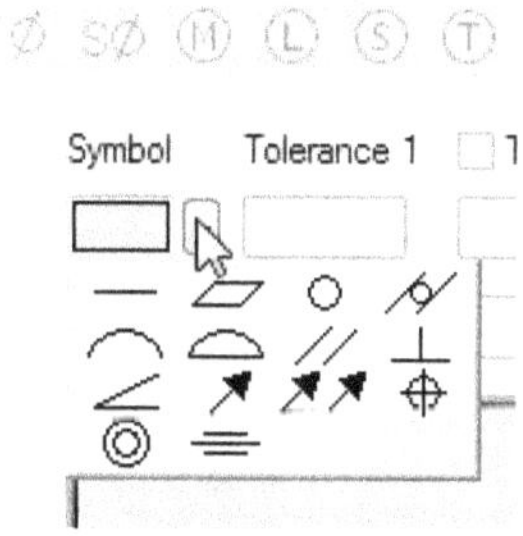

Figure-40. Symbol drop-down

- Click in the **Tolerance** box and specify the deviation.
- Click in the **Primary** box and specify the datum feature label with respect to which the tolerance will be calculated.
- You can specify second tolerance by selecting the **Tolerance 2** check box.
- Click on the **Help** button from the **Properties** dialog box to know more about the box.

Meaning of various geometric tolerance symbols are given in Figure-41.

Figure-42 and Figure-43 shows the use of geometric tolerances in real-world.

Characteristics to be toleranced		Symbols
Form of single features	Straightness	───
	Flatness	▱
	Circularity (roundness)	○
	Cylindricity	⌭
	Profile of any line	⌒
	Profile of any surface	⌓
Orientation of related features	Parallelism	//
	Perpendicularity (squareness)	⊥
	Angularity	∠
Position of related features	Position	⊕
	Concentricity and coaxiality	◎
	Symmetry	=
	Run-out	↗

Figure-41. Meaning of geometric Tolerance symbol

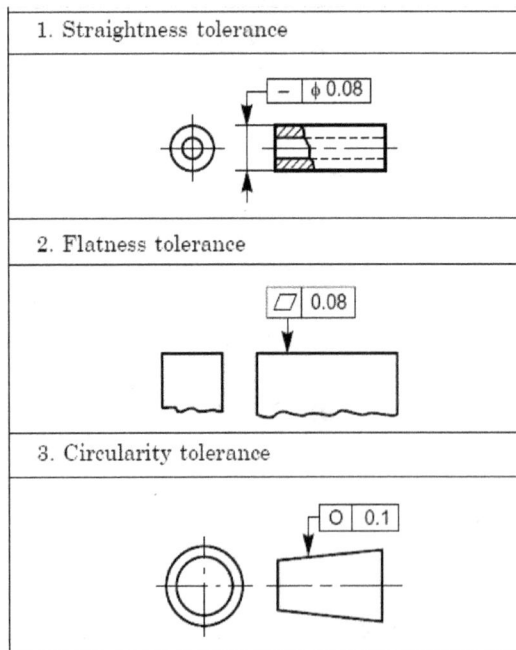

Figure-42. Use of geometric tolerance 1

4. Cylindricity tolerance	8. Concentricity and coaxiality tolerance

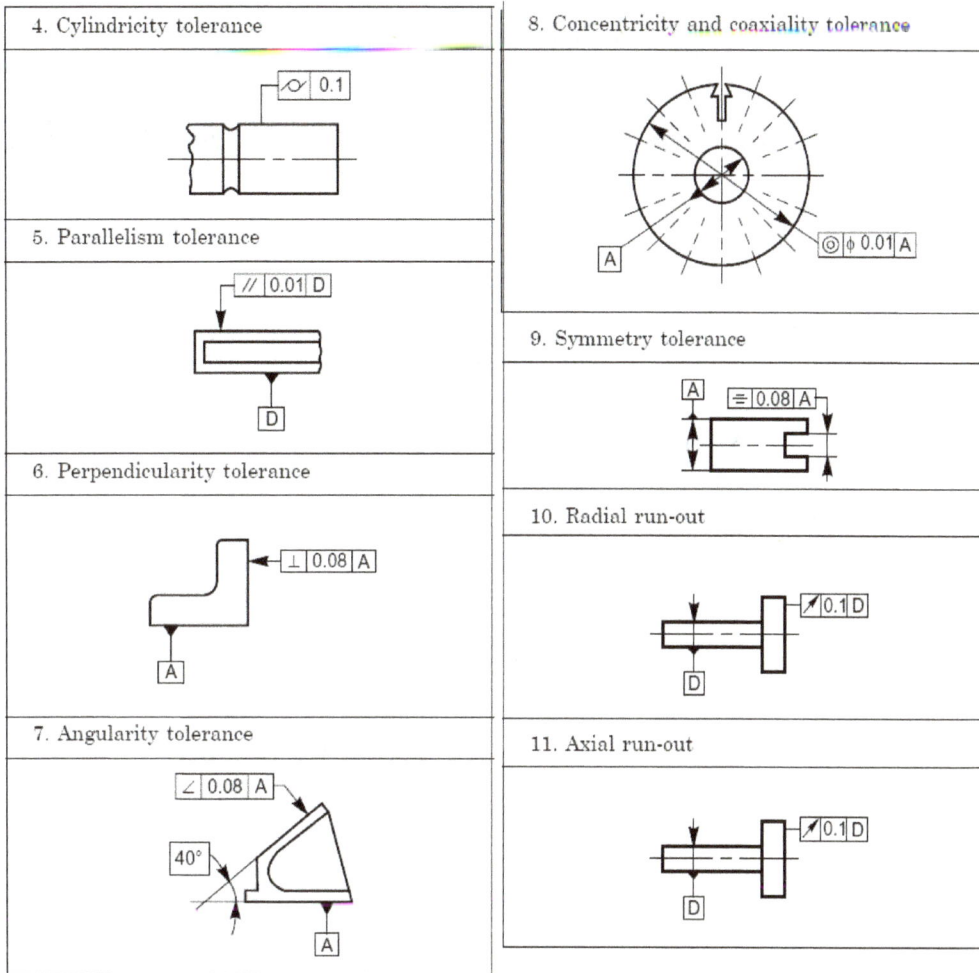

Figure-43. Use of geometric tolerance 2

Surface Finish/Weld Symbol/Hole Callout

The surface finish and the weld symbols are placed in the same way as you place geometric tolerances. Click on the tool, define the parameters and place the symbol by clicking.

To add the hole callout, click on the **Hole Callout** tool and select hole or slot. The callout will be generated automatically. Click to place the callout.

In the same way, you can annotate center line and center mark by selecting the **Centerline** and **Center Mark** tool.

Now, we will add exploded view of the assembly and then we will add bill of material and balloons.

GENERATING EXPLODED VIEW OF ASSEMBLY

To generate exploded view, you must explode the assembly first in the assembly environment. After exploding save the assembly and then follow the steps to generate exploded view.

- Start a new drawing and click on the **Browse** button from the **Model View PropertyManager.**
- Double click on the assembly file that you saved earlier.
- Click on the **Show in exploded state** check box and select the exploded view from the drop-down.
- Select the **Isometric** button from the **Orientation** rollout and click in the drawing area to place the assembly; refer to Figure-44.

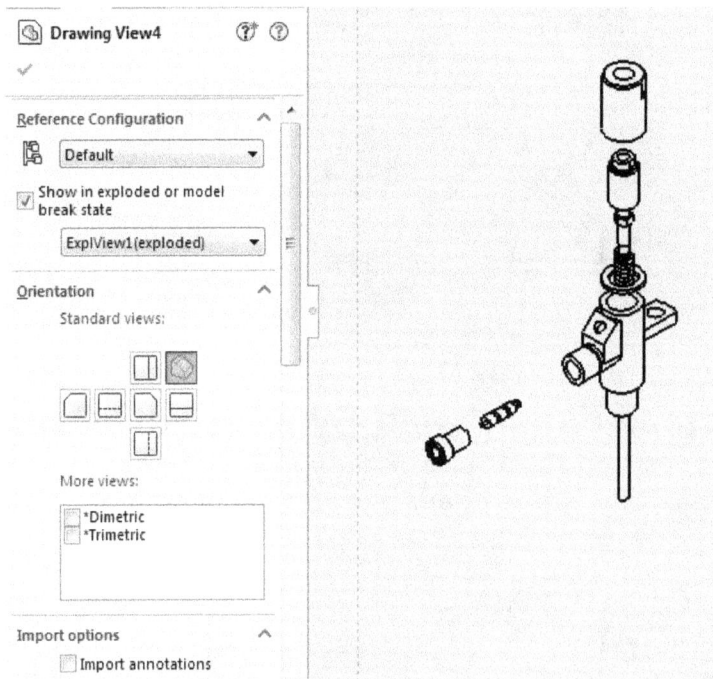

Figure-44. Exploded view of assembly

- You can change the scale as per requirement by using the open in the **PropertyManager.**
- Click on the **OK** button to create the view.

Generating Bill of Material

- Click on the **Annotation** tab and click on the down arrow of **Tables** at the right corner; refer to Figure-45.

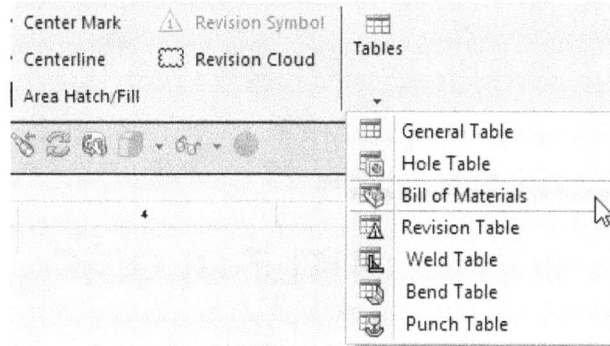

Figure-45. Tables drop-down

- Click on the **Bill of Materials** button. You are asked to select the view.
- Select the exploded view. The **Bill of Materials PropertyManager** will display.
- Click on the **OK** button from the **PropertyManager**. The Bill of Materials table will attach to the cursor.
- Click to place the table; refer to Figure-46.

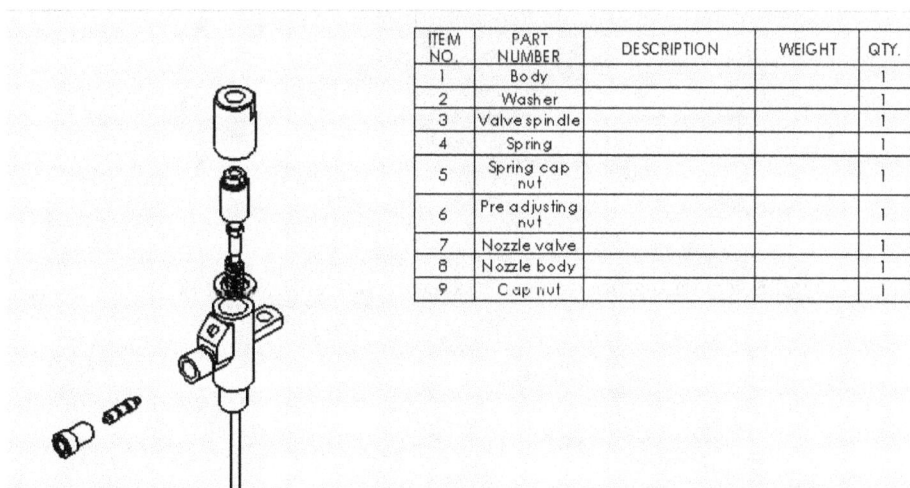

ITEM NO.	PART NUMBER	DESCRIPTION	WEIGHT	QTY.
1	Body			1
2	Washer			1
3	Valve spindle			1
4	Spring			1
5	Spring cap nut			1
6	Pre adjusting nut			1
7	Nozzle valve			1
8	Nozzle body			1
9	Cap nut			1

Figure-46. Bill of Materials

Generating Balloons for Bill of Material

- Click on the **Auto Balloon** button from the **Ribbon** and click on the view. The balloons will be generated automatically.
- Click on the **OK** button from the **PropertyManager** to generate the balloons.
- Drag the balloons to desired positions.

You can use the **Sketch** tab and use the sketcher tools to create custom entities in the drawing.

EDITING TITLE BLOCK

You can edit the title block to specify the information related to designer, part name, versions and so on. To edit the title block, follow the steps given next.

- Right-click on the **Sheet Format** from the **FeatureManager Design Tree**. The shortcut menu will display; refer to Figure-47.

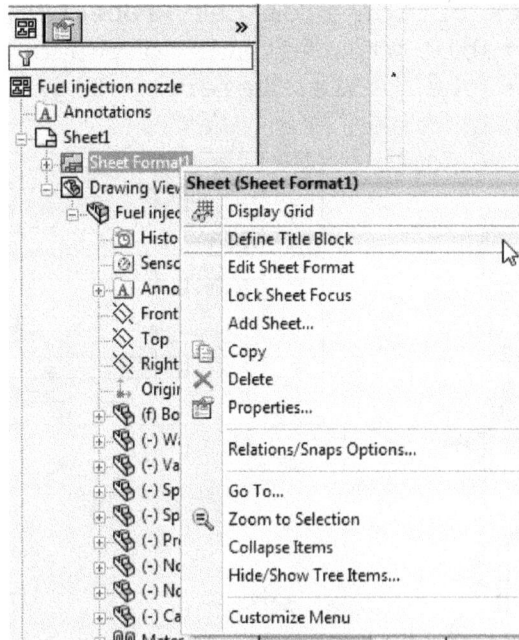

Figure-47. Right click shortcut menu

- Click on the **Edit Sheet Format** button, the title block will become edit able; refer to Figure-48.

- Double-click on the empty boxes in which you want to specify informations.

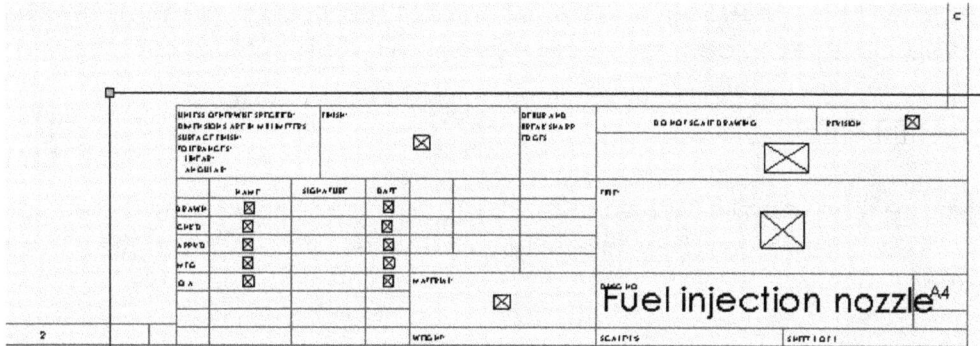

Figure-48. Editable title block

- After specifying the information, click on the **OK** button from the **PropertyManager**.
- Similarly, you can edit the existing information by double-clicking on them.
- Click on the Return button at the top-right corner of the graphic window to return in Drawing mode; refer to Figure-49.

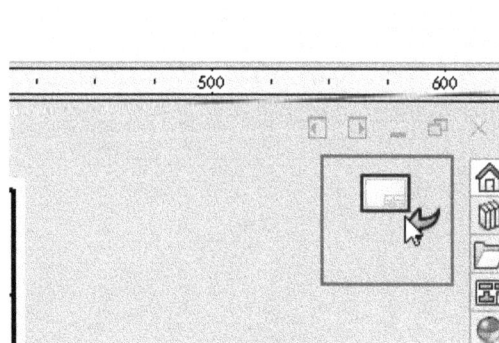

Figure-49. Return to drawing

PRACTICE 1

Generate the drawing views of all the solid models we have created in Practices and Practicals in the previous chapters.

PRACTICE 2

Generate the exploded views, bill of materials and balloons from all the assembly models we have worked on till this chapter.

SELF ASSESSMENT

Q1. In which dialog box, we can set the projection type of views being inserted in drawing?

a. **Sheet Format/Size** dialog box
b. **Sheet Properties** dialog box
c. **New SolidWorks Document** dialog box
d. **Geometric Tolerance** dialog box

Q2. In which of the following tab, the tools to add views in the drawing are available?

a. **View Layout** tab
b. **Annotation** tab
c. **Evaluate** tab
d. **Sheet Format** tab

Q3. We can insert as many model views of a part as required. (T/F)

Q4. We can insert the models views of different parts in same drawing by using the **Model View** tool. (T/F)

Q5. The **Auxiliary View** tool is used to create projected view from selected view by making selected edge of view parallel to screen. (T/F)

Q6. The _____ tool is used to create section in the selected view to display inner detail of the model.

Q7. The _____ tool is used to represent very long objects in the drawing by breaking them at specific span.

Q8. _____ is defined as the maximum permissible overall variation of form or position of a feature.

FOR STUDENT NOTES

Analysis Express

Chapter 10

Topics Covered

The major topics covered in this chapter are:

- *Perform Simulation Xpress Analysis.*
- *Perform Flow Xpress Analysis.*
- *Perform DFM Xpress Analysis.*
- *Perform Costing of manufacturing process*

INTRODUCTION

Analysis Xpress is the combination of tools available in SolidWorks to perform some very useful analyses at a very high pace. There are various types of analyses available in SolidWorks like; SimulationXpress Analysis, FlowXpress Analysis and so on. These analyses are performed for the following functions:

SimulationXpress Analysis: This analysis is used to check whether the component will fail on the specified force/pressure conditions or not. You can generate a report and perform the optimization.

FloXpress Analysis: This analysis is used to check the flow of a fluid through the designed passage.

DFMXpress Analysis: This analysis is used to check whether the created component is manufacturable or not.

StainabilityXpress: This analysis is used to check the impact of environment on the component.

Part Reviewer: This analysis is used to check how the part was created in SolidWorks. This analysis becomes very useful to find error in modeling in SolidWorks.

Along with the above analysis tools, there is one more tool named DriveWorksXpress. This tool is used to increase the speed of designing by making the products formulae based. For example, you can create a formula for Nuts or Bolts and then specify the driving dimensions to create multiple instances of Nuts/Bolts with different sizes.

Note that you need to enter the product codes before using the features discussed in this chapter.

The tools used to perform the named above analyses are discussed next.

SIMULATIONXPRESS ANALYSIS WIZARD

As discussed earlier, the **SimulationXpress Analysis Wizard** tool is used to perform a quick simulation analysis on the model. You can use this tool to perform linear static analysis only. This tool is available in the **Evaluate** tab of the **Ribbon**; refer to Figure-1.

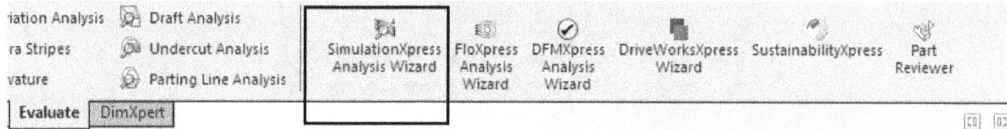

Figure-1 *SimulationXpress_Tool.jpg*

- Click on the **SimulationXpress Analysis Wizard** tool to start SimualtionXpress Analysis, SolidWorks Simulation Xpress will be displayed on the right of the program window; refer to Figure-2. **Make sure that you have opened the solid model for which you want to perform the analysis.**

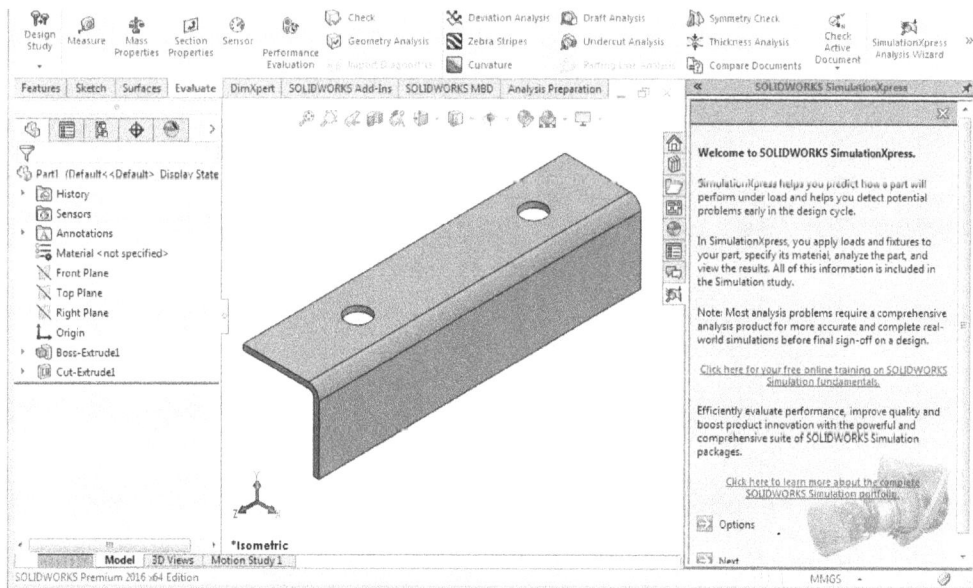

Figure-2 *Interface*

- Click on the **Options** button in the right of the application window; refer to Figure-3. On doing so, the **SimulationXpress Options** dialog box will be displayed as shown in Figure-4.

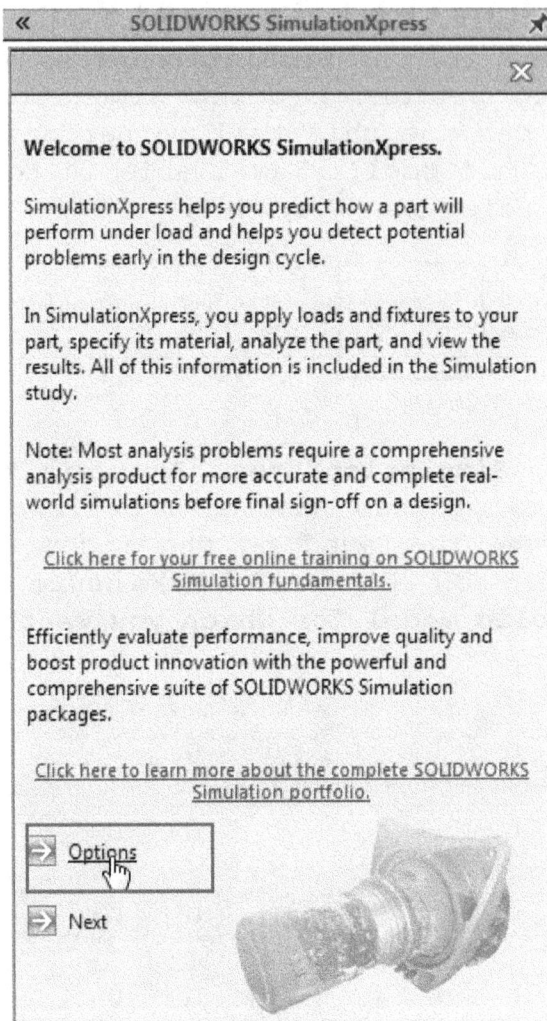

Figure-3 *Options button*

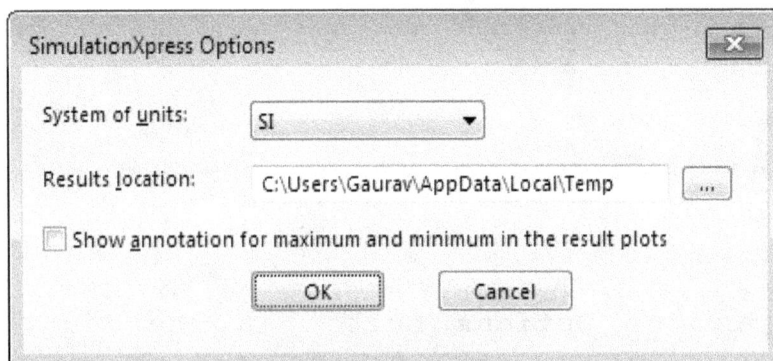

Figure-4 *SimulationXpress Options dialog box*

The options in this dialog box are used to set the unit system of the analysis and save directory for the analysis report.

- Select the **SI** option from the **System of units** drop-down and change the location of result report to the desired one by using the Browse button. [...] After setting all the parameters, select the **OK** button from the dialog box.

FIXTURE SETTING

- Select the **Next** button from the **SolidWorks SimulationXpress** task pane, the **Fixtures page** of SimulationXpress will be displayed as shown in Figure-5.

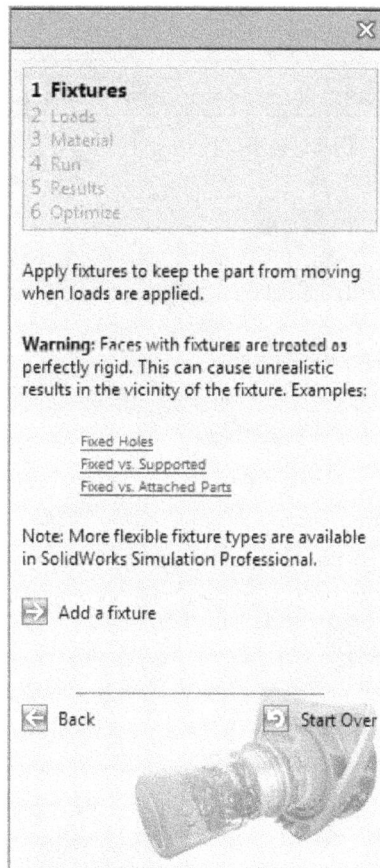

Figure-5 *Fixtures page of SimulationXpress*

1. Click on the **Add a fixture** button to fix a face. On clicking
   on the **Add a fixture** button, the **Fixture FeatureManager**
   will be displayed; refer to Figure-6.

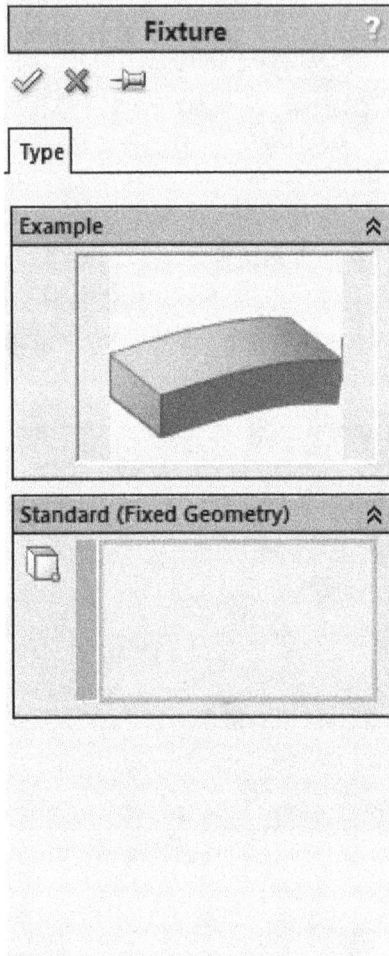

Figure-6 *Fixture PropertyManager*

2. Select the face of the model that you want to be fixed; refer to Figure-7.

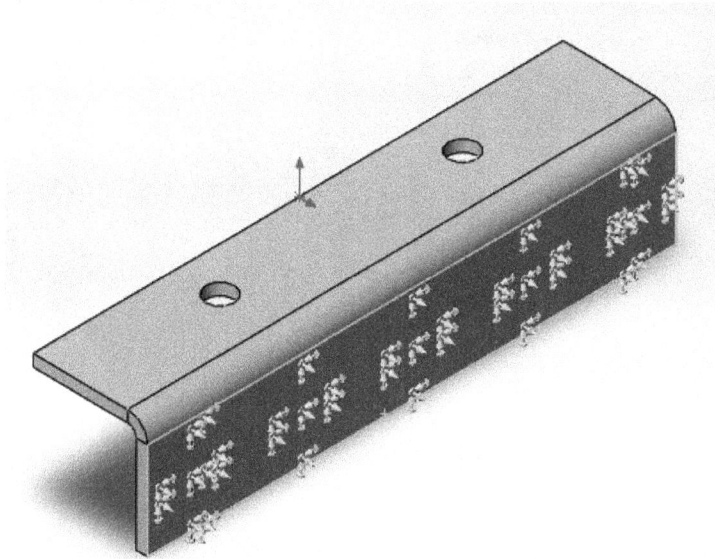

Figure-7 *fixed face*

3. Click on the **OK** button from the **Fixture PropertyManager** and then click on the **Next** link button from the **SimulationXpress** displayed in the right. On doing so, the **Loads** page will be displayed in the right; refer to Figure-8.

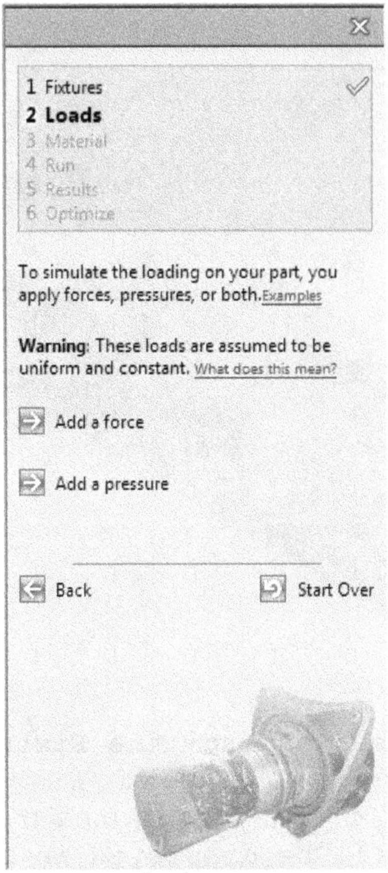

Figure-8 *Loads page*

LOAD SETTING

4. Click on **Add a force** or **Add a pressure** link button from the **SimulationXpress**. In our case, the **Add a pressure** link button is selected. On doing so, the **Pressure PropertyManager** will be displayed as shown in Figure-9.

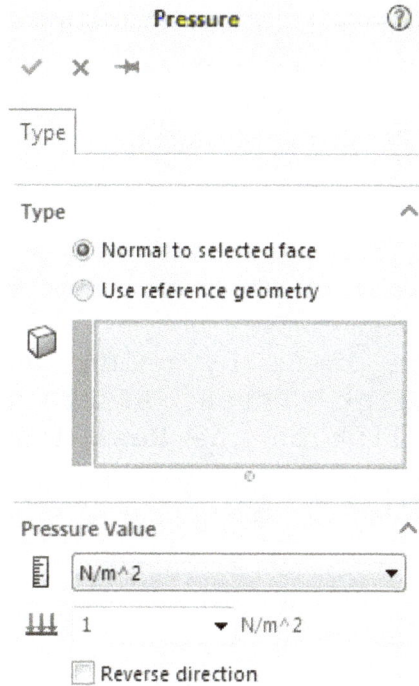

Figure-9 *Pressure PropertyManager*

5. Select the face of the model on which you want to apply the load; refer to Figure-10.

Figure-10 *Face to be selected*

6. Change the unit from N/mm^2 to psi by clicking on the **Unit** drop-down and selecting the **psi** option.

7. Specify the desired pressure value in the **Pressure Value** edit box. After specifying the value, click on the **OK** button from the **PropertyManager**.

You can change the direction of pressure by selecting the Reverse direction check box from the FeatureManager.

8. You can add more loads by using the **Add a force** or **Add a pressure** link button. After specifying all the desired loads, click on the **Next** link button from the **SimulationXpress**. On doing so, the Material page will be displayed; refer to Figure-11.

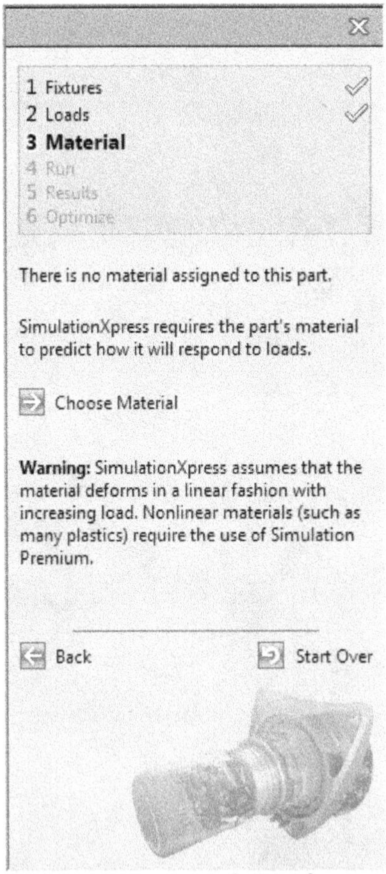

Figure-11 *Material page of SimulationXpress*

MATERIAL SETTING

9. Check the Warning message in the task pane (make sure you understand the conditions of using this analysis) and then click on the **Choose Material** link button from the **SimulationXpress**, the **Materials** dialog box will be displayed as shown in Figure-12.

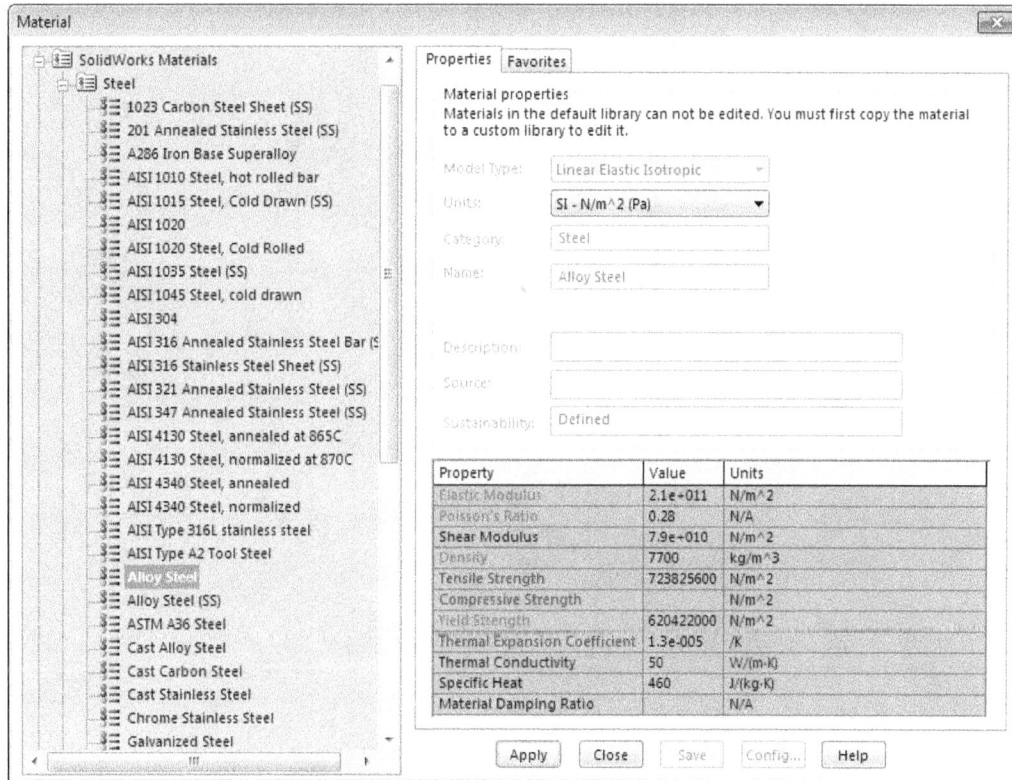

Figure-12 *Materials dialog box*

- Note that some of the properties of the material are highlighted in red color. These highlighted properties are the driving properties for the analysis. Select the desired material from the list in the left and then select the **Apply** button from the dialog box.

- After selecting the **Apply** button, click on the **Close** button to exit the dialog box; the material will be applied and its properties will be displayed in the **SimulationXpress task pane**.

- Select the **Next** button from the **SimulationXpress**. On selecting the button, the **Run** page will be displayed; refer to Figure-13.

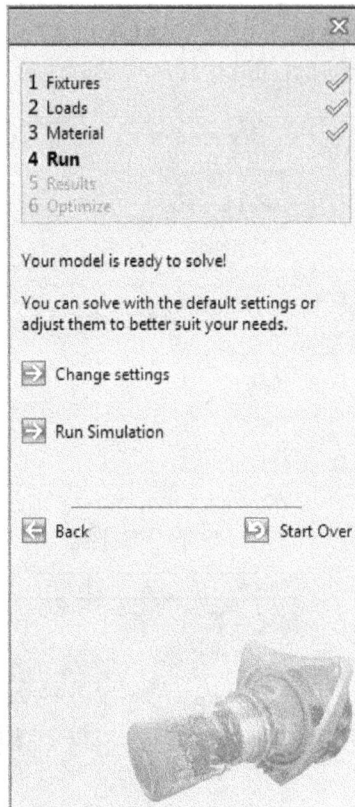

Figure-13 *Run page of SimulationXpress*

RUNNING SIMULATION

- Click on the **Run Simulation** link button from the **SimulationXpress** to check the output. The result will be displayed in the Modeling area and Results page will be displayed as shown in Figure-14.

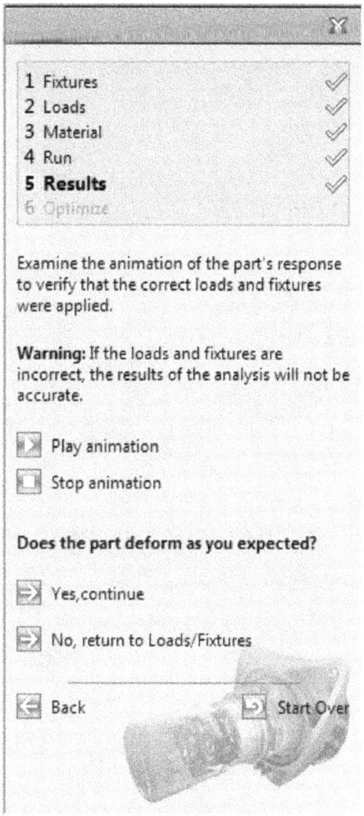

Figure-14 *Results page of*
SimulationXpress

RESULTS

- Select the **Play animation** and **Stop animation** button to start and stop the simulation. If you agree with the result you need to select the **Yes, continue** link button and if you disagree then select the **No, return to Loads/ Fixtures** link button to change the parameters of analysis.

- On selecting the **Yes,continue** button, the modified **Results** page will be displayed as shown in Figure-15.

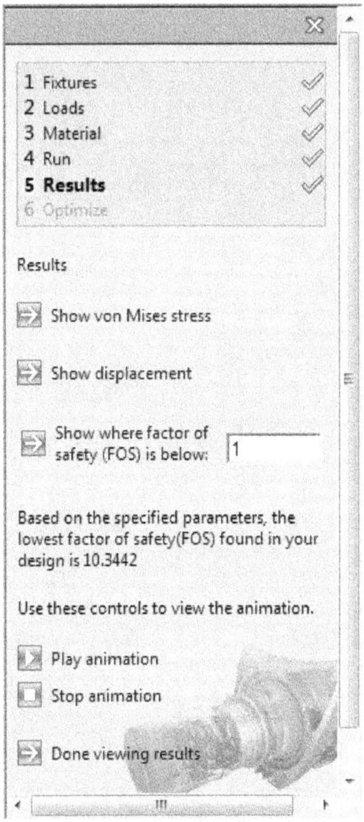

Figure-15 *Modified results page of SimulationXpress*

- You can display von Mises stress or displacement by using the respective button from the SimulationXpress. Also, the suggested FOS will also be displayed in the SimualtionXpress.

- After checking the results, select the **Done viewing results** link button; the modified results page will be displayed as shown in Figure-16. Now, you can generate and eDrawing or you can generate a report by selecting the respective button from the SimualtionXpress.

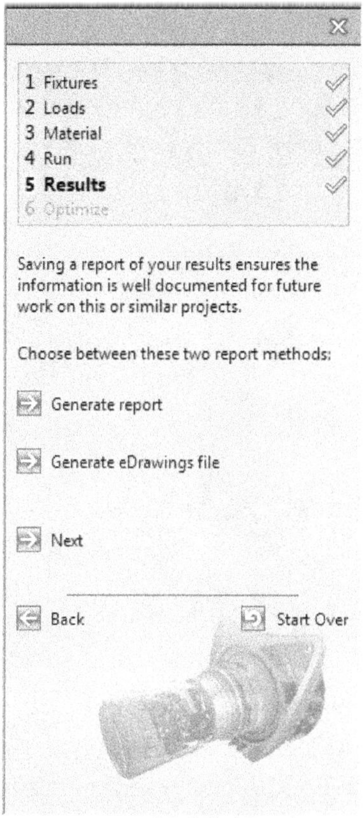

Figure-16 *Modified Results page*

- Select the **Generate report link** button to create the report. On selecting this button, the **Report Settings** dialog box will be displayed as shown in Figure-17.

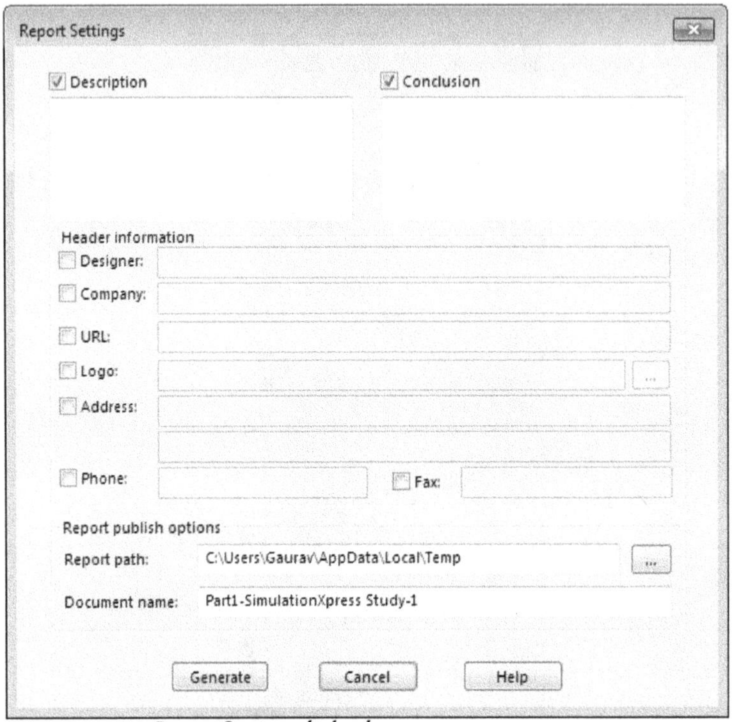

Figure-17 *Report Settings dialog box*

- Select the check boxes to enter information in the desired field in the dialog box. You can enter the path for saving report in the **Report path** field of the dialog box. After specifying the information select the **Generate** button to create the report at the specified path, the **Generating Report** dialog box will be displayed as shown in Figure-18 and after generation gets completed; a word document will open automatically.

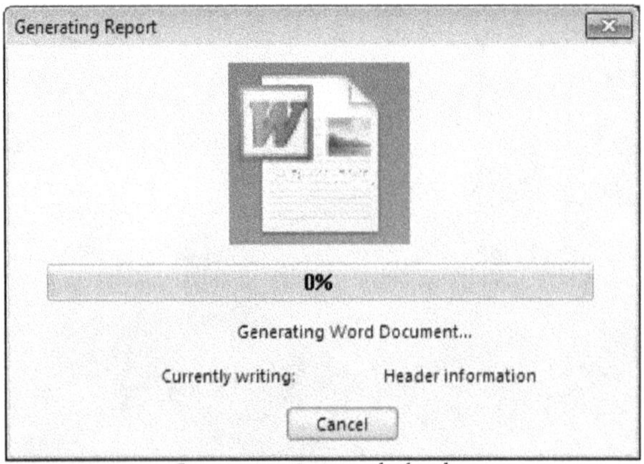

Figure-18 *Generating Report dialog box*

- Similarly, you can generate an eDrawing of the results report by selecting **Generate eDrawings file** link button. The eDrawings can be opened in the eDrawing software provided with the SolidWorks package.

OPTIMIZING

- After checking the results, if you want to optimize the model then select the **Next** link button from the SimualtionXpress, the Optimize page of SimulationXpress will be displayed as shown in Figure-19.

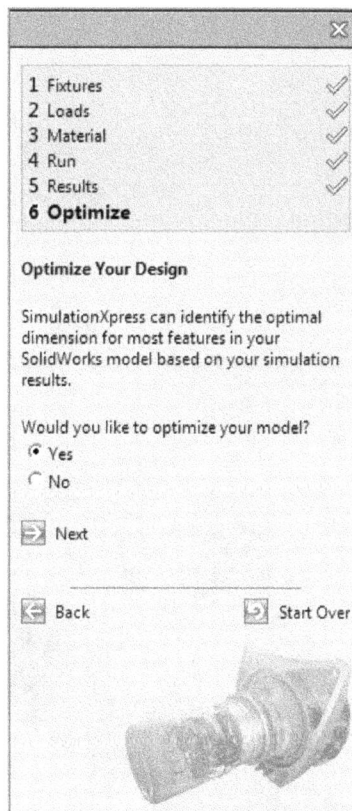

Figure-19 *Optimize page of SimualtionXpress.jpg*

- Make sure that the **Yes** radio button is selected in the page and then select the **Next** button from the SimulationXpress. On doing so, the **Parameters** dialog box will be displayed and the driving dimensions will be displayed in blue color in the modeling area; refer to Figure-20 and Figure-21.

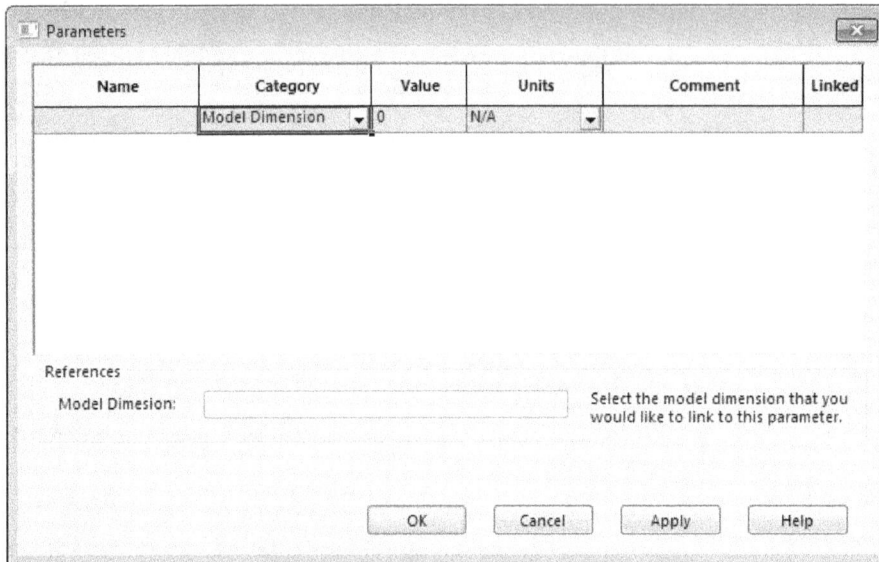

Figure-20 *Parameters dialog box*

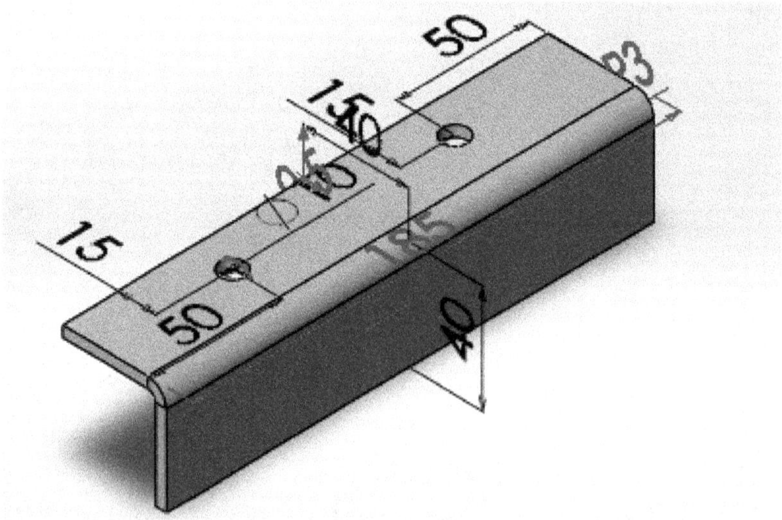

Figure-21 *Driving dimensions of the model*

- Click on one of the driving dimension from the model that you want to change for optimization and then select the **OK** button from the dialog box, the value will be added in the **DesignXpress Study** displayed below the modeling area; refer to Figure-22.

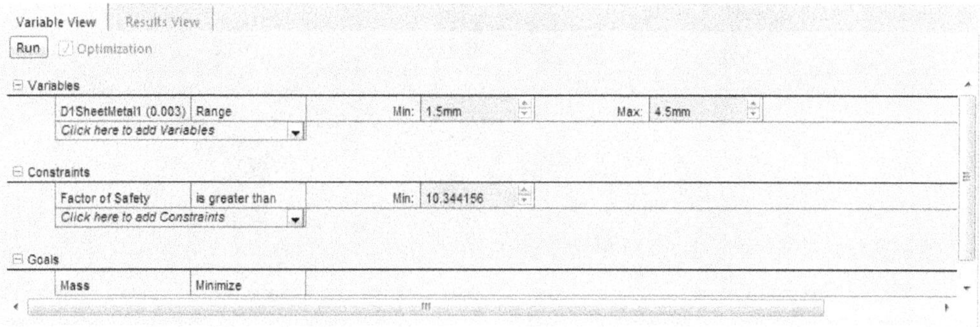

Figure-22 *DesignXpress Study*

- To add more variables for optimization, click on the **Click here to add variables** drop-down under the **Variables** node in the **DesignXpress Study**. On doing so, the Parameters dialog box will be displayed again. Click in the blank Name field and select another dimension from the modeling area to add it in the variables.
- After adding the desired variable, select the OK button from the dialog box. Now, click on the **Click here to add Constraints** drop-down under the **Constraints** node and select the desired constraint from the list. In this case, we have selected **Factor of Safety** as the constraint; refer to Figure-22.
- After setting all the desired parameters; click on the **Run** button displayed at the top in the **DesignXpress Study**. On doing so, the **DesignXpress Study in Progress** dialog box will be displayed; refer to Figure-23.

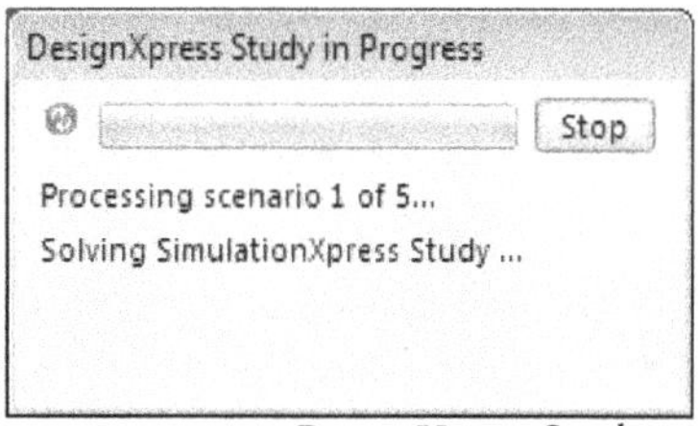

Figure-23 *DesignXpress Study in Progress dialog box*

- After the study is complete, the optimization results will be displayed in the **Results View** tab of the **DesignXpress Study;** refer to Figure-24.

Figure-24 *Results View tab of DesignXpress Study*

- In the **Optimize** page of **SimulationXpress,** the options will be modified as shown in Figure-25.

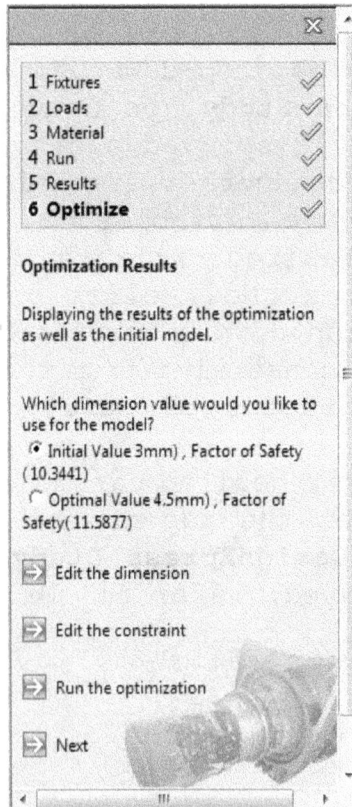

Figure-25 *Modified Optimize page of SimulationXpress*

- Select the radio button for **Optimal** value and then click on the **Run the optimization** link button again to cross check and then click on the **Next** button to apply the optimization.
- Exit the **SimulationXpress** by clicking on the **Close** button at the top-right corner of it and select **Yes** from the mediator dialog boxes displayed.

FLOXPRESS ANALYSIS

FloXpress Analysis is used to check the flow of a fluid through the designed passage. This passage can be in a solid model or it can be in an assembly. This analysis can be performed by using the **FloXpress Analysis Wizard** button available in the **Evaluate** tab of the **Ribbon**; refer to Figure-26.

Figure-26 *FloXpress Analysis Wizard button*

To perform the express flow analysis of a passage, you need to close all the openings in the model with the help of a lid. To do so, perform the following steps.

- Open the model for which you want to perform the flow analysis; refer to Figure-27. Note that using the FloXpress you can check flow of component having only one inlet and one outlet.

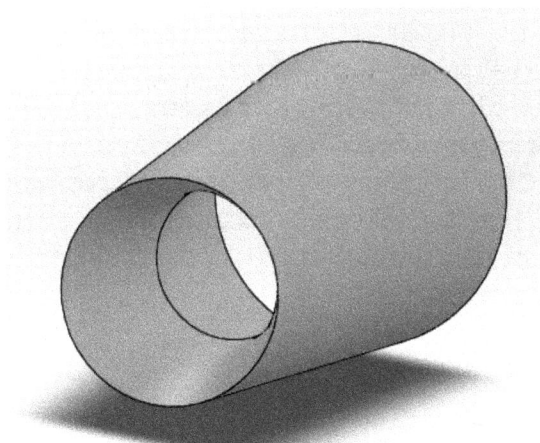

Figure-27 *Model for flow analysis*

PREPARING MODEL

- Now, we need to close both the ends. To do so, select the round edges of one end and then select the **Filled Surface** tool from the **Surfaces** tab in the **Ribbon**; refer to Figure-28. On doing so, the **Filled Surface PropertyManager** will be displayed. Click on the **OK** button from the **PropertyManager**. Similarly, create the filled surface on the other end. The model after creating the filled surfaces will be displayed as shown in Figure-29.

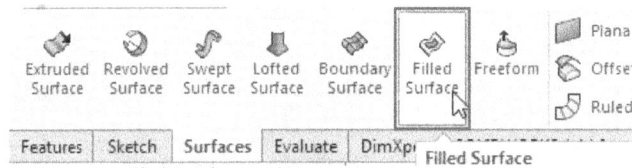

Figure-28 *Filled Surface tool*

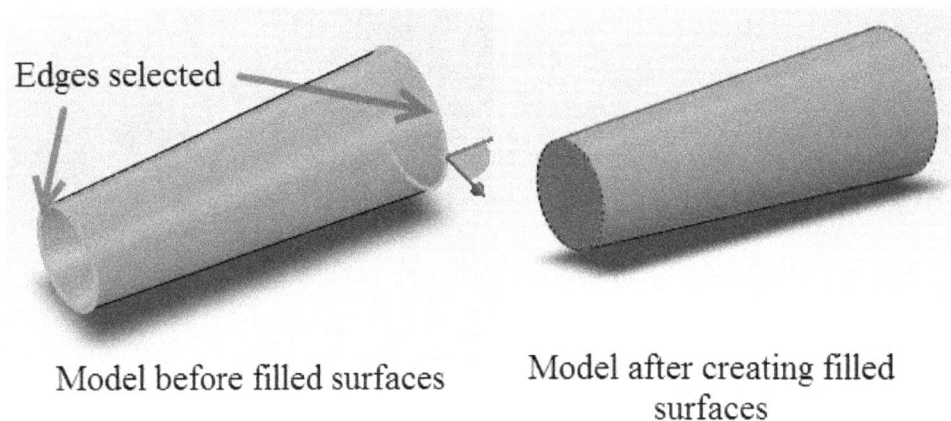

Model before filled surfaces   Model after creating filled surfaces

Figure-29 *Filled surfaces creation*

- Now, we need to thicken these surface to create the lid. To do so, select one of the surfaces and then select the **Thicken** tool from the **Surfaces** tab. On doing so, the **Thicken PropertyManager** will be displayed.
- Select the **OK** button from the **PropertyManager**, the surfaces will be thickened. Similarly, thicken the other surface, the model after thickening will be displayed as shown in Figure-30.

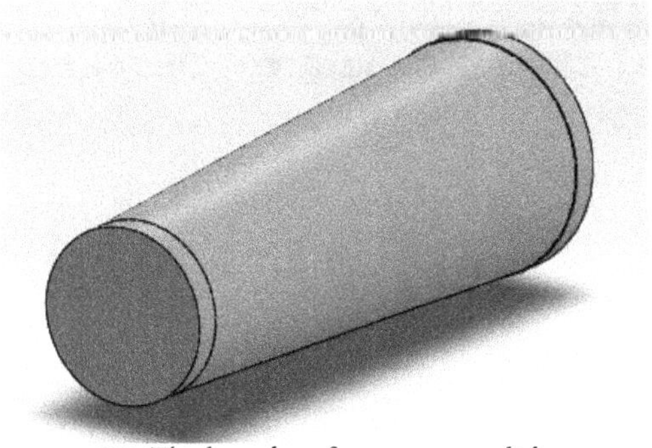

Figure-30 *Thickened surfaces to create lids*

STARTING FLOW ANALYSIS

• Select the **FloXpress Analysis Wizard** tool from the **Evaluate** tab of the **Ribbon**, the **Welcome PropertyManager** will be displayed in the left of the application window; refer to Figure-31.

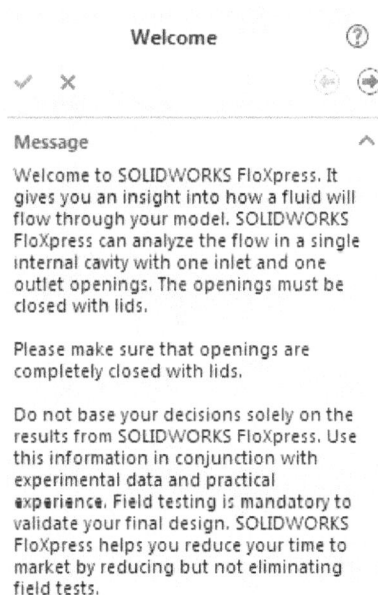

Figure-31 *Welcome PropertyManager*

- Click on the **Next** button from the **PropertyManager**. On doing so, the **Check Geometry PropertyManager** will be displayed as shown in Figure-32.

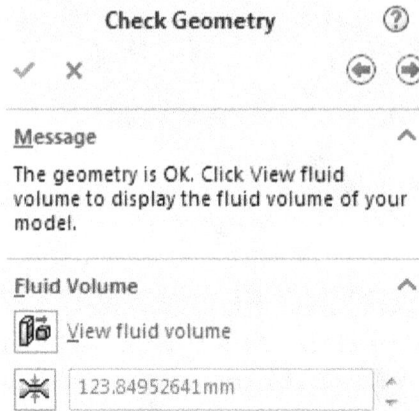

Check Geometry   ⑦

✓ ✗   ⬅ ➡

Message   ⌃

The geometry is OK. Click View fluid volume to display the fluid volume of your model.

Fluid Volume   ⌃

View fluid volume

123.84952641mm

Figure-32 *Check Geometry PropertyManager*

- Click on the **View fluid volume** to check the fluid volume in designed passage. On doing so the model will be displayed as shown in Figure-33.

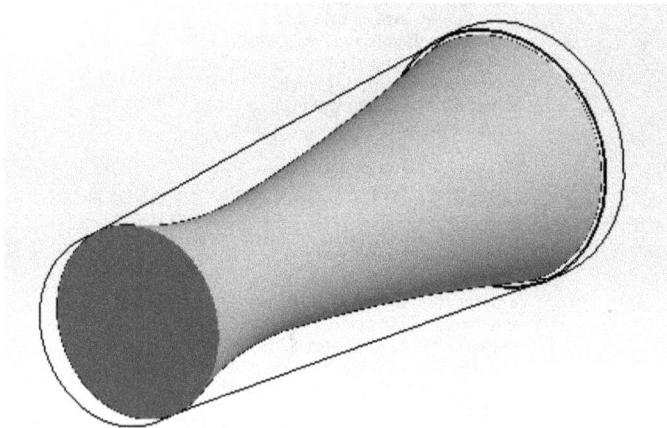

Figure-33 *Fluid Volume in model*

- You can change the smallest flow passage for the model at this stage. To do so, click on the **Smallest Flow Passage** button and then change the value by using the spinner.
- After specify the desired parameters, click on the **Next** button from the **PropertyManager**. On doing so, the **Fluid PropertyManager** will be displayed and you are prompted

to specify the type of fluid for the analysis. You can select either water or air.

- In this case, we select the **Air** radio button. Now, click on the **Next** button from the **PropertyManager**. On doing so, the **Flow Inlet PropertyManager** will be displayed as shown in Figure-34.

Figure-34 *Flow Inlet PropertyManager*

- Select the desired button and then specify the desired values in the edit boxes displayed at the bottom of the **Flow Inlet PropertyManager**.
- Select the inner face of the model as inlet; refer to Figure-35. You might need to right-click on the face and then select the **Select Other** option from the shortcut menu.

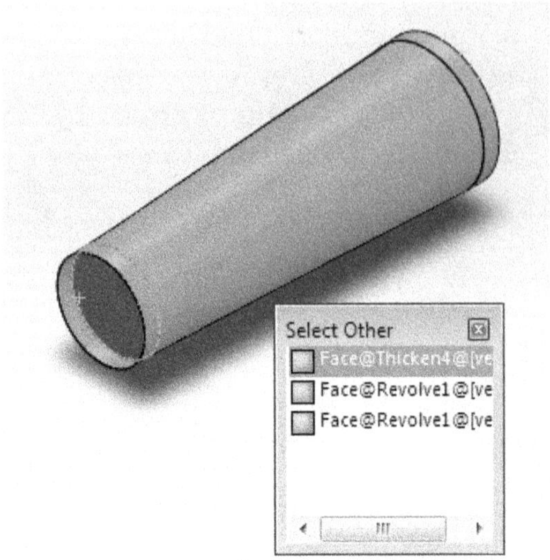

Figure-35 *Face to be selected as Inlet*

- After selecting the face, select the **Next** button from the **PropertyManager**. On doing so, the **Flow Outlet PropertyManager** will be displayed as shown in Figure-36.

Figure-36 *Flow Outlet PropertyManager*

- Now, you need to select the outlet for the model. To do so, move the cursor over the flat face of the outlet lid and then right-click; a shortcut menu will be displayed. Now, select the face as shown in Figure-37.

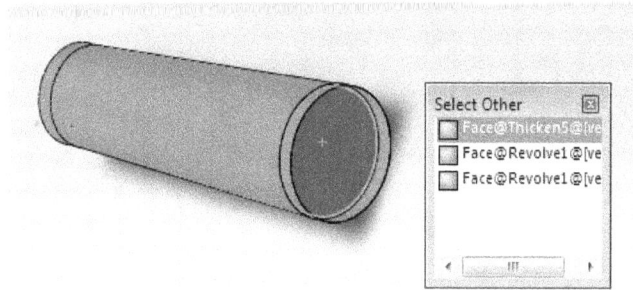

Figure-37 *Face to be selected as Outlet*

- To check the flow fluid, you need to change the display style of the model. To do so, select the **Wireframe** button from the **Display Style** drop-down; refer to Figure-38.

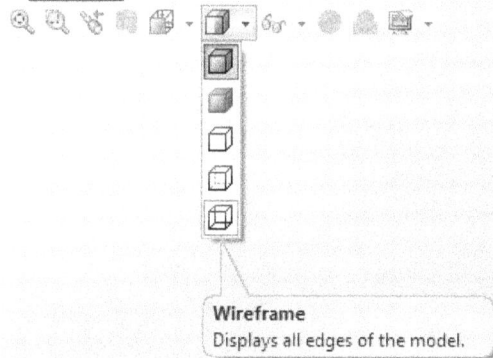

Figure-38 *Wireframe button*

- After changing the display style, the model will be displayed as shown in Figure-39.

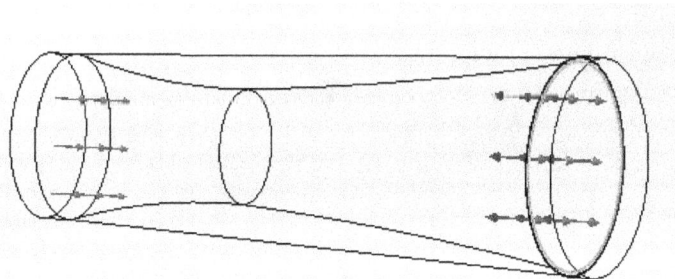

Figure-39 *Model in wireframe style.jpg*

- Click on the **Next** button from the **PropertyManager** to display the **Solve PropertyManager**; refer to Figure-40.

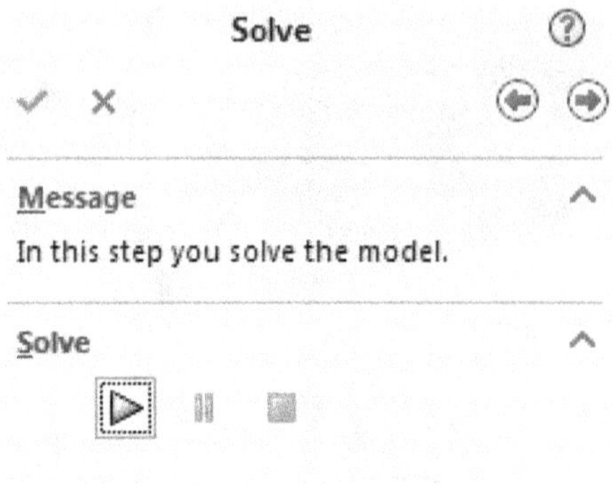

Figure-40 *Solve PropertyManager*

- Click on the **Solve** button to run the analysis. On doing so, the system will start to solve and after the CFD problem is solved, the solution will be displayed.

- You can generate the report of result by using the **Generate Report** button under the **Report** rollout. On doing so, the word document of report will be generated.

DFMXPRESS ANALYSIS

The DFMXpress analysis is used to check whether the model in the Modeling area is manufacturable or not. Using this analysis, you can check a model for its manufacturing by Mill/Drill Manufacturing process, Turn with Mill/drill process, Injection Molding process or Sheetmetal process. To analyze the model for these manufacturing processes, click on the **DFMXpress Analysis Wizard** tool; the DFMXpress task pane will be displayed in the right of the application window; refer to Figure-41.

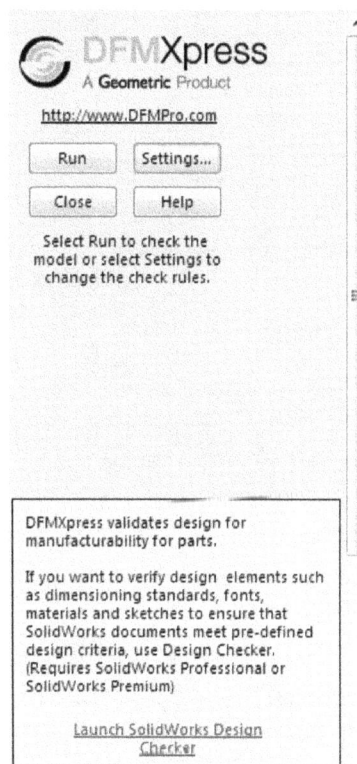

Figure-41 *DFMXpress*

Now, click on the **Settings** button to specify the parameters related to the analysis; refer to Figure-42.

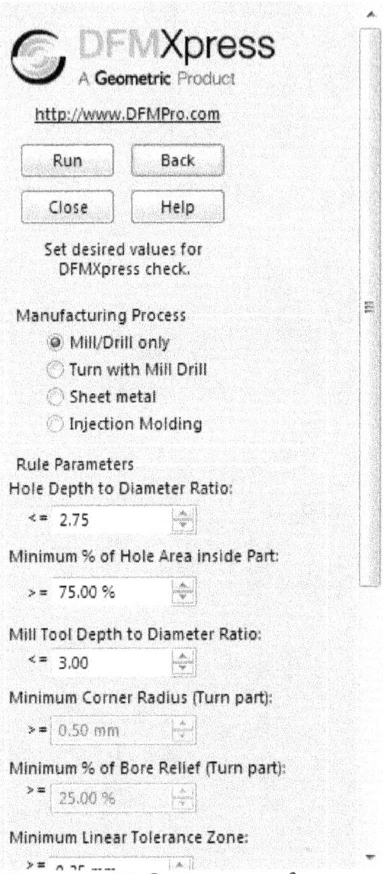

Figure-42 *Settings page of DFMXpress*

Select the desired process and then specify the parameters in the below fields. After specifying all the parameters, click on the **Run** button to run the analysis. After the analysis gets completed, the results will be displayed at the bottom in the DFMXpress.

COSTING

In real world, it is possible to make same part with two or more processes, like part in Figure-43 can be manufactured by milling, shaping, sheetmetal processes or casting process. But in industry, we are also concerned about the cost of manufacturing and quality of product. In this section, we will learn to estimate the cost of manufacturing the model by various processes.

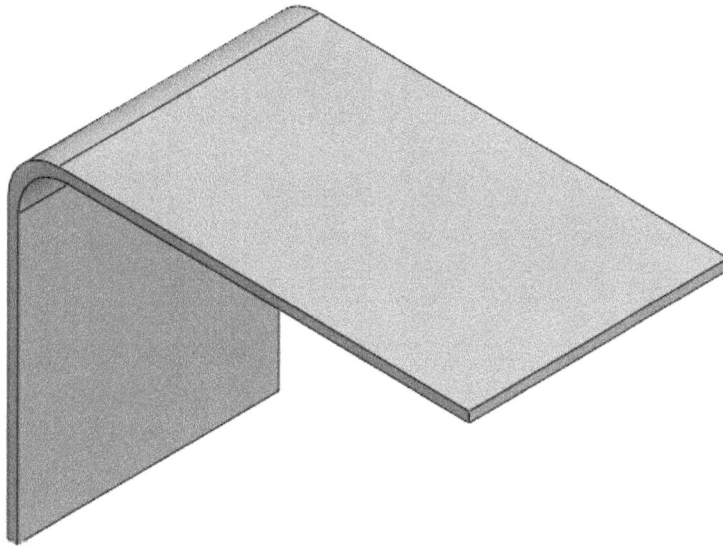

Figure-43 *Model for costing*

- Click on the **Costing** button from the **Evaluate** tab in the **Ribbon**. The Costing interface will be displayed; refer to Figure-44.

Figure-44 *Costing interface*

- In the above figure, cost is estimated on the basis of machining process. You can select the desired process from **Method** drop-down; refer to Figure-45.

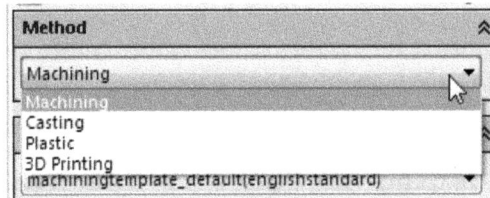

Figure-45 *Method drop-down*

- Click in the **Template** drop-down in **Costing** task pane to select the desired template. Using the **Launch Template Editor** button you can launch the **Costing Template Editor** to edit the template; refer to Figure-46. Note that using the Editor you can change the cost and time of various manufacturing process with respect to the selected material. Like, you can change the feed rate for milling a steel material by clicking on the Mill tab in right of the Costing Template Editor dialog box. Changing the feed rate will change the cycle time and hence the cost of machining.

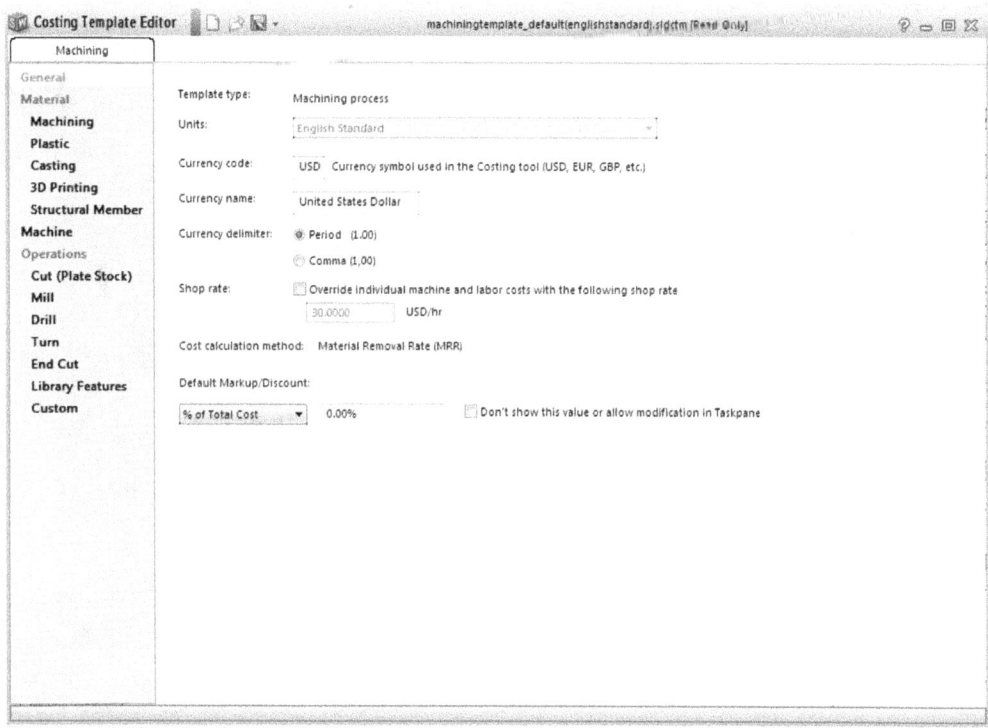

Figure-46 *Costing Template Editor*

- After editing the template, save the file and exit the editor.
- It might be possible that you have some processes which are not having price rate. In that case, you need to manually specify the price. Such processes are displayed under **No Cost Assigned** node; refer to Figure-47.

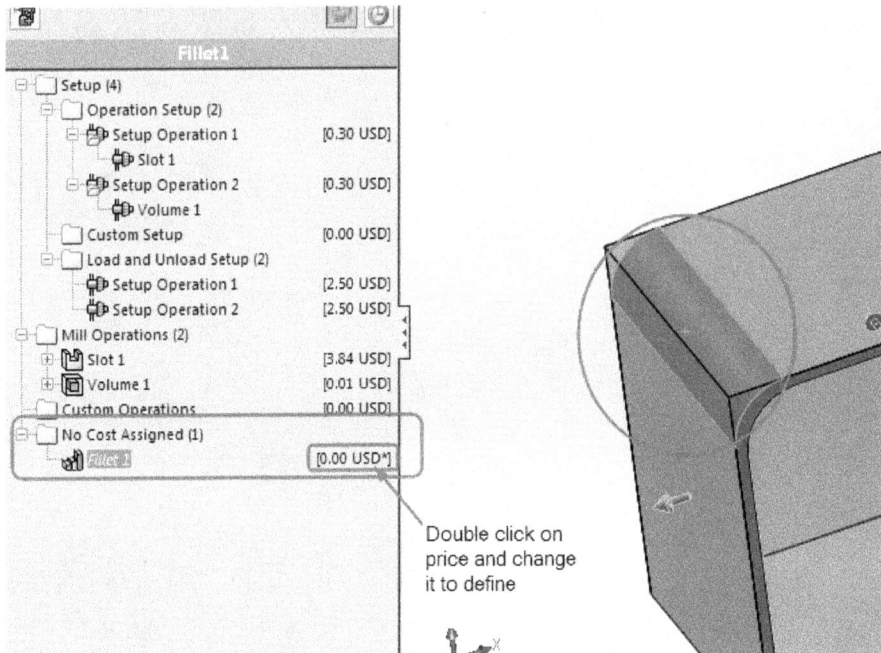

Figure-47 *No cost defined*

- After specifying all the costs, change the material to the desired one and also update the per kg price of material in the **Material** rollout in task pane; refer to Figure-48.

Figure-48 *Material rollout*

- Check the price of manufacturing in the **Estimated Cost per Part** area of **Costing** task pane and get the optimum process for manufacturing your model.

PRACTICE 1

Design a wall bracket to bear a load of 100 kg. There should be two counterbore holes for M10 bolts to mount it on wall. Material for wall bracket is Cast Iron.

SELF-ASSESSMENT

Q1. Which of the following analysis should be used to test the load bearing capacity of a component?

a. FloXpress Analysis
b. SimulationXpress Analysis
c. StainabilityXpress Analysis
d. DFMXpress Analysis

Q2. Which of the following analysis is used to check the flow of a fluid through the designed passage?

a. FloXpress Analysis
b. SimulationXpress Analysis
c. StainabilityXpress Analysis
d. DFMXpress Analysis

Q3. Which of the following analysis is used to check whether the model in the Modeling area is manufacturable or not?

a. FloXpress Analysis
b. SimulationXpress Analysis
c. StainabilityXpress Analysis
d. DFMXpress Analysis

Q4. FloXpress Analysis can be performed with water as well as air. (T/F)

Q5. Different manufacturing processes hardly affect the production of a component in manufacturing facility. (T/F)

FOR STUDENT NOTES

Mold Tools

Chapter 11

Topics Covered

The major topics covered in this chapter are:

- *Starting the Mold tools*
- *Analyzing the model for molding*
- *Preparing the model for mold*
- *Starting the Mold project*
- *Creating the parting line*
- *Creating the Shutoff surfaces*
- *Creating the Parting surface*
- *Creating the Splitting the Core and Cavity from Tooling*

STARTING THE MOLD TOOLS

To start with the mold tools, first you need to open or import a model. To open/ import a model follow steps given next.

- Go to the **File menu** and click on the **Open** option or click on the **Open** button from the **Menu Bar** or click on the **Open a Document** link in the **SolidWorks Resources Task Pane**; refer to Figure 1. The **Open** dialog box is being displayed as Figure 2.

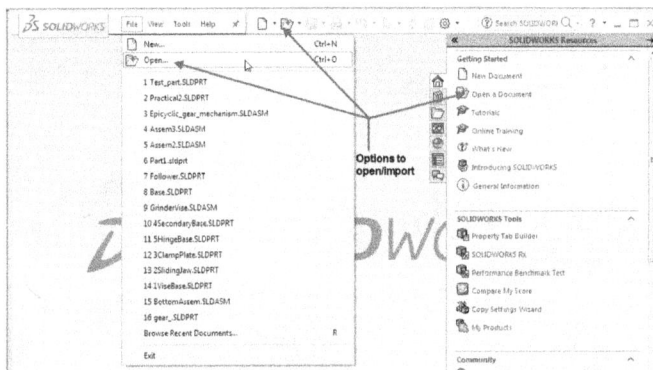

Figure 1. Open or Import methods

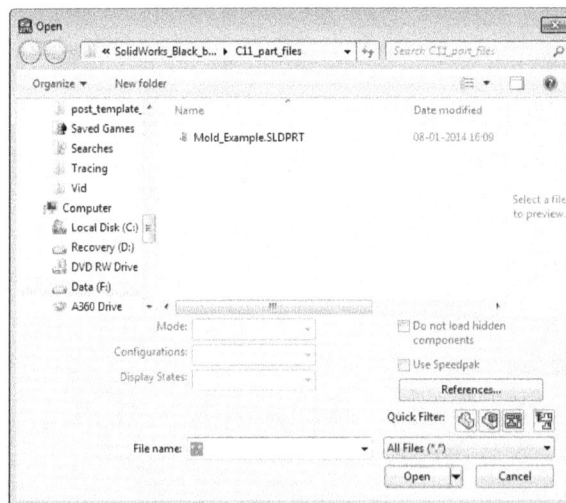

Figure 2. Open dialog box

- Click on the **SolidWorks Files** drop-down displayed at the bottom right in the dialog box. List of file types supported by SolidWorks will be displayed; refer to Figure 3.

SolidWorks Files (*.sldprt; *.sldasm; *.slddr
Part (*.prt;*.sldprt)
Assembly (*.asm;*.sldasm)
Drawing (*.drw;*.slddrw)
DXF (*.dxf)
DWG (*.dwg)
Adobe Photoshop Files (*.psd)
Adobe Illustrator Files (*.ai)
Lib Feat Part (*.lfp;*.sldlfp)
Template (*.prtdot;*.asmdot;*.drwdot)
Parasolid (*.x_t;*.x_b;*.xmt_txt;*.xmt_bin)
IGES (*.igs;*.iges)
STEP AP203/214 (*.step;*.stp)
IFC 2x3 (*.ifc)
ACIS (*.sat)
VDAFS (*.vda)
VRML (*.wrl)
STL (*.stl)
CATIA Graphics (*.cgr)
CATIA V5 (*.catpart;*.catproduct)
ProE/Creo Part (*.prt;*.prt.*;*.xpr)
ProE/Creo Assembly (*.asm;*.asm.*;*.xas)
Unigraphics/NX (*.prt)
Inventor Part (*.ipt)
Inventor Assembly (*.iam)
Solid Edge Part (*.par;*.psm)
Solid Edge Assembly (*.asm)
CADKEY (*.prt;*.ckd)
Add-Ins (*.dll)
IDF (*.emn;*.brd;*.bdf;*.idb)
CADKEY (*.prt;*.ckd)

Figure 3. File types supported

- Select the type for your file from the list and browse to the folder in which you have placed the file.
- Double-click on the file, you want to open or import. The file will open in the application; refer to Figure 4.

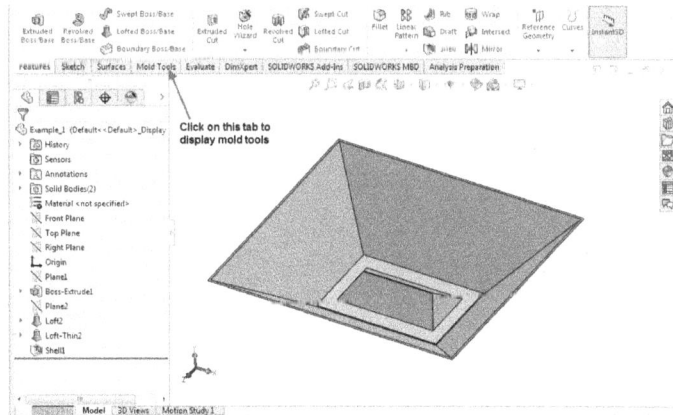

Figure 4. Opened file

- Click on **Mold Tools** tab of **Ribbon** to display tools of molding. If the **Mold Tools** tab is not displaying by default, then right-click on any of the tab displaying in the **Ribbon**. On right-clicking a menu will be displayed; refer to Figure 5.

Figure 5. Tabs menu

- Click on the **Mold Tools** option from this list, the **Mold Tools** tab will get added in the Tab Bar. On clicking the **Mold Tools** tab, the Mold Tools will be displayed as shown in Figure 6.

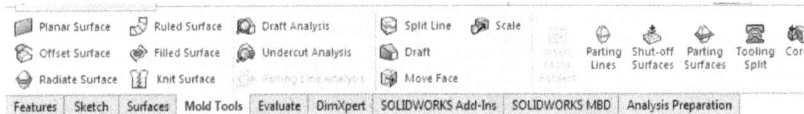

Figure 6. Mold tools tab

- Most of the tools have already been discussed in the book. Now, you will learn about the tools that are specifically used for making mold.

Now, we have the model opened and we need to analyze it for the possibility of its mold design.

ANALYZING THE MODEL

There are three tools available in the **Mold Tools** tab to analyze the model for molding: **Draft Analysis**, **Undercut Analysis** and **Parting Line Analysis**. These tools are discussed next.

Draft Analysis

The **Draft Analysis** tool is used to check the draft angles of various faces in the model. Draft is an important requirement of Molding. Draft is the taper angle given to various faces of the mold part for easy and safe ejection from the mold tooling (core and cavity). The angle value of draft depends on the material and geometry of the mold part. Typically 1° to 3° of draft is given on all faces of the mold part. If you have steps at parting line, then 5° to 7° of draft is required for shutoff. (**Shutoff** is the surface where core and cavity meet each other.) To perform the draft analysis, follow the steps given next.

* Click on the **Draft Analysis** tool in the tab, **Draft Analysis PropertyManager** will be displayed in the left of the screen; refer to Figure 7.

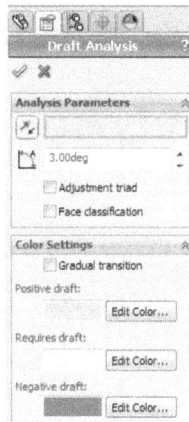

Figure 7. Draft analysis property-manager

* A box is highlighted in blue color in the **Analysis Parameters** rollout of the **PropertyManager.** You are prompted to select a face with respect to which the angles will be measured and the pull direction will be defined.
* Select a flat face from the model. Note that this face will become the mating face of core and cavity later. Figure 8 shows the face selected as neutral face.

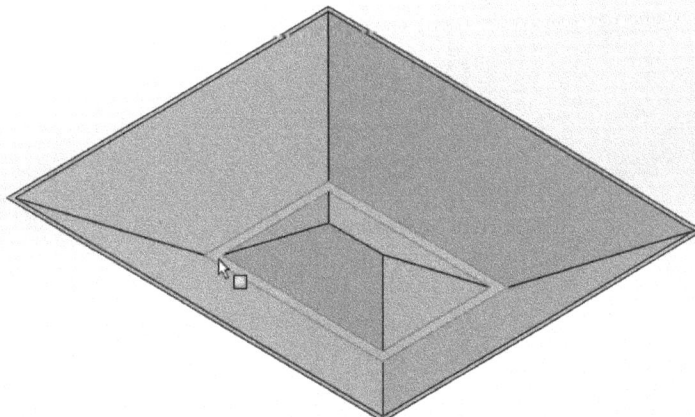

Figure 8. Face to be selected

- As you select the face, all the faces of the model will be painted with the colors specified in the lower area of the **PropertyManager**.
- You can change the colors of the faces as per your requirement by selecting the **Edit Color** buttons displayed in the lower area of the **PropertyManager**.
- Note that when you move cursor over the faces of the model, the draft angle value of the current face will be displayed along with the cursor; refer to Figure 9.

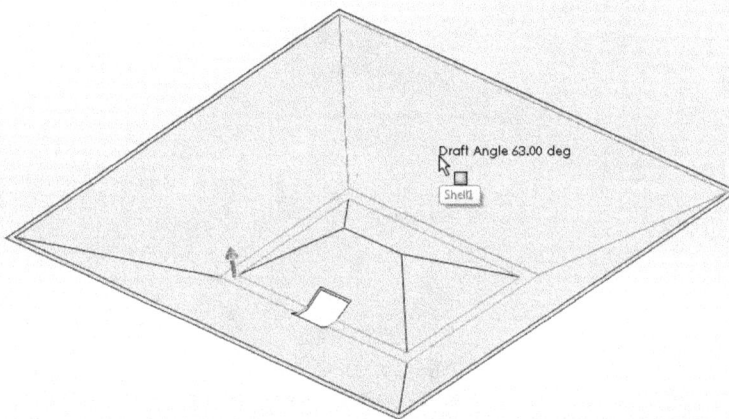

Figure 9. Draft angle over the faces

- If you need to adjust the pull direction, then select the Adjustment Triad check box. On doing so, the triad will be displayed around the pull direction arrow; refer to Figure 10. Using this triad, you can change the direction of pull.

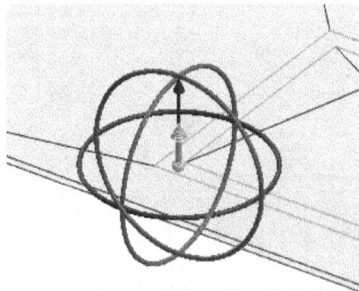

Figure 10. Adjustment triad

- Now, check the colors of the model. By default, **Green** color denotes the positive draft which means that the faces are going to be in the **Core**. **Red** color denotes negative draft which means that the faces are going to be in the **Cavity**.

Yellow color denotes that the faces do not have draft angle applied. **The faces that are painted yellow need to be worked on.** You need to apply the desired draft angle to these faces. The method to apply draft will be discussed later in this chapter.

Undercut Analysis

The **Undercut Analysis** tool is used to check the faces of the mold part that behave as undercuts. SolidWorks classify the undercut faces in five categories: **Direction1 undercut, Direction2 undercut, Occluded undercut, Straddle undercut,** and **No undercut.**

Direction1 undercut: The faces in this category are imprinted on Core tooling.

Direction2 undercut: The faces in this category are imprinted on Cavity tooling.

Occluded undercut: The faces in this category are those faces which can neither be included in core nor in cavity. These faces require sliders. In SolidWorks, sometimes these faces are shifted to No undercut category.

Straddle undercut: The faces in this category are those which you can put in any of the two: the core steel or the cavity steel.

No undercut: The faces in this category are those which are not counted as undercut. But in SolidWorks, sometimes these are the occluded faces. Most of the time, you need to split the face to transfer it in core/cavity steel.

Now, we will perform the Undercut Analysis on the mold part. The steps are given next.

- Exit all the other analyses if still active. Now, click on the **Undercut Analysis** tool. **Undercut Analysis PropertyManager** will be displayed; refer to Figure 11.

Figure 11. Undercut
analysis PropertyManager

- You can select a pull face or you can select the parting line (if available).
- On selecting the pull face, the mold part will be displayed; refer to Figure 12.

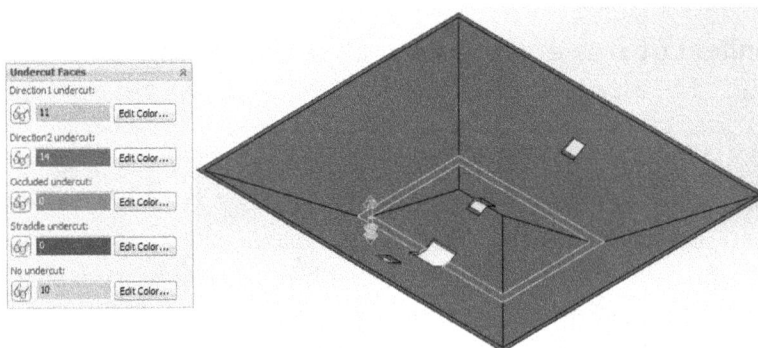

Figure 12. Model with undercut report

- You can see from the model and report that there are 10 faces that come in **No undercut** category. If you want to hide all the other faces and want to display only No undercut faces, then click on the **Show/Hide** button 👁 next to all other categories in the PropertyManager; refer to Figure 13.

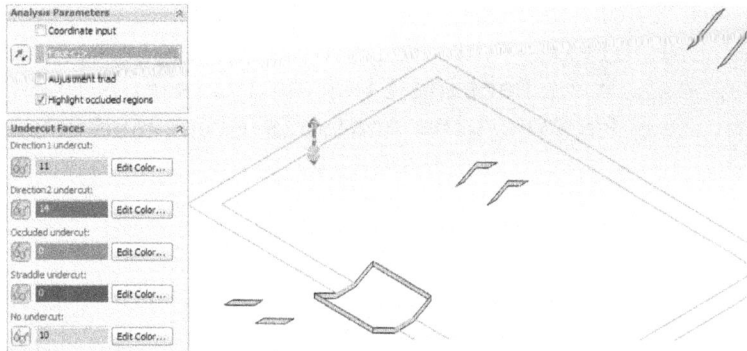
Figure 13. Mold part after hiding other categories

- Sometimes, we need to split these faces to include them in their respective die steel. You will learn about the splitting later in this chapter.

Parting Line Analysis

The **Parting Line** tool is used to check the possible parting line for the mold part. Using this tool, you can check the parting line for multiple pull directions. The steps to perform Parting Line Analysis are given next.

- Check the part given in Figure 14 and try to find out the parting lines manually (Parting line is the line at which the core and cavity meet.)

Figure 14. Part for parting line check

- Now, we will check what should be the parting lines by which we can divide the mold part into core and cavity. To do so, click on **Parting Line Analysis** tool to start the Analysis. The **Parting Line Analysis PropertyManager** will be displayed as shown in Figure 15.

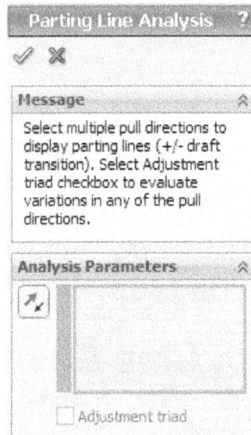

Figure 15. Parting Line Analysis PropertyManager

- Select the faces perpendicular to which the pull directions are two be defined. Refer to Figure 16.

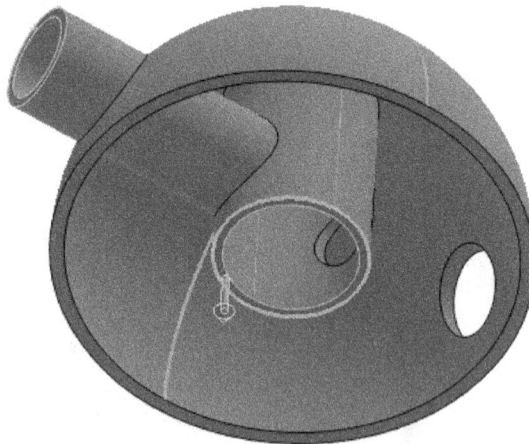

Figure 16. faces selected for parting line analysis

- Check the dark and light lines. In the above figure, there are three types of lines: dark lines, dashed lines, and dotted lines. Dark lines denote main core and cavity. Dashed lines and dotted lines denote sub-inserts for the mold or the outline where the mold part needs to be divided. (Sub-Inserts are the parts that are assembled in the main Core/Cavity to create a mold part.)

PREPARING MODEL FOR MOLD

After performing the above three analyses, we need to modify the mold part so that molding becomes feasible. There are four options available in SolidWorks to prepare the model for molding: **Split Line**, **Draft**, **Move Face**, and **Scale**. We will explain these tools and their uses one by one.

Splitting Faces using Split Line tool

In some of the cases, a face of mold part cannot be completely allotted to core or cavity. In those cases, you need to divide that face into two or more parts that can fit for core and cavity. The steps given next explain the use of splitting for mold.

- Perform the **Undercut Analysis** to find out the areas where you need to split the faces to accommodate the face in core/cavity.
- After finding the areas that are required for splitting, click on the **Split Line** tool from the **Ribbon**. The **Split Line PropertyManager** will be displayed as shown in Figure 17.

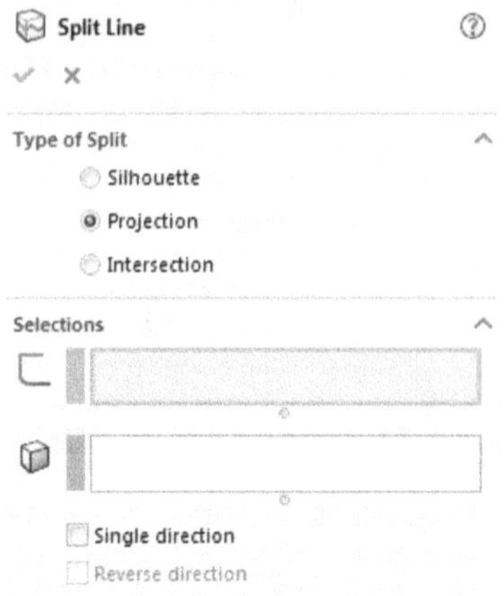

Figure 17. Split Line PropertyManager

- By default, **Projection** radio button is selected and you are supposed to select a sketch for dividing the face. Figure 18 shows a mold part that need to be divided from mid.

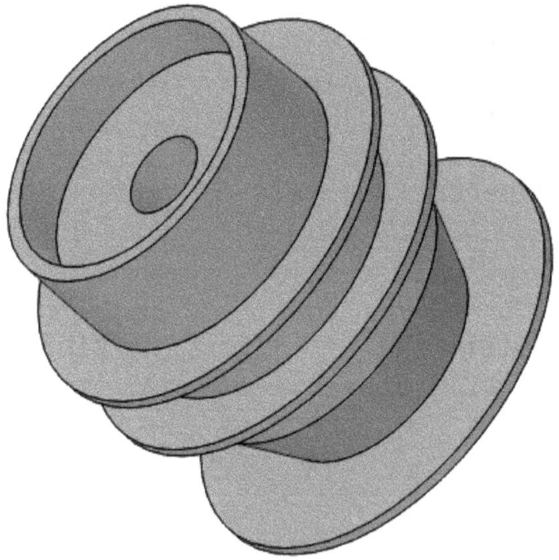

Figure 18. mold part to be splitted

- A line sketch is drawn at the center of the mold part so that it extends beyond the part at top and bottom; refer to Figure 19.

Note that to draw the sketch, you need to cancel the **Split Line** tool, select the **Sketch** tab from the **Ribbon** and after you draw the sketch, you need to select the **Split Line** tool again.

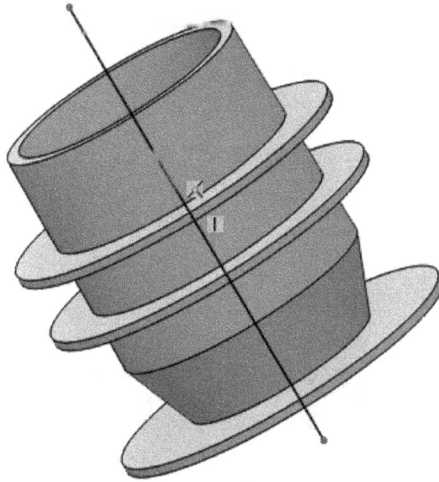

Figure 19. Sketched line drawn at the center

- Select the line sketch, it will be displayed in the first selection box in the **Split Line PropertyManager**.
- Click in the next selection box. When you hover the cursor in this selection box; the **Faces to Split** call-out is displayed.
- Select all the faces of the mold part that you want to split by using the projection of sketch; refer to Figure 20.
- Select the **OK** button from the **PropertyManager**, the mold part will be displayed as shown in Figure 21.

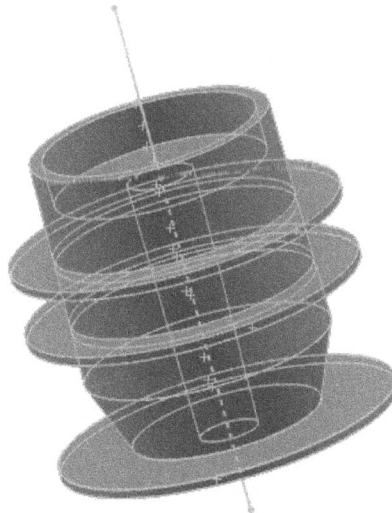

Figure 20. faces selected

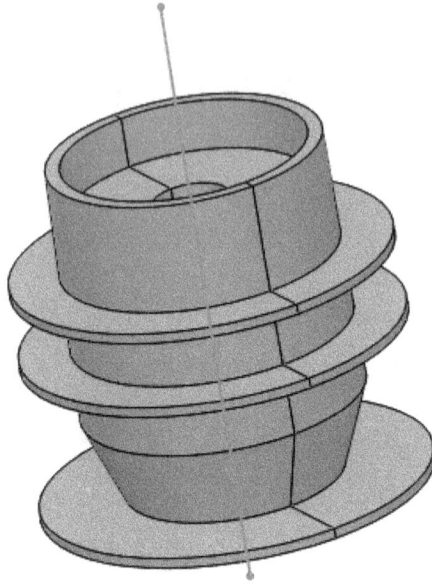

Figure 21. Mold part after splitting

- You can also split this part by using the **Silhouette** radio button. Select this radio button, the first selection box will change with **Direction of Pull** selection box.
- Select a plane or face or edge that defines the direction of pull; refer to Figure 22.
- Click in the next selection box and select all the required faces of the mold part that you want to split.

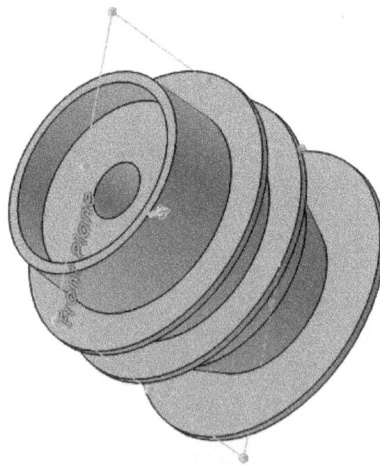

Figure 22. Direction of pull

- Click on the **OK** button, the part will split by the plane; refer to Figure 23.

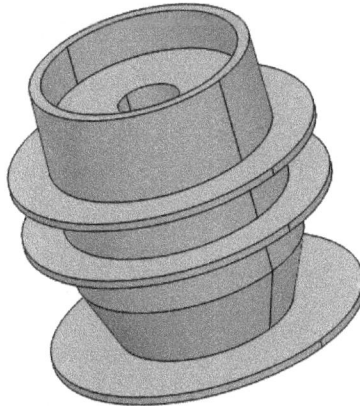

Figure 23. Silhouette split

- In the same way, you can split the mold part by using the **Intersection** radio button.

Applying draft using Draft tool

After you perform the draft analysis, some of the faces of the mold part will be displayed in yellow color which means that they require the draft. In such cases, we use the **Draft** tool to apply draft angle at those faces. The following steps explain the procedure to apply draft angle.

- Click on the **Draft** tool to apply draft. The **DraftXpert PropertyManager** will be displayed as shown in Figure 24.

Figure 24. DraftXpert PropertyManager 1

- Select the plane which will act as neutral plane for the draft angle and then select the faces on which you want to apply draft angle; refer Figure 25.

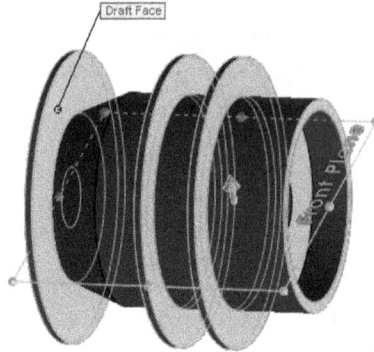

Figure 25. faces selected for applying draft

- Change the angle value by using the spinner in the **PropertyManager** and then click on the **OK** button to apply the draft.

Increasing/Decreasing thickness of walls using the Move Face tool

Since it is recommended to have a uniform thickness in the mold part, at some places you may require to change the thickness of part. The steps to change the thickness of wall are given next.

- Click on the **Move Face** tool from the **Ribbon**. The **Move Face PropertyManager** will be displayed as shown in Figure 26.

Figure 26. Move Face PropertyManager

- Select the face that you want to move and then specify the distance value in the **ΔX, ΔY,** or **ΔZ** spinners as per your requirement.
- If you want to rotate the faces then click on the **Rotate** radio button and then specify the angle values in the respective spinners at the bottom of the **PropertyManager**.

Figure 27 shows a model while increasing its wall thickness.

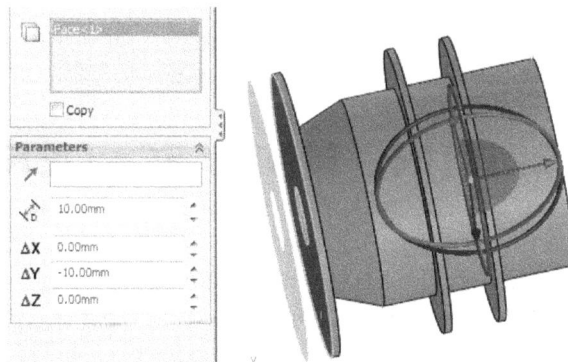

Figure 27. Model while increasing wall thickness

Scaling the model to allow shrinkage in part

While creating mold for a part, we need to increase the size of model by certain percentage so that it do not get undersized when it comes out of mold after cooling. (During the cooling of mold part, its plastic shrinks by a certain amount). The steps to scale a mold part are given next.

- Click on the **Scale** tool, the **Scale PropertyManager** will be displayed as shown in Figure 28.

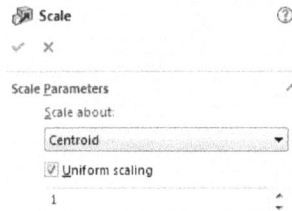

Figure 28. Scale PropertyManager

- By default, **Centroid** is selected in the **Scale about** drop-down. You can select **Origin** or a **Coordinate System** as reference for scaling by using respective option from the **PropertyManager**.
- After selecting the desired option from the drop-down, specify the value of scale in the spinner.
- Shrinkage value is given by mold designer or material supplier. You need to add the shrinkage value supplied to you in 1. For example, the shrinkage value for ABS plastic is 0.004 supplied to you. In this case, specify the scale factor as 1.004 in the spinner.

After performing the analyses and doing the required operation, we are ready to start the mold project.

INSERTING MOLD FOLDER

This is the first step when you start a mold project in SolidWorks. To start the project, click on the **Insert Mold Folders** button; the folders that are required for various components of mold will be created in the current project.

PARTING LINE

In this step, we design the parting line by using the standard identification of SolidWorks based on draft analysis. Follow the steps given below to create parting line.

- Click on the **Parting Line** tool from the **Ribbon**. The **Parting Line PropertyManager** will display; refer to Figure 29.

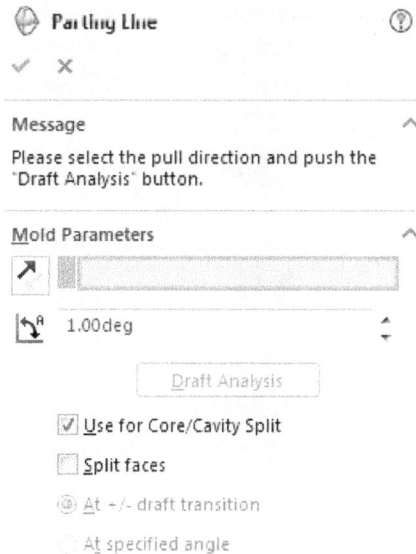

Figure 29. Parting Line PropertyManager

- Click on the neutral plane i.e. a flat face with respect to which the draft angles will be measured.
- Set the draft value in the spinner and click on the **Draft Analysis** button. The model will be colored for core and cavity. Also, the parting line will display in violet color; refer to Figure 30.

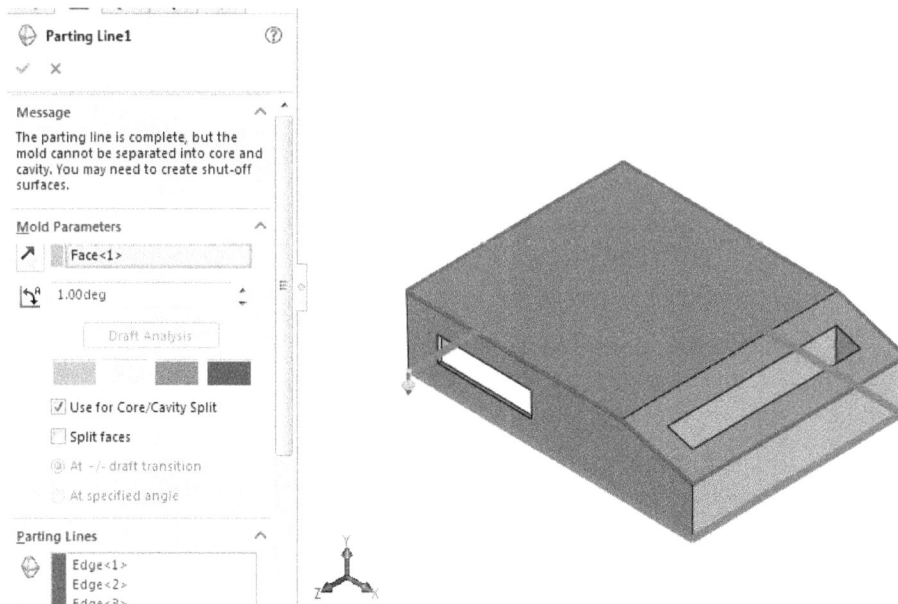

Figure 30. Parting line preview

- Click on the **OK** button from the **Parting Line PropertyManager** to create the parting line.

SHUT-OFF SURFACES

Any opening in the model must be closed by a surface so that the core steel and cavity steel meet at defined surface. The **Shut-off Surfaces** tool performs this job for us. The steps to use this tool are given next.

- Click on the **Shut-off Surfaces** tool from the **Ribbon**. The **Shut-off Surface PropertyManager** will display and the open loop will be selected automatically; refer to Figure 31.

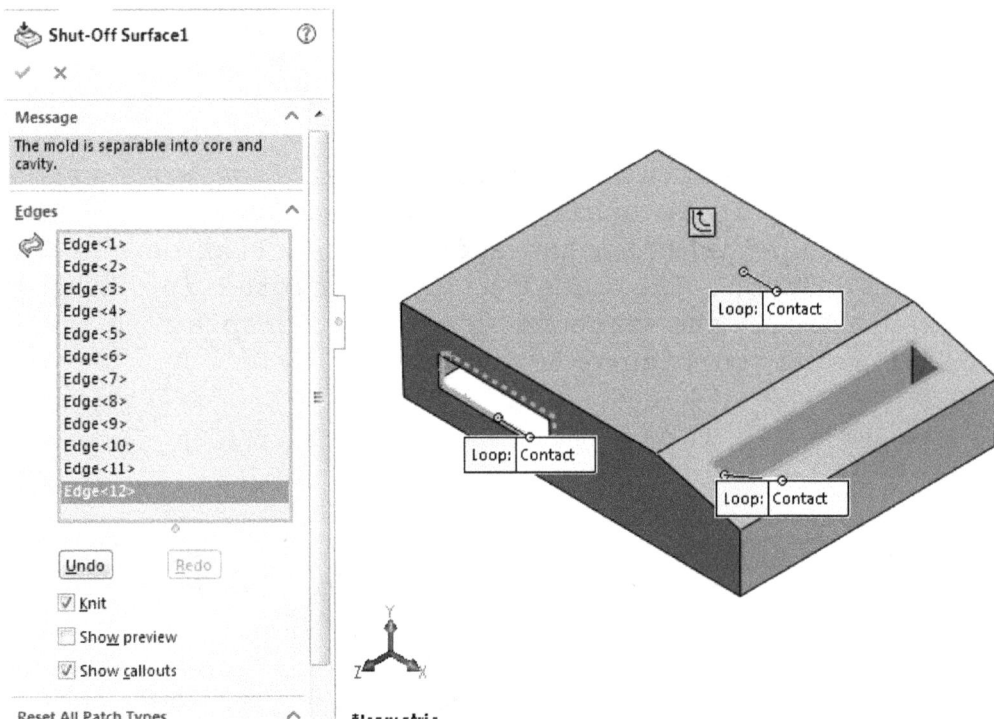

Figure 31. Shut-Off Surface PropertyManager

- Click on the **All Tangent** button from the **Reset All Patch Types** rollout, to create straight patches.
- Click on the **OK** button to create the shut-off surfaces.
- Note that in the above figure, you need to deselect all the edges that are vertical. One the flat face patch should be created; refer to Figure 32.

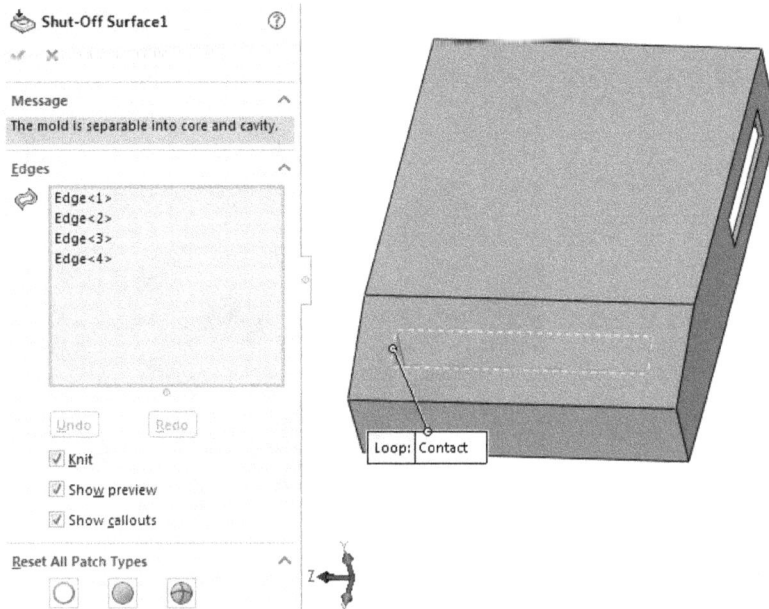

Figure 32. Shut-off surface to be created for this case

PARTING SURFACES

Parting surface is the surface by which the core and cavity is separated. Parting surfaces are created by the use of parting lines. Note that parting surface is always a continuous surface. The steps to create parting surface are given next.

• Click on the **Parting Surfaces** tool from the **Ribbon**. The **Parting Surface PropertyManager** will display. Also, the preview of parting surface will display; refer to Figure 33.

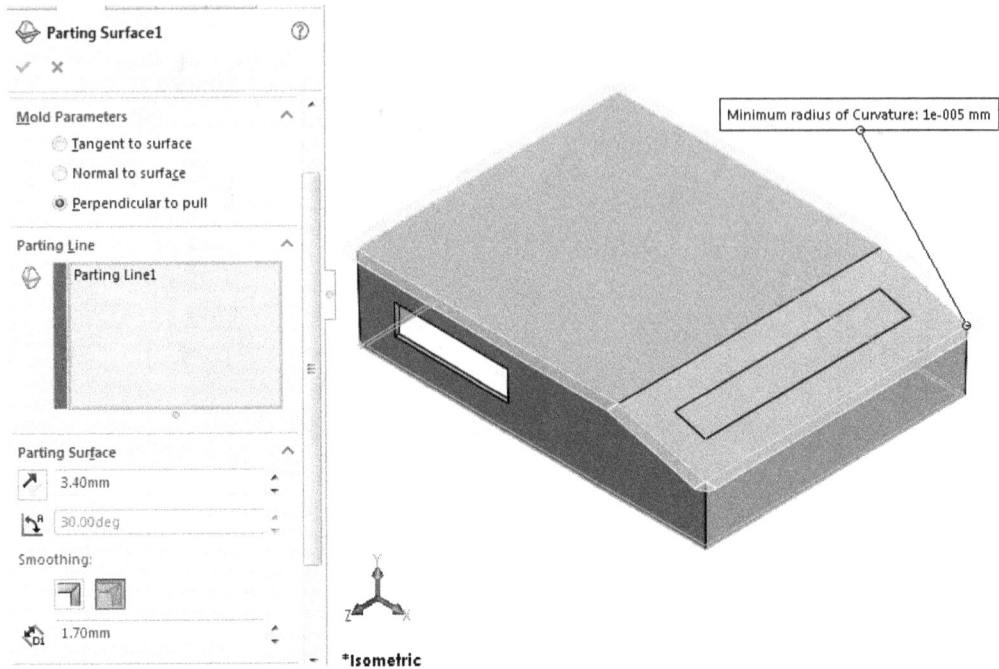

Figure 33. Parting surface

- Increase the length of parting surface by using the first spinner in the **Parting Surface** rollout in the **PropertyManager.**
- You can change the direction of parting surface by using the radio buttons in the **Mold Parameters** rollout.
- If the **Perpendicular to pull** radio button is selected then click on the **Manual Mode** check box to manually change the direction of surface by dragging the key points; refer to Figure 34.
- Click on the **OK** button from the **PropertyManager.**

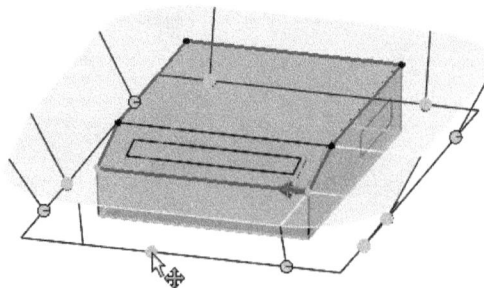

Figure 34. Manual editing of parting surface

TOOLING SPLIT

Now, we want to extract core and cavity from the model by using the parting surface. To split the tooling to generate core and cavity, follow the steps given below.

- Click on the **Tooling Split** tool from the **Ribbon**. You are asked to create sketch for the tooling from which core and cavity will be extracted.
- Select a plane parallel to the parting surface to create tooling; refer to Figure 35.

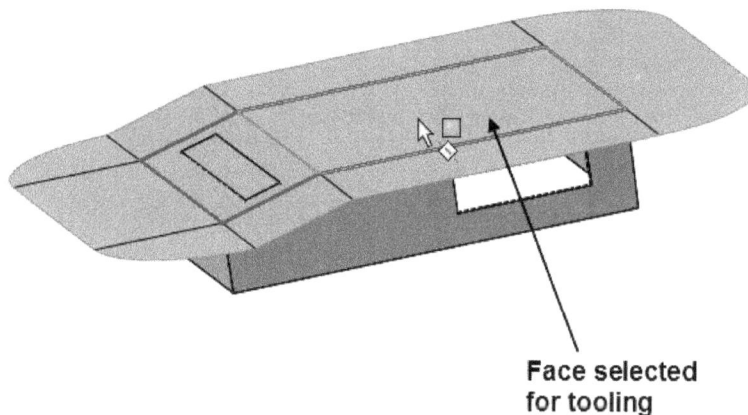

Face selected
for tooling

Figure 35. Face selected for tooling

- On selecting the face, the sketching environment will display.
- Create the sketch for the tooling. You can change the display style to **Hidden Lines Visible** for clear view; refer to Figure 36.

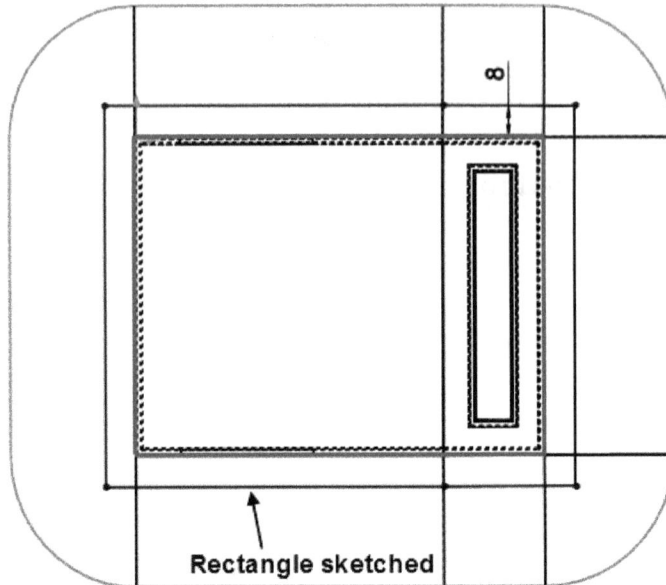

Figure 36. Sketch for tooling

- Exit the sketch environment and change the display style to shaded and orientation to isometric; refer to Figure 37.

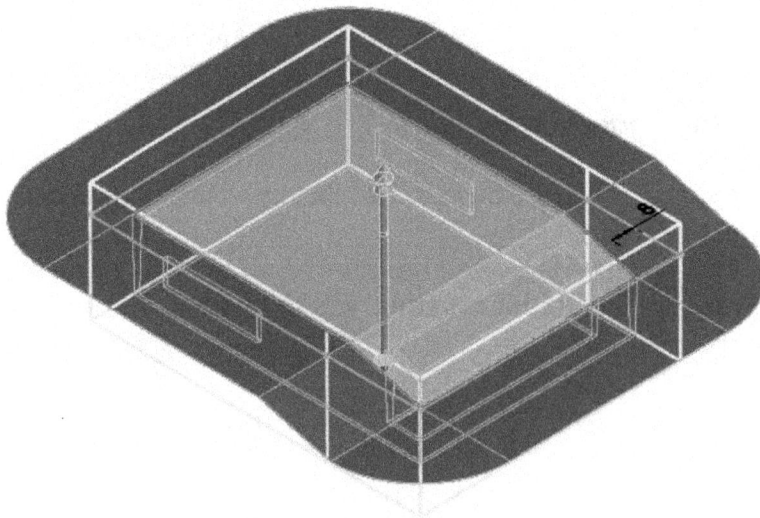

Figure 37. Extrusion tooling split

- Specify the desired extrusion height in the spinners.
- Click **OK** from the **PropertyManager** to creating the tooling split.

- The core and cavity are added in the **Solid Bodies** folder of **FeatureManager Design Tree**; refer to Figure 38.

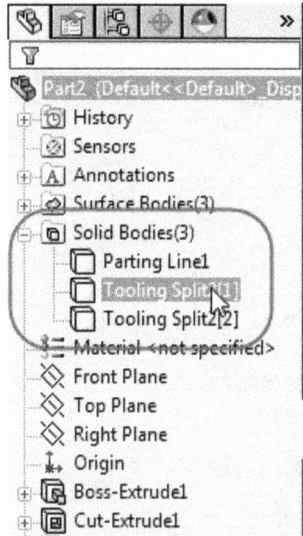

Figure 38. Core and cavity in Solid Bodies folder

Right-click on any of the tooling split and select the **Isolate** option from the menu to check the part separately.

CORE

There are a few portions in the core or cavity that are to be inserted from sides. Such parts are called sliders in engineering. To create sliders, follow the steps given below.

- Isolate the tooling from which you want to extract sliders.
- Click on the **Core** tool from the **Ribbon.**
- Select the side face of the tooling from which you want to create slider. Change the display style to **Hidden Lines Visible** for clear view; refer to Figure 39.

**Face selected for
creating slider**

Figure 39. Face selection for slider

- Create sketch for the slider; refer to Figure 40. You can use offset tool for easy creation of sketch.

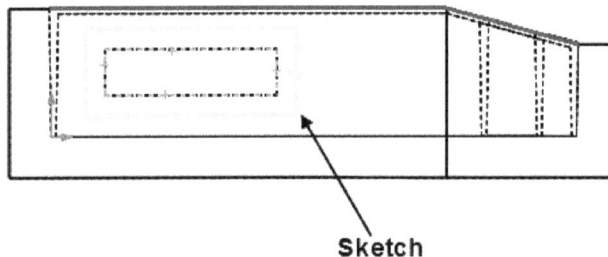

Sketch

Figure 40. Sketch for slider

- Exit the sketch environment and extrude the sketch tool the face you want to include in slider.
- Click on the **OK** button from the **PropertyManager** to create the slider; refer to Figure 41.

To move the individual parts of mold, click on the **Move/ Copy** option from the **Features** cascading menu of **Insert** menu; refer to Figure 42. The **Move/Copy Body PropertyManager** will display. Select the part that you want to move. Triad will be displayed providing you the options to move the part. Move the selected part to desired location and click **OK** from the **PropertyManager.** Similarly, you can move other parts of the tool split. Refer to Figure 43.

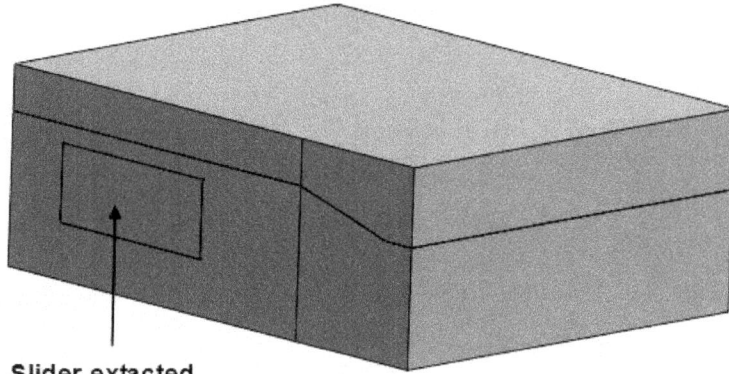

Slider extacted

Figure 41. Slider created

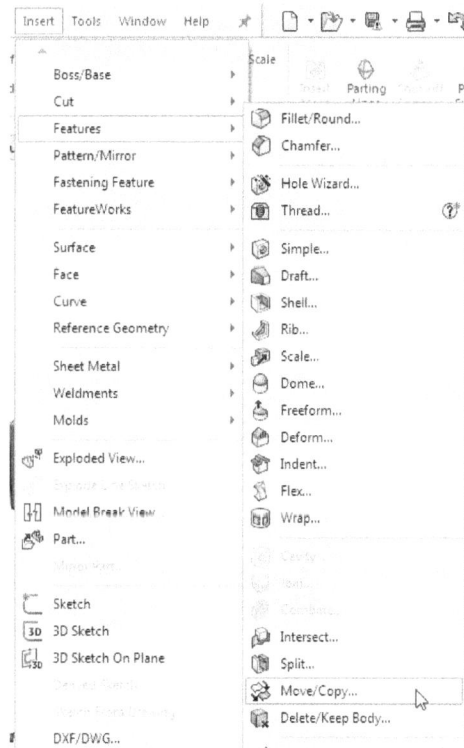

Figure 42. Move or Copy tool

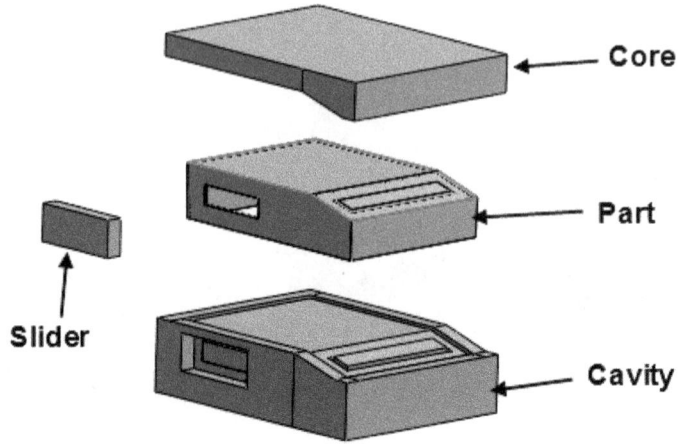

Figure 43. Tool split parts after moving

PRACTICE 1

Make a mold core and cavity of the model shown in Figure 44 out of a 380x440x100 mm^3 block. The part file is available in resources of the book.

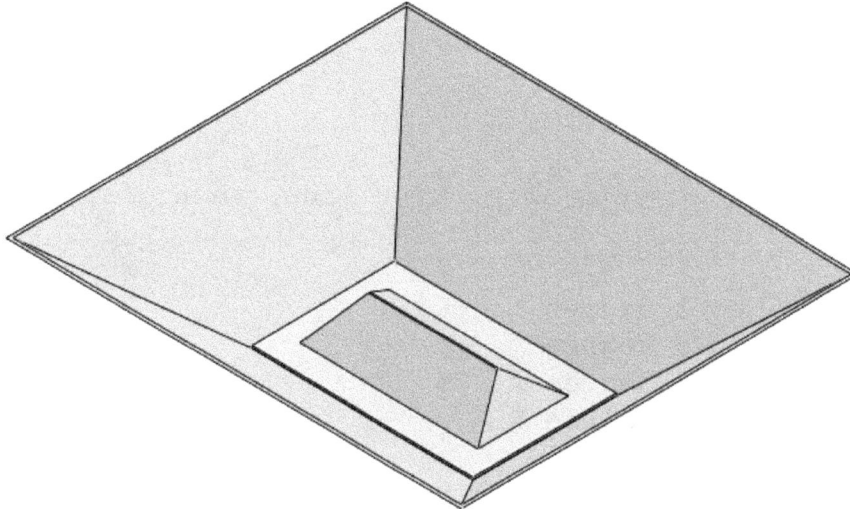

Figure 44. Practice 1 model

SELF ASSESSMENT

Q1. Which of the following analysis is not performed on the part for checking possibility of mold design?

a. Draft Analysis
b. Stress Analysis
c. Undercut Analysis
d. Parting Line Analysis

Q2. If there are steps at parting line, then draft required for shutoff is

a. 1 degree to 3 degree
b. 3 degree to 5 degree
c. 5 degree to 7 degree
d. 7 degree to 9 degree

Q3. To compensate for shrinkage of molded part, we need to use _____ tool.

a. **Draft**
b. **Scale**
c. **Move Face**
d. **Split Line**

Q4. Any opening in the model can be closed by _____ tool so that the core steel and cavity steel meet at defined surface.

a. **Shut-off Surfaces**
b. **Parting Surfaces**
c. **Move Face**
d. **Draft**

Q5. _____ is the surface by which the core and cavity is separated.

a. Shut-off surface
b. Parting surface
c. Ruled surface
d. Offset surface

FOR STUDENT NOTES

FOR STUDENT NOTES

Sheet metal and Practice

Chapter 12

Topics Covered

The major topics covered in this chapter are:

- *Sheet metal Introduction.*
- *Sheet metal Creation Tools.*
- *Tools for applying cut.*
- *Corners Modification*
- *Flat Pattern.*
- *Practice.*

SHEET METAL INTRODUCTION

Sheet metal is used when you need a component of thickness in the range of 0.16 mm to 12.70 mm and we do not require conventional cutting machines. The components that can be created by Punch-press and bending machines are designed in Sheet Metal environment. In SolidWorks, there is a separate tab to design sheet metal components named **Sheet Metal**; refer to Figure-1.

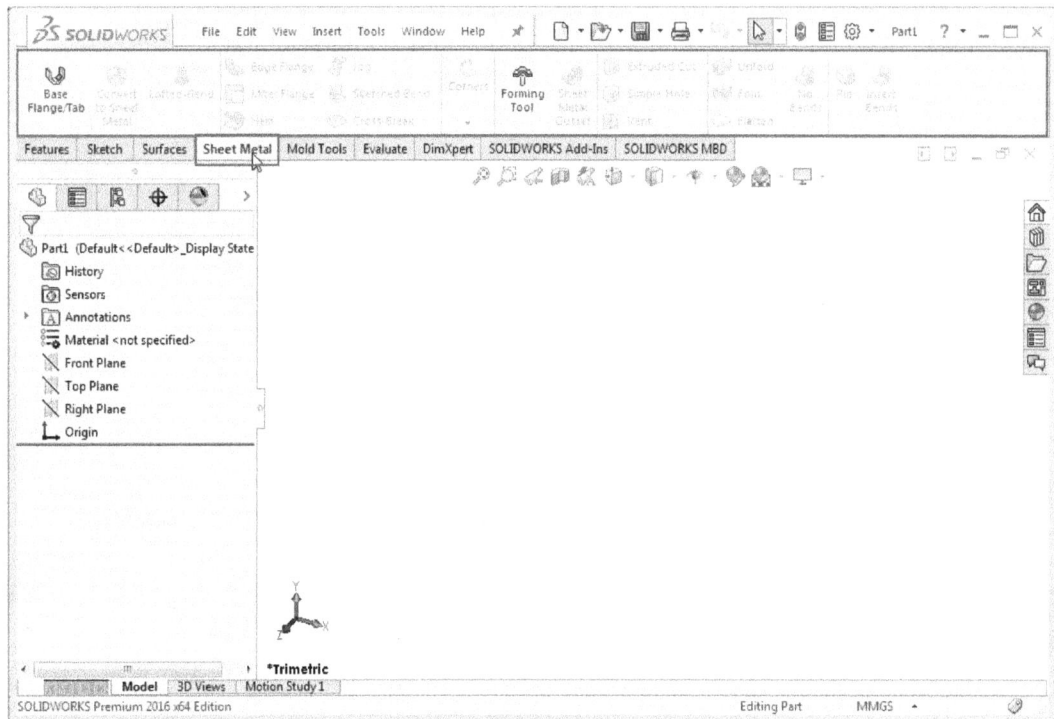

Figure-1. Sheet metal environment

The tools to create sheet metal designs are explained next.

BASE FLANGE/TAB

Base Flange/Tab tool is used to create base feature of the sheet metal component. All the other features will be created on this base flange. The steps to create base flange/tab are given next.

- Click on the **Base Flange/Tab** tool. You are asked to select the sketching plane to draw sketch of the base flange.

- Select a plane and draw the sketch.
- Exit the sketch environment. The **Base Flange PropertyManager** will display; refer to Figure-2 and Figure-3.

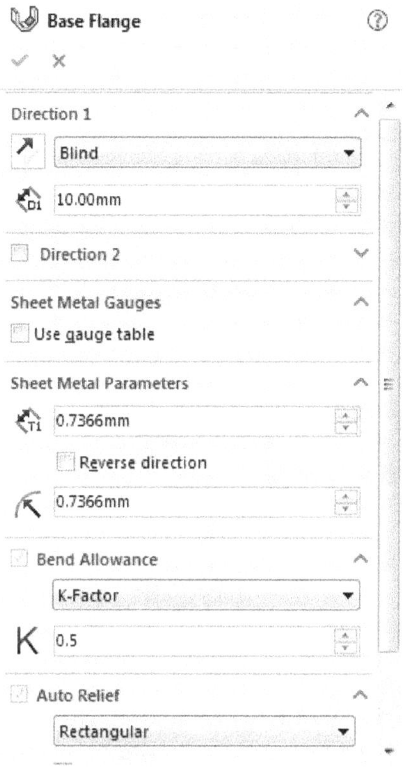

Figure-2. Base Flange PropertyManager
with open sketch selected

Figure-3. Base Flange Property-
Manager with closed sketch

Setting Parameters for Base Flange/Tab
with Open sketch

- Specify the desired length of the flange/tab in direction 1 and direction 2 by using the options in the **Direction 1** and **Direction 2** rollouts in the **PropertyManager**.
- Set the desired thickness and bend radius in the edit boxes available in Sheet Metal Parameters or you can use the gauge table for defining parameters.
- To use the gauge table, select the **Use gauge table** check box from the **Sheet Metal Gauges** rollout. You are asked to select a sample table.
- Select the desired sample table from the drop-down; refer to Figure-4.

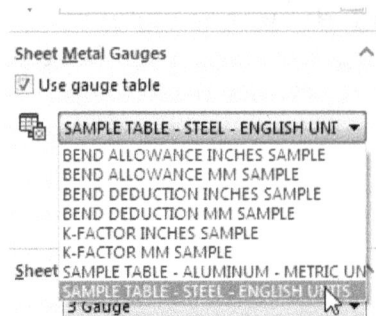

Figure-4. Sample table selected

- Select the desired gauge of sheet from the **Gauge** drop-down in the **Sheet Metal Parameters** rollout.
- Click on the **OK** button from the **PropertyManager** to create the base flange/tab.

Setting Parameters for Base Flange/Tab with Close sketch

- Set the desired thickness for the flange/tab in the edit box in the **Sheet Metal Parameters** rollout.
- You can also set the parameters by using the gauge table as discussed earlier.

LOFTED-BEND

Lofted-Bend tool is used to create sheet metal component by joining two or more sketch sections. This tool works in the same way as the **Lofted Surfaces** or **Lofted Boss/Base** tool work. Figure-5 shows a lofted bend. Procedure to use this tool is given next.

- Make sure you have two or more open sketches in the graphics area and then click on the **Lofted-Bend** tool from the **Ribbon**. The **Lofted-Bends PropertyManager** will be displayed; refer to Figure-6.

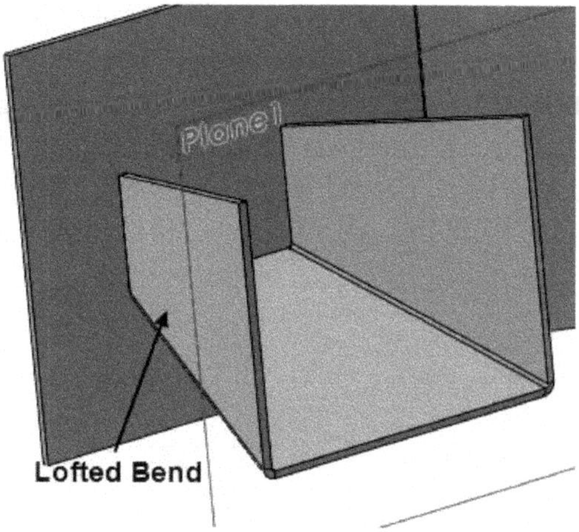

Figure-5. Lofted bend

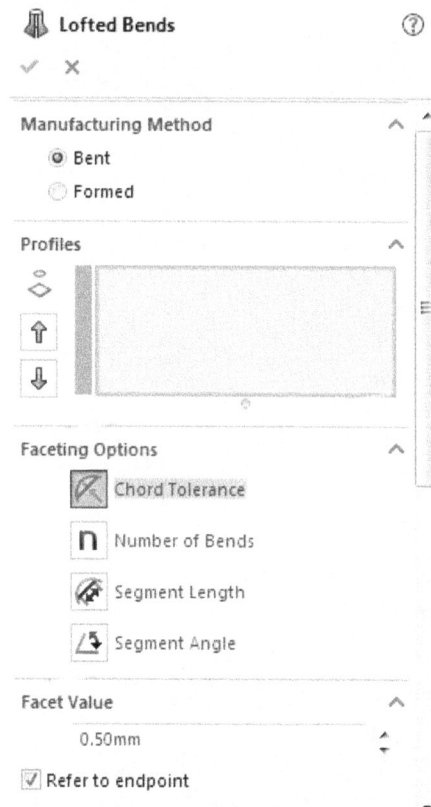

Figure-6. Lofted-Bends PropertyManager

- Select the **Bent** radio button if you want to create real physical bends, rather than formed geometry and approximated bend lines in a flat pattern. Bent lofted bends form a realistic transition between two profiles to facilitate instructions for press brake manufacturing.

- Select the **Formed** radio button if you want to create a formed transition between two profiles assuming that a forming tool is used to create lofted bend. In this case, you cannot have sharp corners in the profiles; refer to Figure-7.

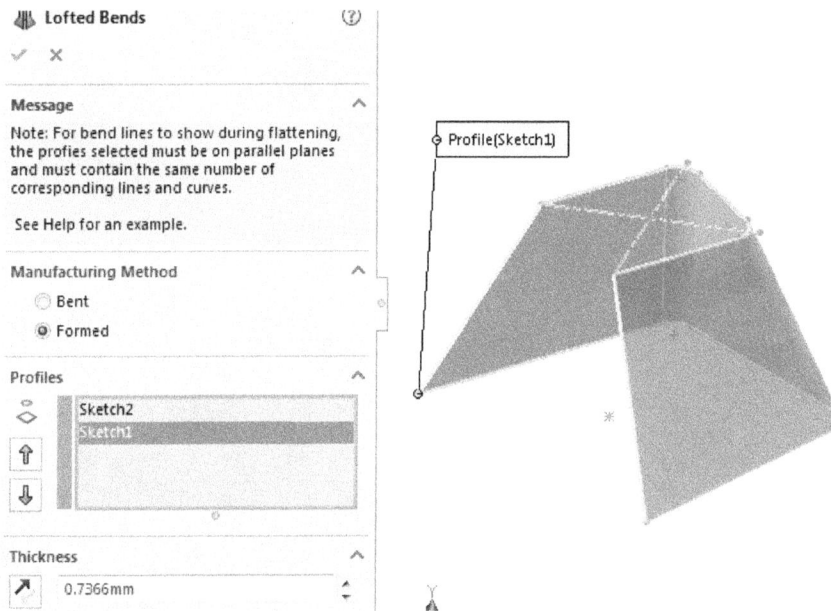

Figure-7. Formed lofted bend

- In most of the cases, you will be using the **Bent** radio button. Select the **Bent** radio button and then select the profiles.
- Select the **Chord Tolerance** button from the **Faceting Options** rollout and specify the desired tolerance for cord created during bend transition from point to arc of sections; refer to Figure-8. Similarly, you can use the **Number of Bends**, **Segment Length**, and **Segment Angle** buttons in the same way.
- Specify the sheetmetal parameter in the **Sheet Metal Parameters** rollout as discussed earlier.

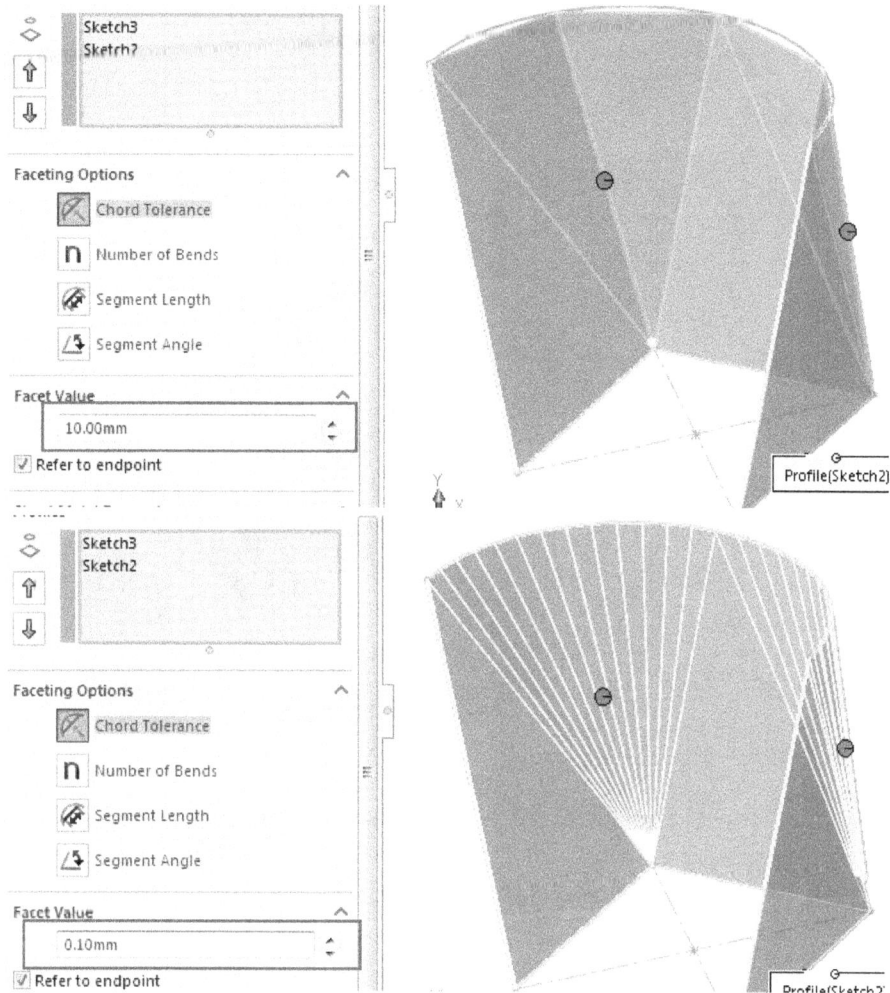

Figure-8. Variation due to chord tolerance

EDGE FLANGE

The **Edge Flange** tool is used to create a flange by using selected edge. The steps to do so are given next.

- Click on the **Edge Flange** tool. The **Edge Flange PropertyManager** will display.
- Select an edge using which you want to create the flange wall. The flange end will get attached to cursor; refer to Figure-9.

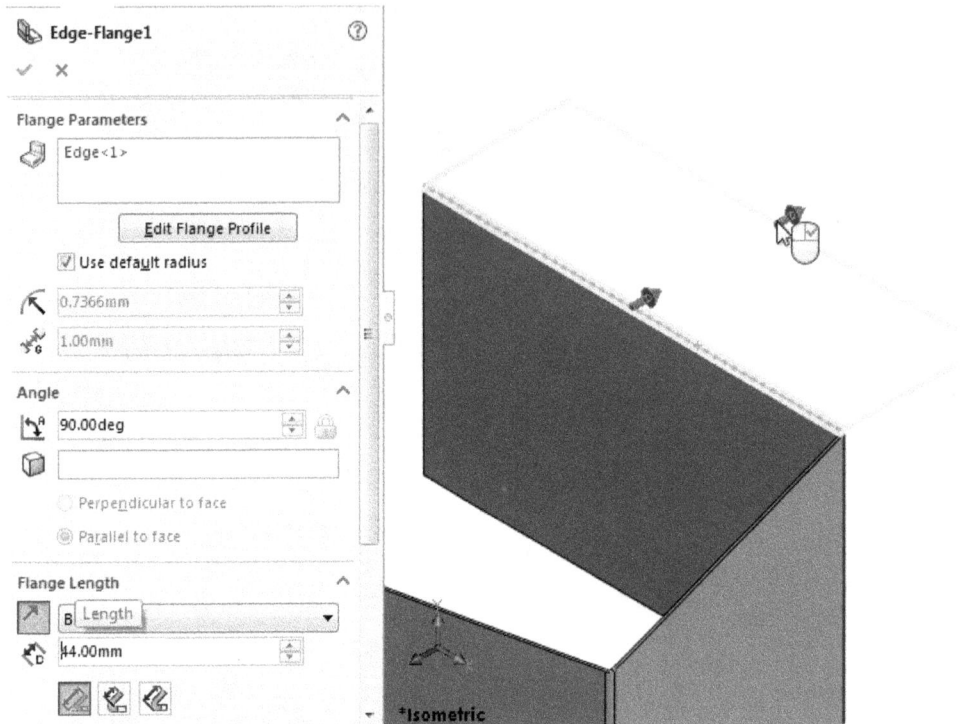

Figure-9. Edge Flange PropertyManager and preview of flange

- Click to specify the end point of the flange.
- You can change the length of the flange by using the spinner in the **Flange Length** rollout.

Some of the important options in the **PropertyManager** are explained next.

Flange Parameters Rollout

The options in the **Flange Parameters** rollout are used to define the edge reference to be used for creating the edge flange, the bending radius, and the profile of the edge flange. These options are discussed next.

Edge

The **Edge selection** box is used to select the edges to create the edge flange.

Edit Flange Profile

The **Edit Flange Profile** button is chosen to edit the profile of the edge flange. By default, the edge flange is created along the entire length of the selected edge. To edit the profile of the edge flange, choose the **Edit Flange Profile** button; the **Profile Sketch** dialog box will be displayed informing you that the sketch is valid. Also, the sketching environment will be invoked in the background. Edit the sketch of the profile of the edge flange using the sketching tools.

You will also notice that while editing the sketch of the edge flange, the **Profile Sketch** dialog box informs you whether the sketch is valid for creating the edge flange or not. If the status of the sketch is shown valid in the **Profile Sketch** dialog box, the preview of the flange will be displayed in the drawing area. After editing the profile, choose the **Finish** button from the **Profile Sketch** dialog box; the flange will be created and the **Edge-Flange PropertyManager** will be automatically closed. Note that if you want to modify the other parameters of the flange, choose the **Back** button from the **Profile Sketch** dialog box. Figure 5 shows the edge flange created along the entire length of the selected edge. Figure 6 shows the edited sketch of the edge flange and Figure 7 shows the resulting edge flange.

Angle Rollout

The **Angle** rollout is used to define the angle of the flange. The default angle of the flange is 90 degrees. You can define any other angle of the flange by using the **Flange Angle** spinner. The angle of the edge flange can be greater than 0-degree and less than 180 degrees. You can also select a face and specify whether the resulting flange will be parallel or normal to it. Figure 8 shows an edge flange created at an angle of 45 degrees. Figure 9 shows an edge flange created at an angle of 135 degrees.

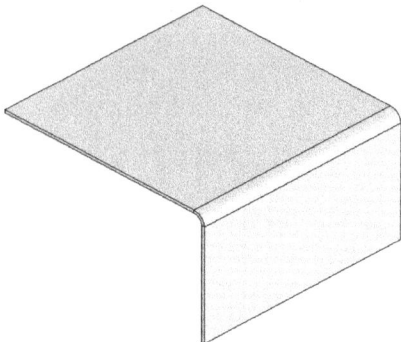

Figure-10. Edge flange created along the entire length of the edge

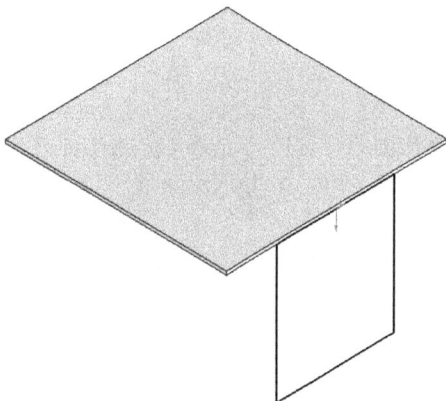

Figure-11. Edited sketch of the edge flange

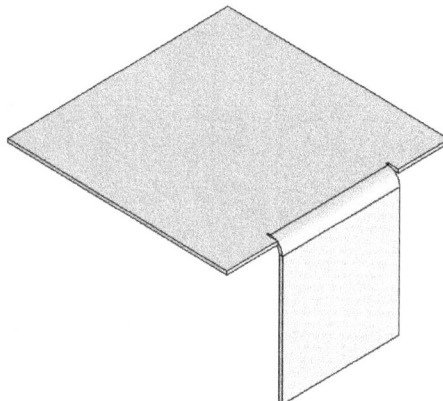

Figure-12. Resulting edge flange

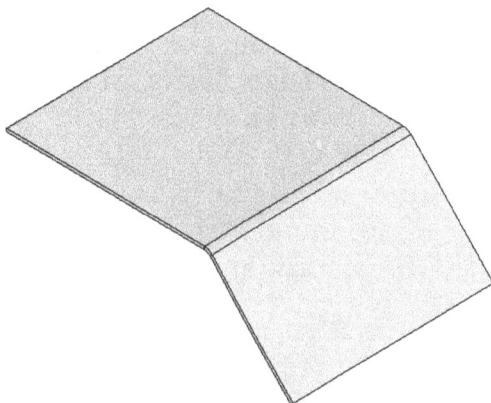

Figure-13. Edge flange created at an angle of 45 degrees

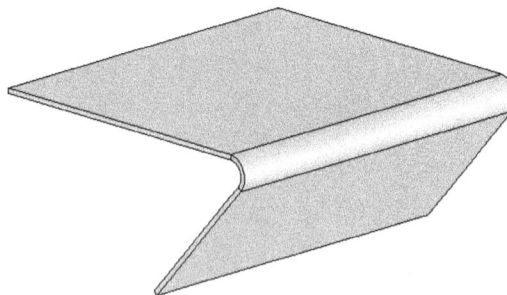

Figure-14. Edge flange created at an angle of 135 degrees

Flange Length Rollout

The **Flange Length** rollout is used to define the length of the flange. In other words, the options for feature termination are available in this rollout. These options are the same as discussed earlier. The other two options provided in this rollout are discussed next.

Outer Virtual Sharp

The **Outer Virtual Sharp** button is used to define the length of the flange from the outer virtual sharp. The outer virtual sharp is an imaginary vertex created by extending the tangent lines virtually from the outer radius of the bend, as shown in Figure 10.

Inner Virtual Sharp

The **Inner Virtual Sharp** button is chosen by default and is used to define the length of the flange from the inner virtual sharp. The inner virtual sharp is an imaginary vertex created by extending the tangent lines virtually from the inner radius of the bend, as shown in Figure 10.

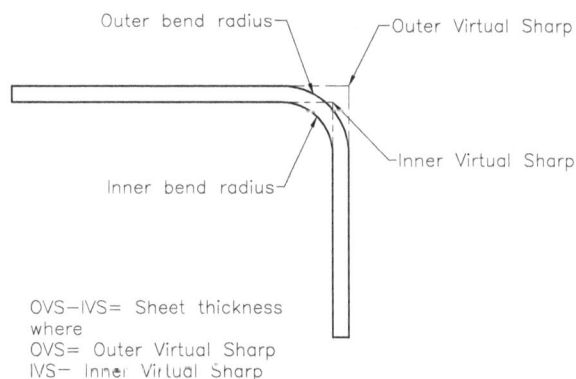

Figure-15. Outer Virtual Sharp and Inner Virtual Sharp

Tangent Bend

The **Tangent Bend** button is used to define the length of the flange from the imaginary line that is created by extending the tangent line from the outer radius of the bend and parallel to the end edge of the flange to be created, refer to Figure 11. This button will be available only for the flange to be created whose bend radius is greater than 90-degree.

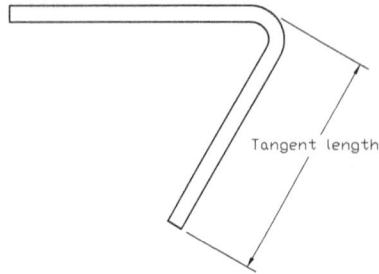

Figure-16. Tangent length of
***Tangent Bend** in base flange*

Flange Position Rollout

The **Flange Position** rollout is used to define the position of the flange on an edge. The options in this rollout are discussed next.

Material Inside

The **Material Inside** button is used to create the edge flange in such a way that the material of the flange after the bend lies inside the maximum limit of sheet. Figure 12 shows the edge flange created with the Material Inside button chosen.

Material Outside

The **Material Outside** button is chosen by default and the edge flange is created such that the material of the flange after the bend lies outside the maximum limit of the sheet. Figure 13 shows the edge flange created with the Material Outside button chosen.

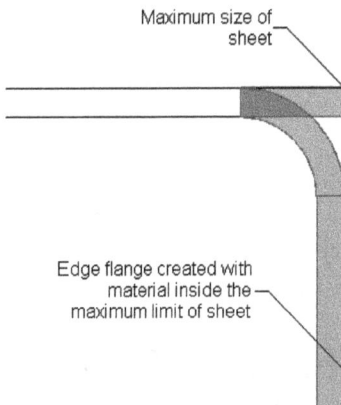

Figure-17. Edge flange created with the
***Material Inside** button chosen*

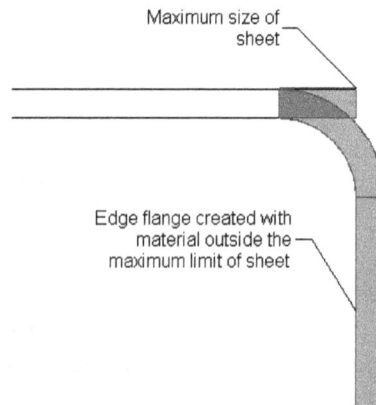

Figure-18. Edge flange created with the
***Material Outside** button chosen*

Bend Outside

The **Bend Outside** button is used to create an edge flange such that the bending of the sheet starts from the point that is beyond the maximum limit of the sheet, as shown in Figure 14.

Bend from Virtual Sharp

The **Bend from Virtual Sharp** button is used to create an edge flange with the bending of the sheet starting from the virtual sharp. The position of the flange depends on whether you choose the **Outer Virtual Sharp** button, **Inner Virtual Sharp** button, or the **Tangent Bend** button from the **Flange Length** rollout. Figure 15 shows the edge flange created with the **Inner Virtual Sharp** and **Bend from Virtual Sharp** buttons chosen.

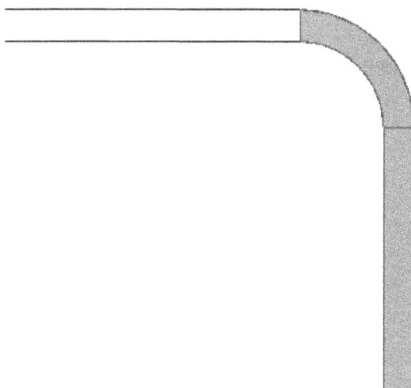

Maximum size of
sheet

Virtual
sharp

Bending of sheet started from
the virtual sharp

*Figure-19. Edge flange created with the **Bend Outside** button chosen*

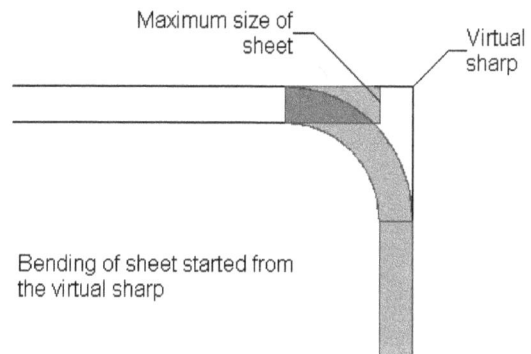

*Figure-20. Edge flange created with the **Bend from Virtual Sharp** button chosen*

Tangent to Bend

The **Tangent to Bend** button is used to create the edge flange in such a way that the material of the flange after bending lies tangent to the maximum limit of the sheet, refer to Figure 16. Note that this option is not valid for the bend angle less than 90 degree.

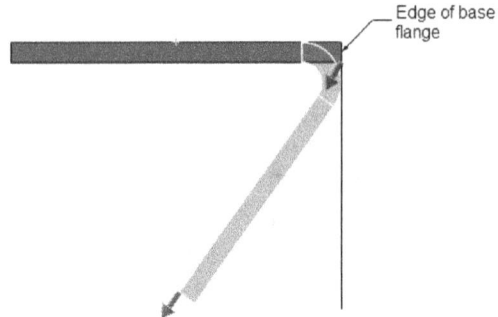

Edge of base
flange

Figure-21. Edge flange created with the
Tangent to Bend *button chosen*

Trim side bends

Select the **Trim side bends** check box to trim extra materials
in the bends surrounding the current edge flange. By default,
this check box is not selected. Figure 17 shows the edge
flange created with the **Trim side bends** check box cleared.
Figure 18 shows the edge flange created with the **Trim side
bends** check box selected.

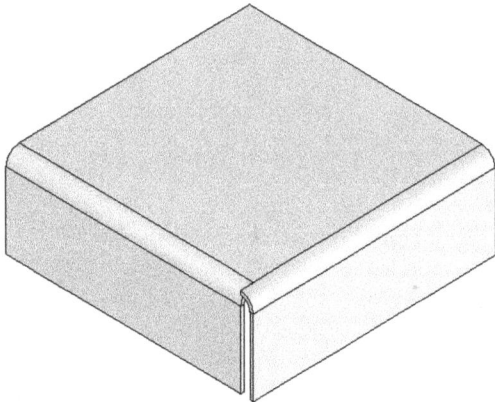

Figure-22. Edge flange created with the
Trim side bends *check box cleared*

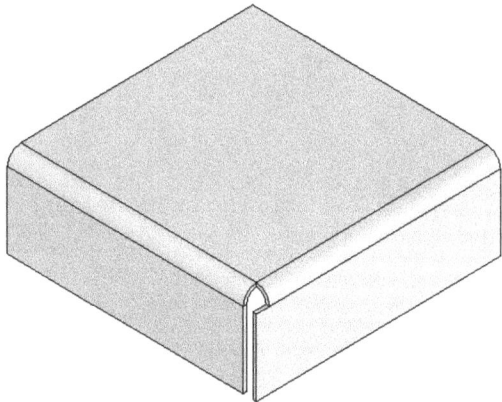

Figure-23. Edge flange created with the
Trim side bends *check box selected*

Offset

The **Offset** check box is available only when you create an
edge flange using the **Material Inside, Material Outside, Bend
Outside,** or **Tangent to Bend** options. This check box is used
to create an edge flange at an offset distance from the
selected edge reference. On selecting the **Offset** check box,

the **Offset End Condition** drop-down and the **Offset Distance** spinner will be displayed. Specify the offset distance using the spinner. Figure 19 shows the edge flange created with the **Offset** check box cleared. Figure 20 shows the edge flange created with the **Offset** check box selected and the offset distance specified in the **Offset Distance** spinner.

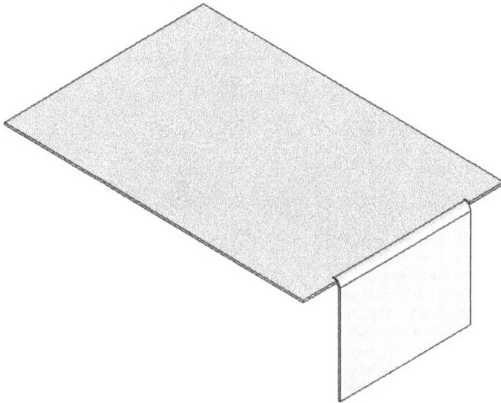

Figure-24. Edge flange created with the ***Offset*** *check box cleared*

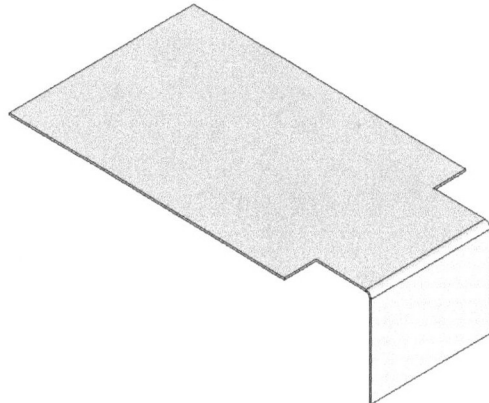

Figure-25. Edge flange created with the ***Offset*** *check box selected*

Custom Bend Allowance Rollout

The **Custom Bend Allowance** rollout is used to define the bend allowance other than the default bend allowance that you defined while creating the base flange. To apply the custom bend allowance, expand this rollout by selecting the **Custom Bend Allowance** check box. Then use the options in this rollout to define the bend allowance for the current bend as discussed earlier.

Custom Relief Type Rollout

The **Custom Relief Type** rollout is used to define the type of relief other than the default that was defined while creating the base flange. To apply the custom relief, expand this rollout by selecting the check box in the title bar of the **Custom Relief Type** rollout, as shown in Figure 21.

The types of reliefs that can be defined for a sheet metal component are discussed next.

Obround Relief

The **Obround** option is used to provide the obround relief such that the edges of the relief merging with the sheet are rounded. The **Use relief ratio** check box is selected by default. Therefore, you can modify the value of the relief ratio by setting the value in the **Relief Ratio** spinner. If you clear the **Use relief ratio** check box, the **Relief Width** and **Relief Depth** spinners will be displayed, as shown in Figure 22. You can modify the relief width and relief depth individually by using these two spinners.

Figure 23 shows the edge flange created by providing the obround relief with the default relief ratio. Figure 24 shows the edge flange created by providing obround relief after modifying the relief ratio.

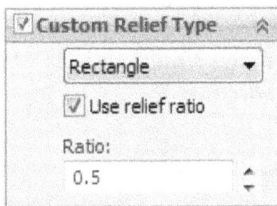

*Figure-26. The **Custom Relief Type** rollout*

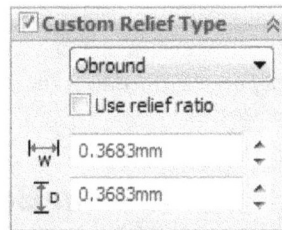

*Figure-27. The **Relief Width** and **Relief Depth** spinners displayed in the **Custom Relief Type** rollout*

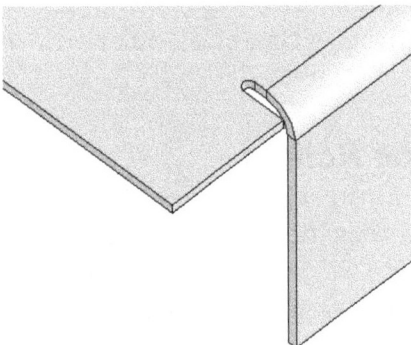

Figure-28. Edge flange created with the default relief ratio

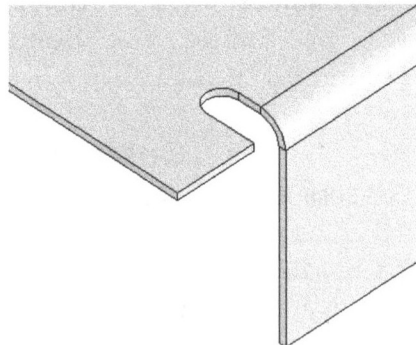

Figure-29. Edge flange created after modifying the relief ratio

Rectangle Relief

The **Rectangle** option is selected by default in this rollout. This option is used to provide the rectangular relief to the sheet metal components. The options for defining the rectangular relief are the same as discussed in the previous paragraph. Figure 25 shows an edge flange created by providing the rectangular relief with the default relief ratio. Figure 26 shows an edge flange created by providing rectangular relief after modifying the relief ratio.

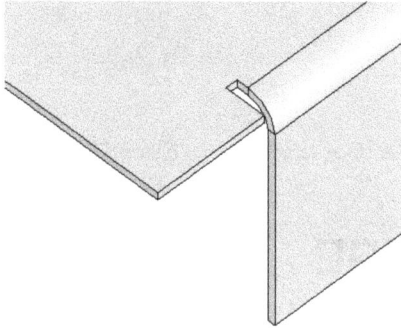

Figure-30. Edge flange created by providing the rectangular relief with default relief ratio

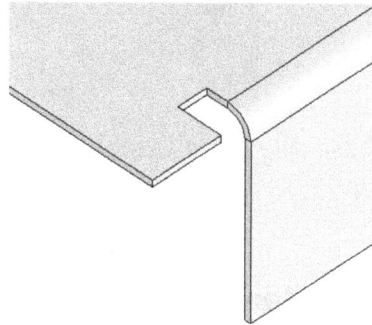

Figure-31. Edge flange created by providing the rectangular relief after modifying the relief ratio

Tear Relief

You can provide the tear relief to an edge flange by using the **Tear** option. The tear relief will tear the sheet in order to accommodate the bending of the sheet. When you select the **Tear** option from the **Relief Type** drop-down list, all the other options are replaced by the **Rip** and **Extend** buttons, as shown in Figure 27.

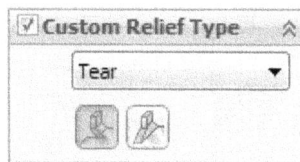

*Figure-32. The **Custom Relief Type** rollout with the **Tear** option selected from the **Relief Type** drop-down list*

The **Rip** button is chosen by default. This option rips or tears the sheet to accommodate the bending of the sheet, as shown in Figure 28. When the **Extend** button is chosen, the outer faces of the bend will be extended to the outer faces of the sheet on which you create the edge flange, as shown in Figure 29.

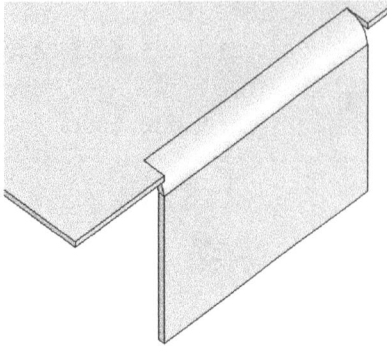

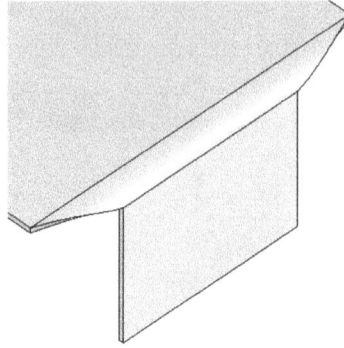

*Figure-33. Tear relief with the **Rip** button chosen* *Figure-34. Tear relief with the **Extend** button chosen*

MITER FLANGE

The **Miter Flange** tool is used to create a flange of specified shape. This type of flange is best used in creating tray type shapes. The steps to create miter flange are given next.

* Click on the **Miter Flange** tool from the **Ribbon.** You are asked to select a plane/edge to define sketching plane.
* Select the edge on which you want to create flange. The sketching environment will display.
* Create an open sketch to define shape of flange; refer to Figure-35 and exit the sketching environment. Preview of flange will display; refer to Figure-36.

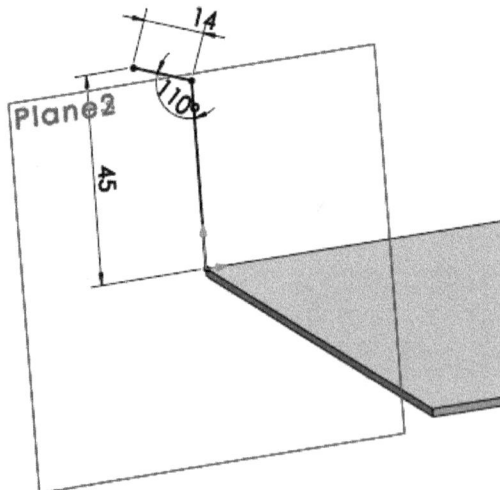

Figure-35. Sketch for miter flange

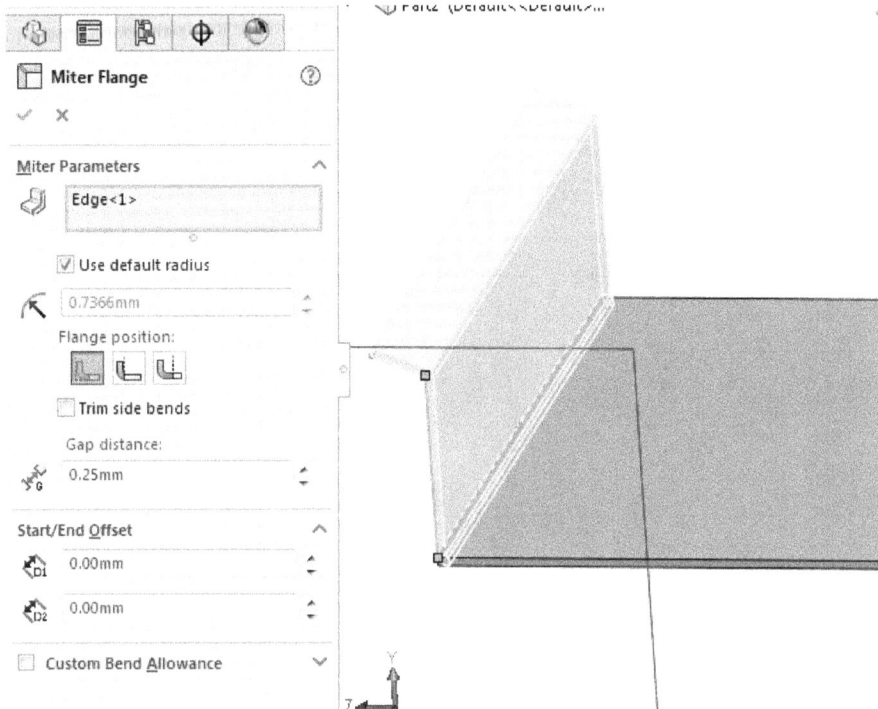

Figure-36. Preview of miter flange

- If you want to create tray then select all the edges of the current loop; refer to Figure-37.

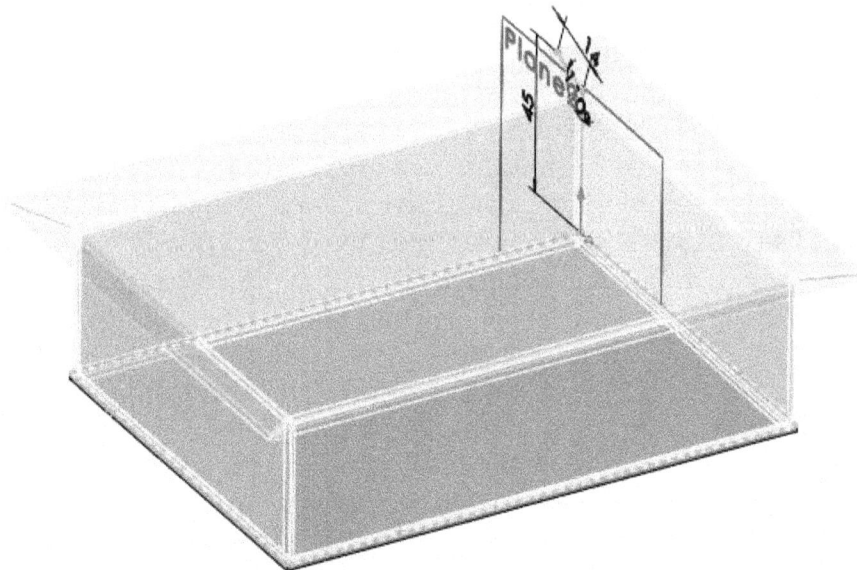

Figure-37. Closed loop of miter flange

• Click on the **OK** button to create the flange.

Some of the important options in the **PropertyManager** are explained next.

Gap distance Area

The spinner in this area is used to define the rip distance between two consecutive flanges. Set the value in the **Rip Gap** spinner to modify the distance value of the rip. Figure 33 shows the miter flange created using the default distance value and Figure 34 shows the miter flange created using the modified rip distance. In case of error enter the higher gap value.

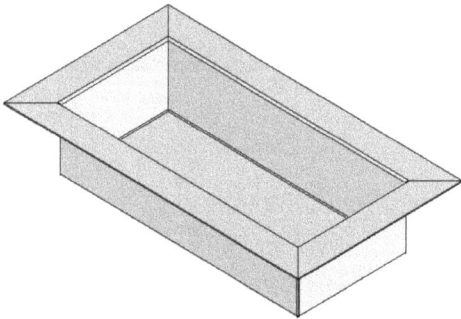

Figure-38. Miter flange with the default rip distance

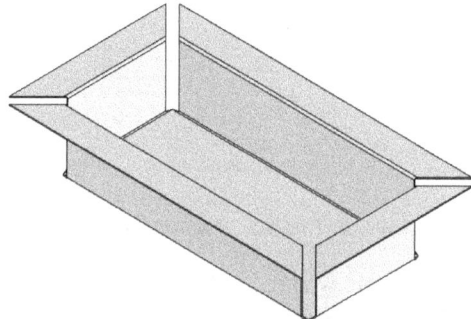

Figure-39. Miter flange with the modified rip distance

Start/End Offset Rollout

You can specify the start and end offset distances of the miter flange by using the options in the **Start/End Offset** rollout. The **Start Offset Distance** spinner is used to specify the offset distance from the start face of the miter flange. The **End Offset Distance** spinner is used to specify the offset distance from the end face of the miter flange. If the start and end offset distances are applied to the miter flange created on the continuous edges of the base flange, the start offset distance will be applied to the first edge and the end offset distance will be applied to the edge selected at last. Figure 35 shows the miter flange created on a single edge with the start and end offsets. Figure 36 shows the offsets applied to the miter flange created by selecting all the edges of the base flange.

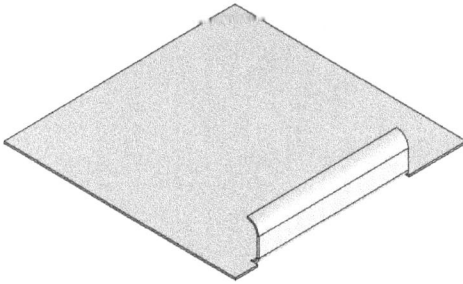

Figure-40. Miter flange created at an offset
distance on a single edge

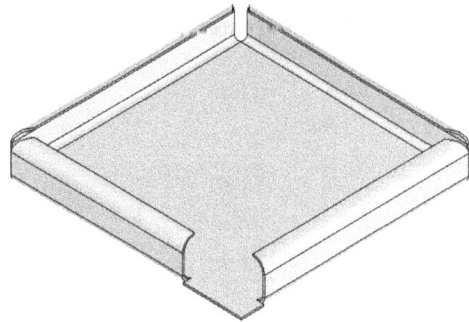

Figure-41. Miter flange created at an offset
distance on all edges

HEM

The **Hem** tool is used to create bend at the end edge of the
sheet. The steps to create hems are given next.

*   Click on the **Hem** tool from the **Ribbon**. The **Hem PropertyManager**
    will display; refer to Figure-42.

Figure-42. Hem PropertyManager

*   Select the edge on which you want to create the hem.
    Preview of hem will display; refer to Figure-43.

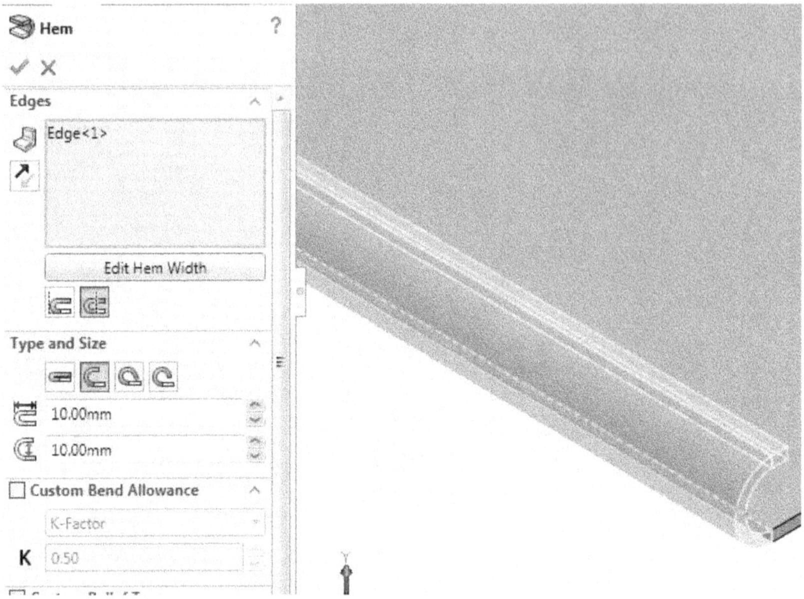

Figure-43. Preview of hem

- Select the desired shape using the **Closed, Open, Tear Drop,** or **Rolled** button from the **Type and Size** rollout.
- Specify the size and other parameters and click on the **OK** button from the **PropertyManager** to create the hem.

JOG

The **Jog** tool can be used to create double bend in the sheet. The steps to create jog are given next.

- Click on the **Jog** tool from the **Ribbon.** You are asked to select a plane for creating sketch.
- Sketch the bend line on the sheet metal face and exit the sketching environment. You are asked to select a face that you want to be fixed.
- Select the fixed face. The preview of jog will display; refer to Figure-44.

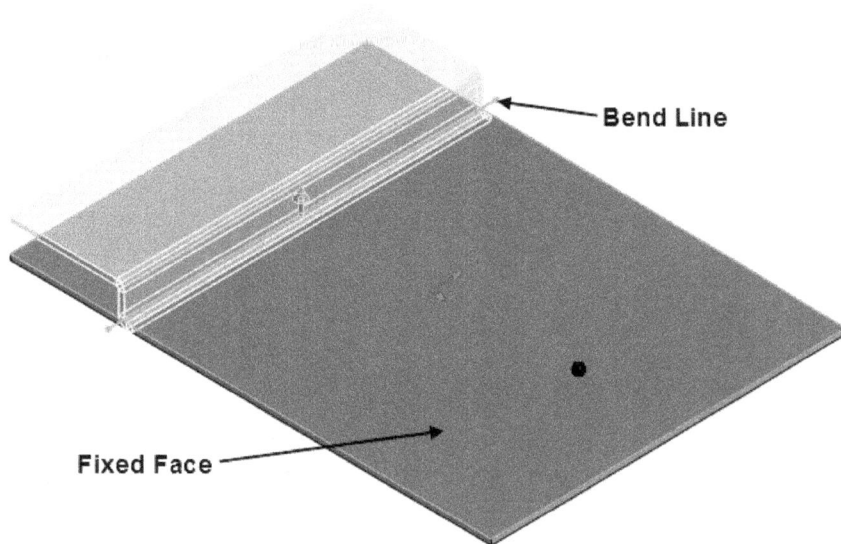

Figure-44. Preview of Jog

- Specify the desired parameters in the **PropertyManager** displayed and click on the **OK** button to create the bend.

SKETCHED BEND

The **Sketched Bend** tool is used to bend a sheet metal part by specified angle at selected bend line. The steps to bend a part are given next.

- Click on the **Sketched Bend** tool from the **Ribbon**. You are asked to select a flat face to bend.
- Select the face of sheet metal part. The sketching environment will display.
- Draw the bend line in such a way that it intersect to the edges of the sheet metal part's face.
- Exit the sketch and specify the parameters in the **PropertyManager**.
- Click on the **OK** button from the **PropertyManager**. The bend will be created with specified parameters; refer to Figure-45.

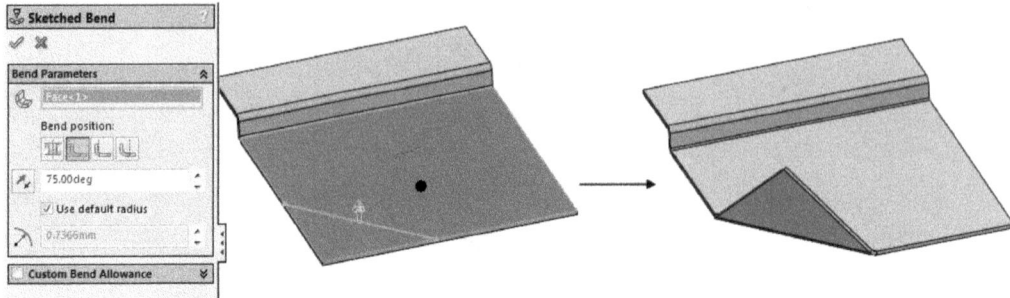

Figure-45. Sketched Bend

CROSS-BREAK

The **Cross-Break** tool is used to create cross-break in the sheet metal part. In HVAC or other duct works, cross-breaks are provided to stiffen the sheet. The procedure to use this tool is given next.

* Click on the **Cross-Break** tool from the **Sheet Metal** tab in the **Ribbon**. The **Cross Break PropertyManager** will be displayed; refer to Figure-46.

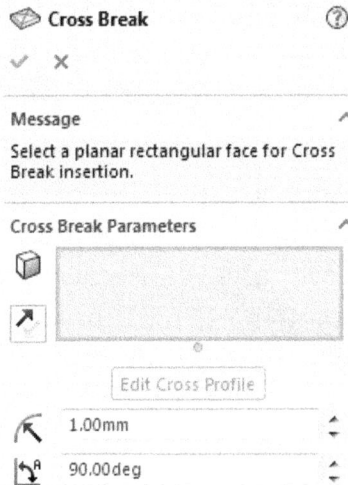

Figure-46. Cross Break Property-Manager

* Select the face on which you want to apply the cross-break. Preview of the cross-break will be displayed.
* Set the desired parameters and click on the **OK** button from the **PropertyManager**.

CLOSE CORNER

The **Close Corner** tool is used to change the closing of flange walls at corners. The steps to use this tool are given next.

• Click on the **Closed Corner** tool from **Corners** drop-down in the **Ribbon**. The **Closed Corner PropertyManager** will display.
• Select the face at corner that you want to close. Preview will be displayed; refer to Figure-47.

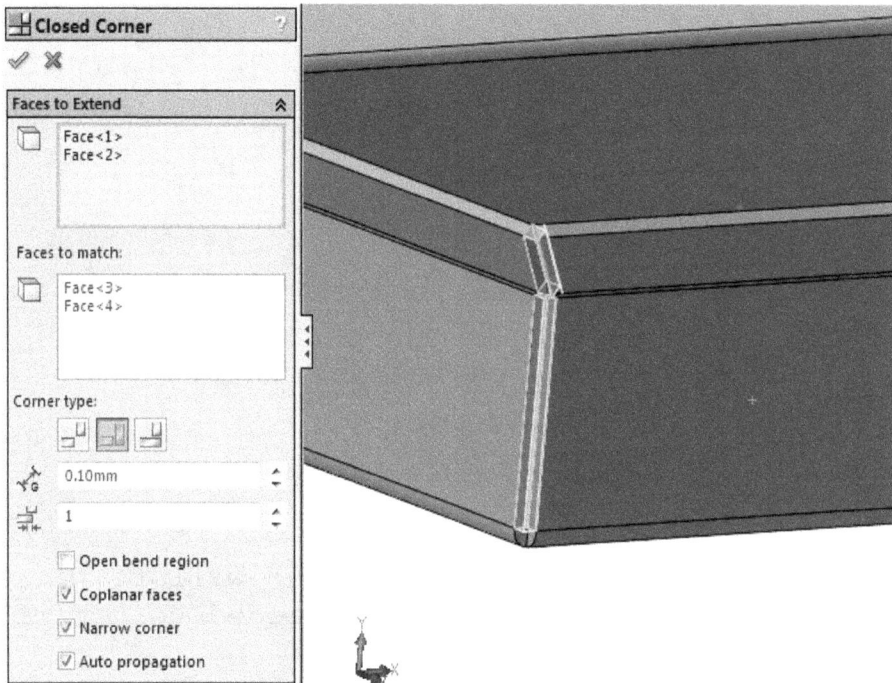

Figure-47. Preview of closed corner

• Set the desired options in the **PropertyManager** and click on the **OK** button to create closed corner.

WELDED CORNER

The **Welded Corner** tool is used to weld the gap between two adjacent walls. The procedure to use this tool is similar to **Close Corner** tool. The procedure is discussed next.

• Click on the **Welded Corner** tool from the **Corners** drop-down in the **Ribbon**; refer to Figure-48. The **Welded Corner PropertyManager** will be displayed; refer to Figure-49.

Figure-48. Corner drop-down

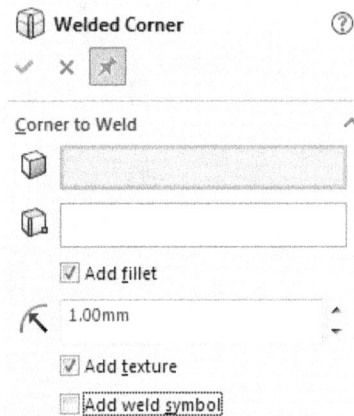

*Figure-49. Welded Corner Proper-
tyManager*

- Select the side face of sheet metal wall to be closed by welding bead. Preview of the welded corner will be displayed; refer to Figure-50.

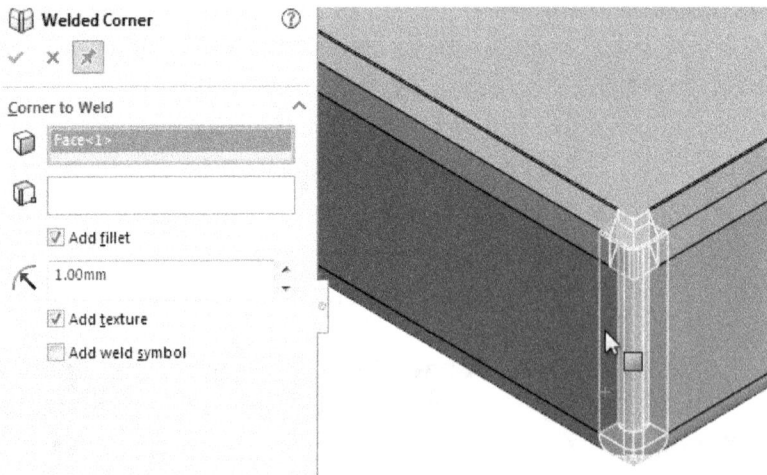

Figure-50. Preview of welded corner

- Set the fillet radius and other parameters in the **PropertyManager** and click on the **OK** button to create the welded corner.

BREAK-CORNER/CORNER-TRIM

The **Break-Corner/Corner-Trim** tool is used to chip-off the sharp corners of the sheet metal component. The procedure to use this tool is given next.

- Click on the **Break-Corner/Corner-Trim** tool from the **Corners** drop-down in the **Ribbon**. The **Break Corner PropertyManager** will be displayed; refer to Figure-51.

Figure-51. Break Corner PropertyManager

- Select the sharp edge of the sheet metal component. Preview of break corner will be displayed; refer to Figure-52.

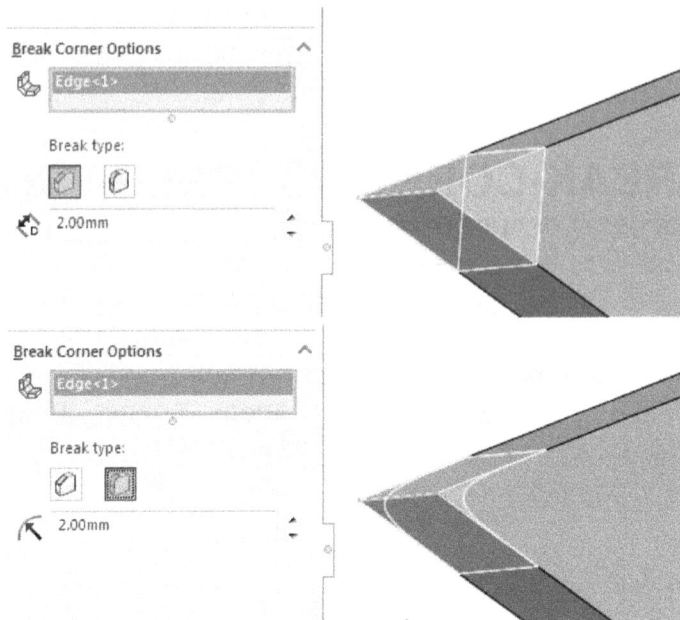

Figure-52. Preview of break corner

- Set the desired parameters and click on the **OK** button from the **PropertyManager** to create the corner break.

CORNER RELIEF

The **Corner Relief** tool is used to apply bend relief at the corners of sheet metal parts. The procedure to use this tool is given next.

- Click on the **Corner Relief** tool from the **Corners** drop-down in the **Ribbon**. The **Corner Relief PropertyManager** will be displayed; refer to Figure-53.
- Click on the **Collect all corners** button from the **PropertyManager**. All the corners in the sheet metal model will be displayed in the selection box in **PropertyManager**.
- Select one or all the corners from the selection box and select the desired button from the **Relief Options** rollout in the **PropertyManager**.

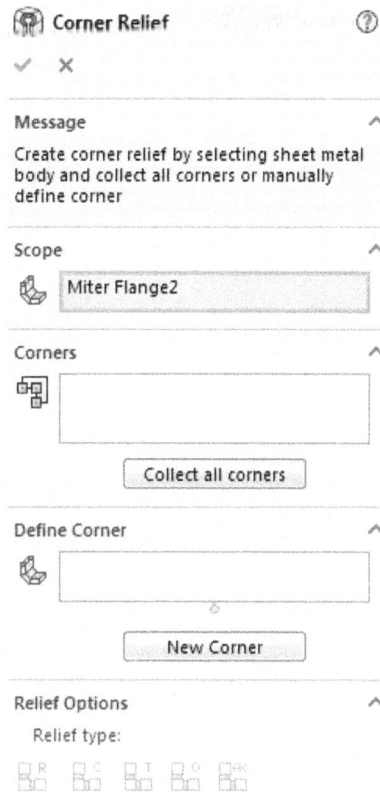

Figure-53. Corner Relief Property-Manager

SHEET METAL GUSSET

The **Sheet Metal Gusset** tool is used to add rib to support any sheet metal flange. The steps to create sheet metal gusset are given next.

• Click on the **Sheet Metal Gusset** tool from the **Ribbon**. The **Sheet Metal Gusset PropertyManager** will display; refer to Figure-54.

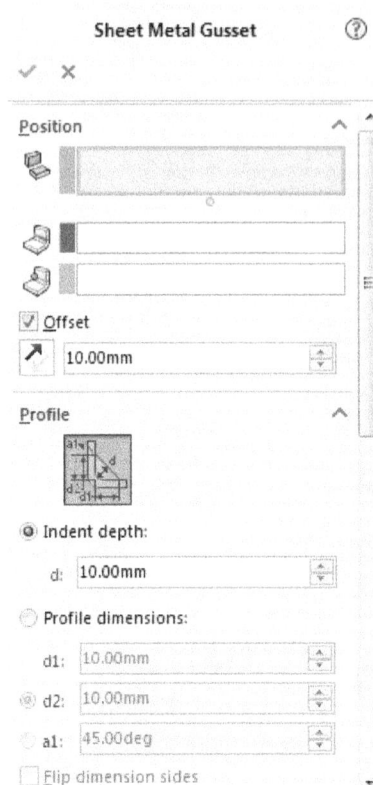

Figure-54. Sheet Metal Gusset
PropertyManager

- Select the two faces (one of base and other of flange). The preview of gusset will display; refer to Figure-55.
- Set the desired parameters and click on the **OK** button from the **PropertyManager** to create the gusset.

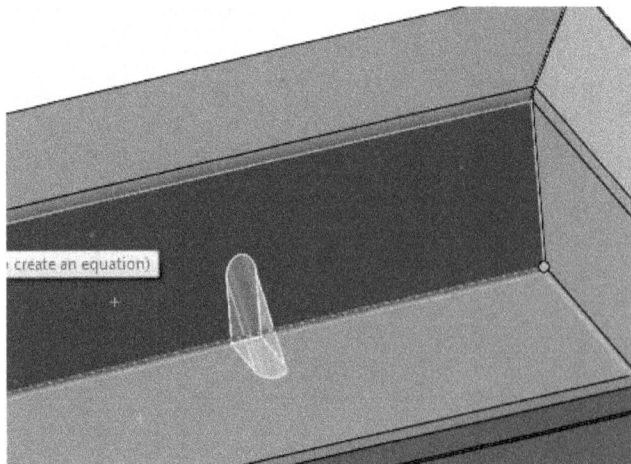

Figure-55. Preview of gusset

EXTRUDE CUT

The **Extrude Cut** tool is used to create cuts in the sheet metal part by extrusion. The steps to create extruded cut are given next.

- Click on the **Extruded Cut** tool from the **Ribbon**. You are asked to select the face of sheet metal part.
- Click on the face and draw the sketch of the cut
- Exit the sketch and click on the **OK** button to create the cut.

FLATTEN

The **Flatten** tool is used to create flat pattern of the sheet metal part. Click on the tool from the **Ribbon** and Flat pattern will be displayed. In some cases, you may be asked to specify the base. In those cases, select the flat face that you want to be fixed. Figure-56 shows a sheet metal part and its flat pattern.

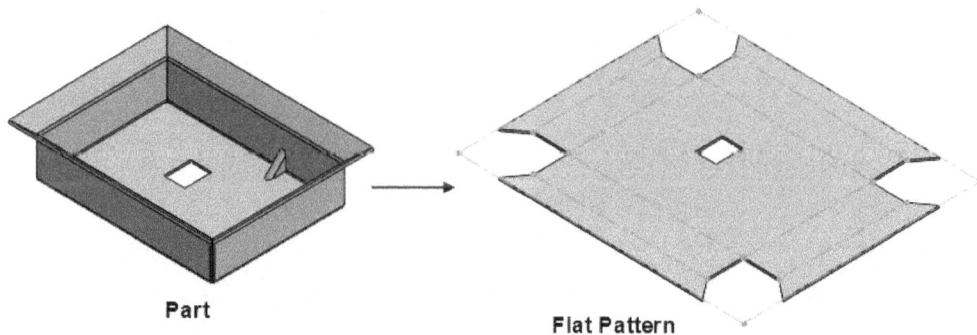

Part   Flat Pattern

Figure-56. Part and its flat pattern

INSERTING FLAT PATTERN IN DRAWING

- Start the drawing by using the **Make Drawing from Part/ Assembly** button from the **New** drop-down of **Menu Bar**; refer to Figure-57. The drawing mode will open.

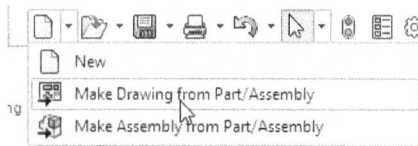

Figure-57. New drop-down

- Select the sheet size from the list and click on the **OK** button from the dialog box displayed. The **View Palette** will be displayed in the right of the drawing; refer to Figure-58.

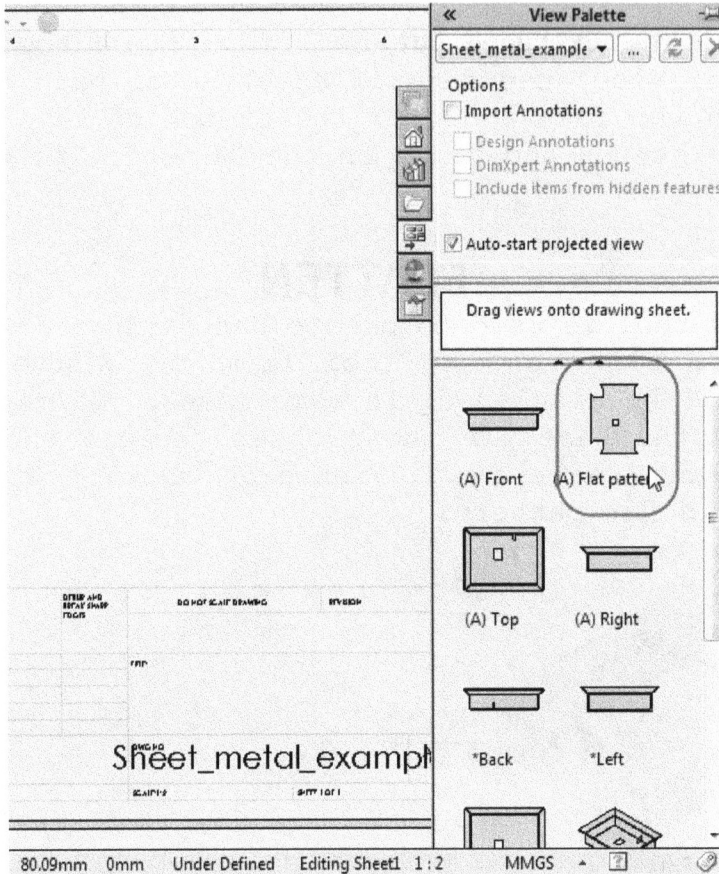

Figure-58. View Palette

- Drag the **Flat pattern** view from the palette and place it at desired location in the drawing; refer to Figure-59.

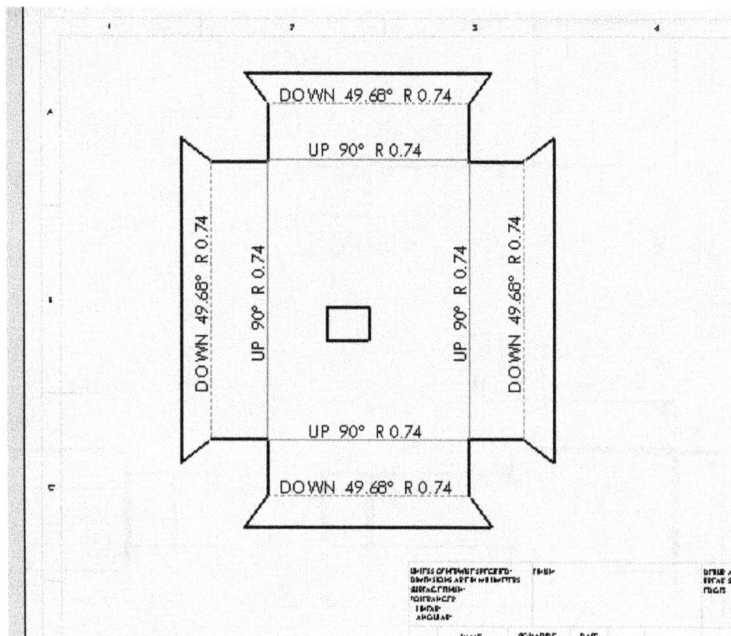

Figure-59. Flat pattern view

PRACTICE 1

Create the sheet metal model as shown in Figure-60. Dimensions are given in Figure-61.

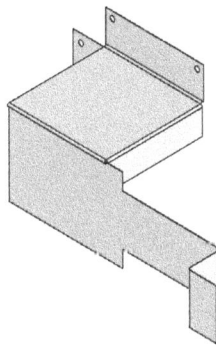

Figure-60. Sheet metal model

Detail A

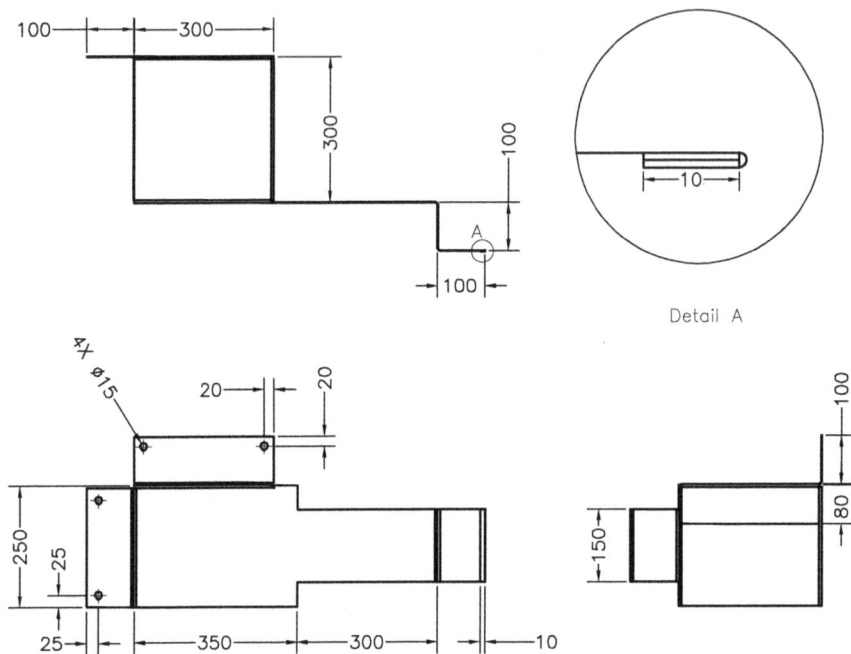

Thickness of sheet= 1mm
Bend Radius= 5mm

Figure-61. *Drawing views*

To get more drawings for practice, write us at **cadcamcaeworks@ gmail.com**

SELF ASSESSMENT

Q1. The _____ tool is used to create base feature of the sheet metal component.

Q2. The _____ tool is used to create sheet metal component by joining two or more sketch sections.

Q3. The _____ relief will tear the sheet in order to accommodate the bending of the sheet.

Q4. The _____ tool is used to create a flange of desired shape.

Q5. The _____ tool is used to create bend at the end edge of the sheet.

Q6. The **Jog** tool can be used to create double bend in the sheet. (T/F)

Q7. In HVAC or other duct works, cross-breaks are provided to stiffen the sheet. (T/F)

Q8. The **Close Corner** tool is used to make all the corners of a sheet metal part closer. (T/F)

FOR STUDENT NOTES

Weldments

Chapter 13

Topics Covered

The major topics covered in this chapter are:

- *Introduction.*
- *Weldment tool.*
- *Structural members.*
- *End Cap.*
- *Weld Beads.*
- *Welding Symbols in Drawing*

INTRODUCTION

This chapter is dedicated to welding joints or so called weldments. Welding is a method to permanently join parts with the help of welding beads. In SolidWorks, we can represent the welding beads in model as well as in drawing. In model, we can display a solid bead of weld around the selected faces/edges. We can also attach the welding bead symbol there. In drawing, we can attach the welding symbol to the affected edges/faces. We can also insert the cut list to tabulate the components used in the model. But, before we start using SolidWorks for weldments, its important to revise some basics of welding.

WELDING SYMBOLS AND REPRESENTATION IN DRAWING

The symbols to represent various type of welds are given next.

Butt/Groove Weld Symbols

Various symbols that come under this category are given next. Refer to Figure-1 and Figure-2.

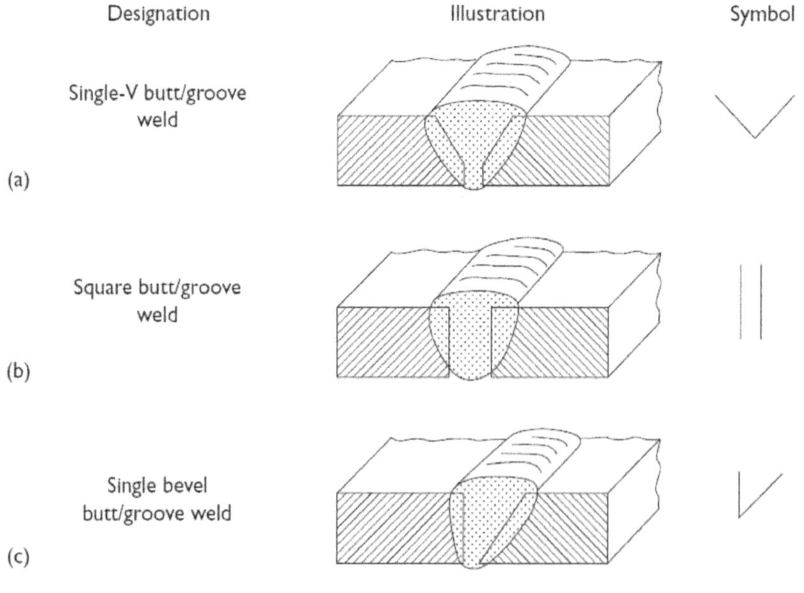

Designation	Illustration	Symbol
Single-V butt/groove weld (a)		
Square butt/groove weld (b)		
Single bevel butt/groove weld (c)		

Figure-1. Welding symbols list 1

(d) Single-U butt/groove weld

(e) Single-J butt/groove weld

(f) Butt weld between plates with raised edges (ISO)
Edge weld on a flanged groove joint (AWS)

ISO

AWS

(g) Single-V butt weld with broad root face

(h) Single bevel butt weld with broad root face

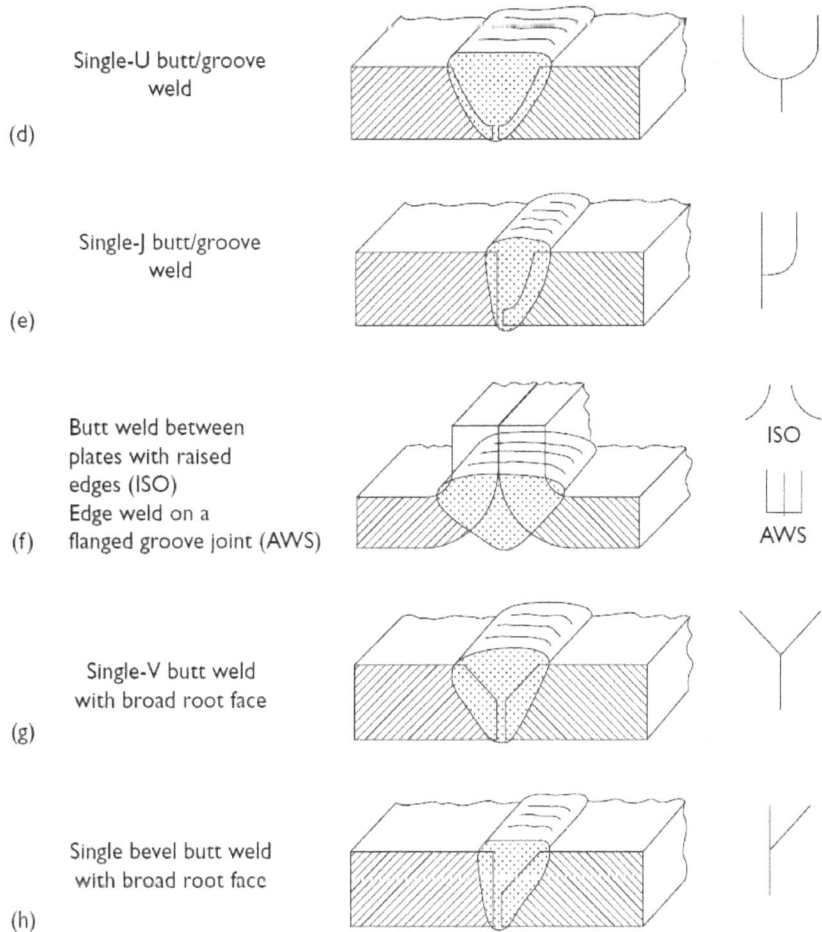

Figure-2. Welding symbols list 2

Fillet and Edge Weld Symbols

Various symbols that come under this category are given next. Refer to Figure-3.

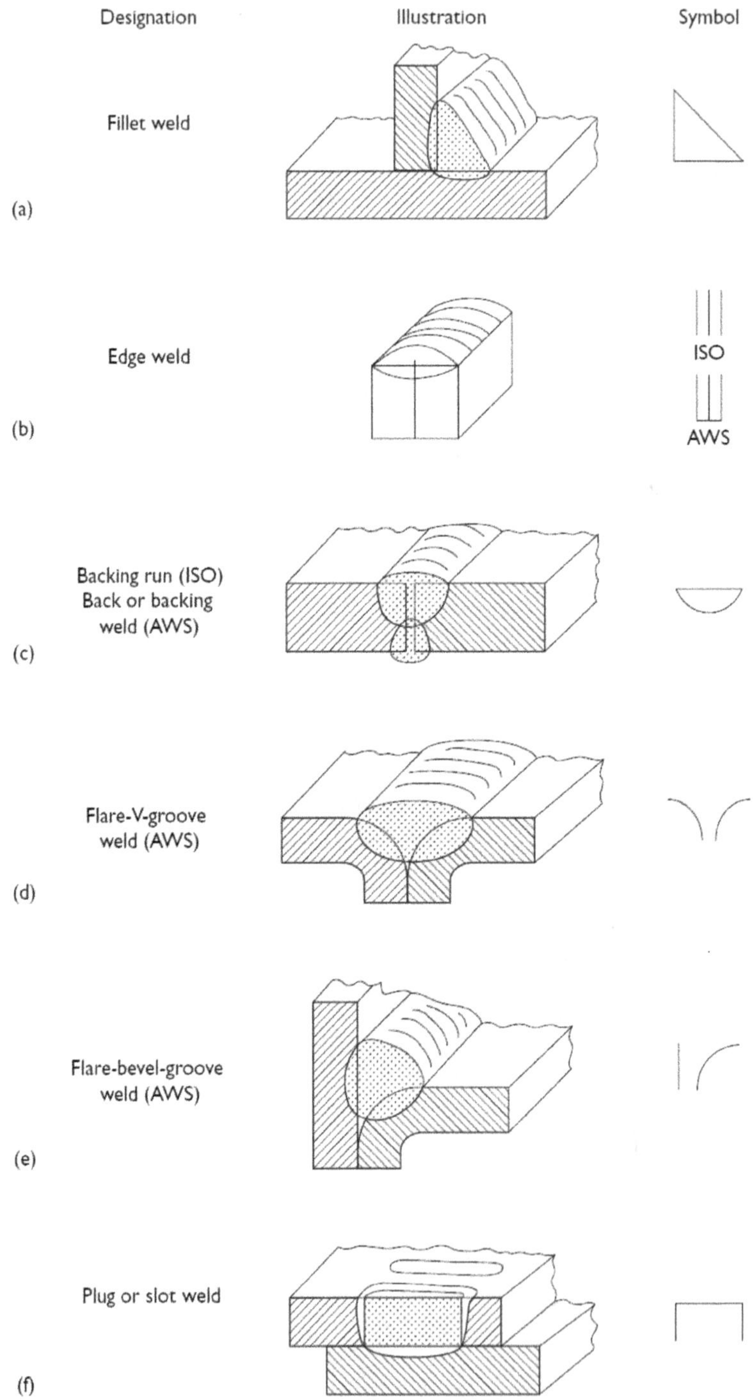

Designation	Illustration	Symbol
Fillet weld (a)		
Edge weld (b)		ISO AWS
Backing run (ISO) Back or backing weld (AWS) (c)		
Flare-V-groove weld (AWS) (d)		
Flare-bevel-groove weld (AWS) (e)		
Plug or slot weld (f)		

Figure-3. Welding symbols list 3

Miscellaneous Weld Symbols

Various symbols that come under this category are given next. Refer to Figure-4.

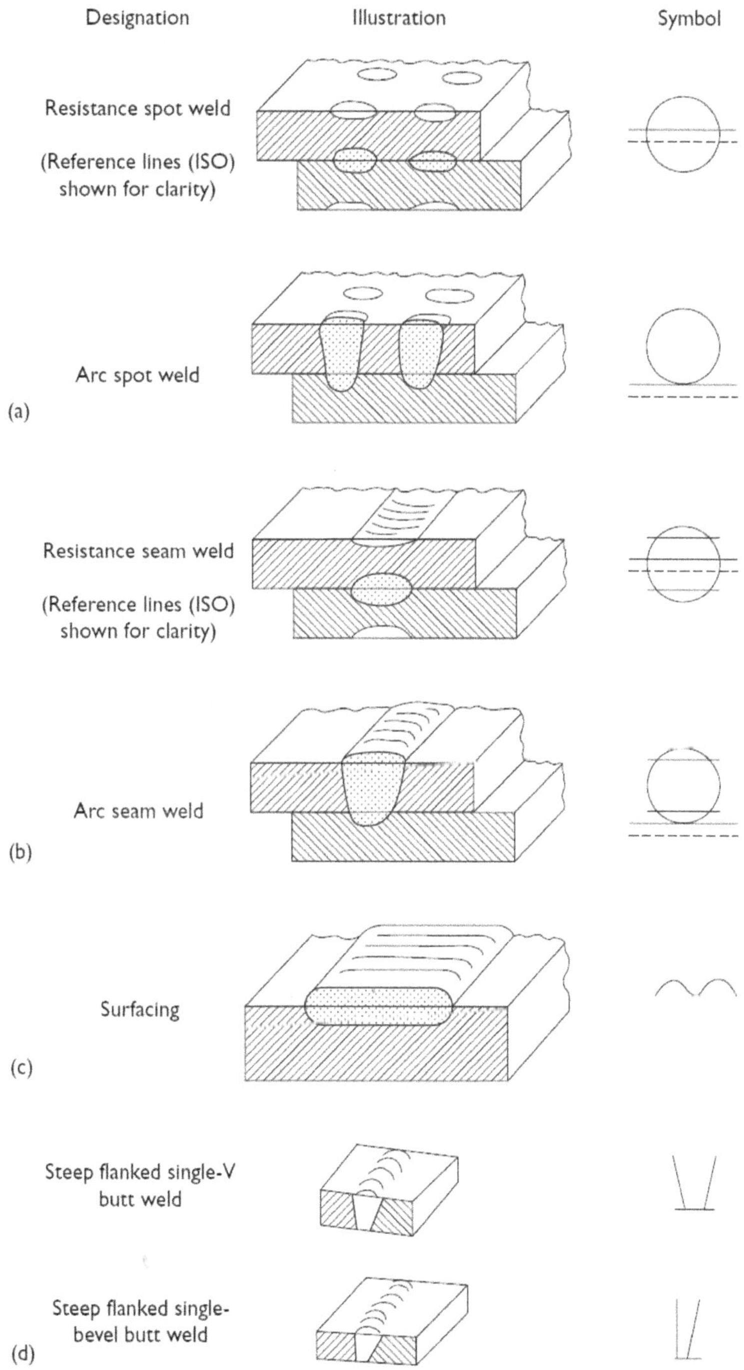

Designation	Illustration	Symbol

Resistance spot weld

(Reference lines (ISO) shown for clarity)

Arc spot weld

(a)

Resistance seam weld

(Reference lines (ISO) shown for clarity)

Arc seam weld

(b)

Surfacing

(c)

Steep flanked single-V butt weld

Steep flanked single-bevel butt weld

(d)

Figure-4. Welding symbols list 4

Now, we know various symbols used in welding drawings but keep a note that placement of welding symbol along the arrow decides the side on which the welding will be done on the object; refer to Figure-5.

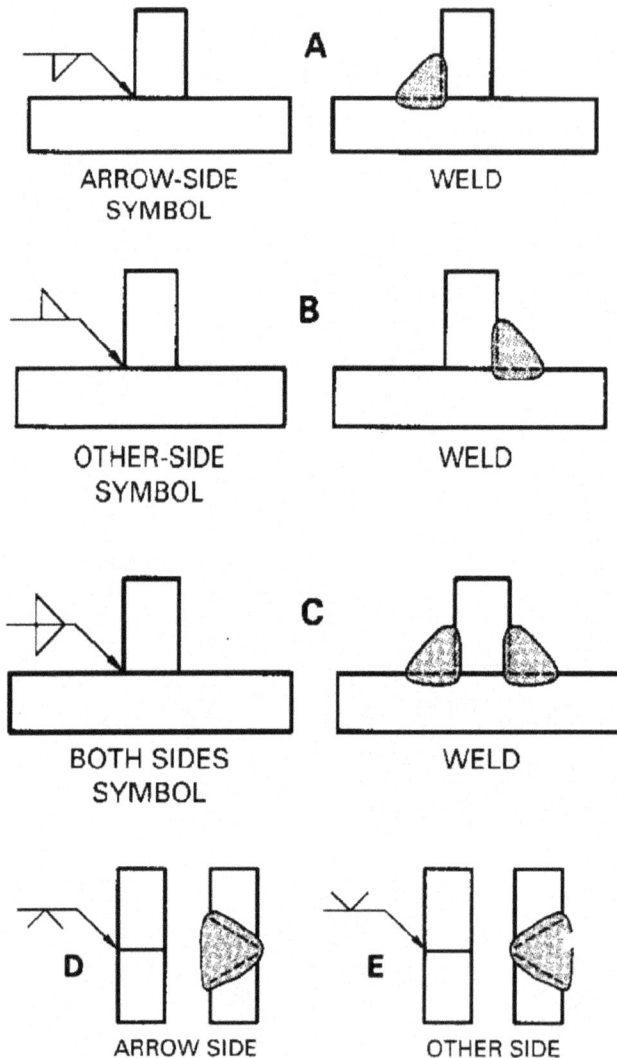

Figure-5. Deciding weld bead side

Dimensioning a weld bead

Alike the other measurements, weld is also measured with respect to various references so that we can control its quality. Figure-6 shows the information required for dimensioning a weld bead.

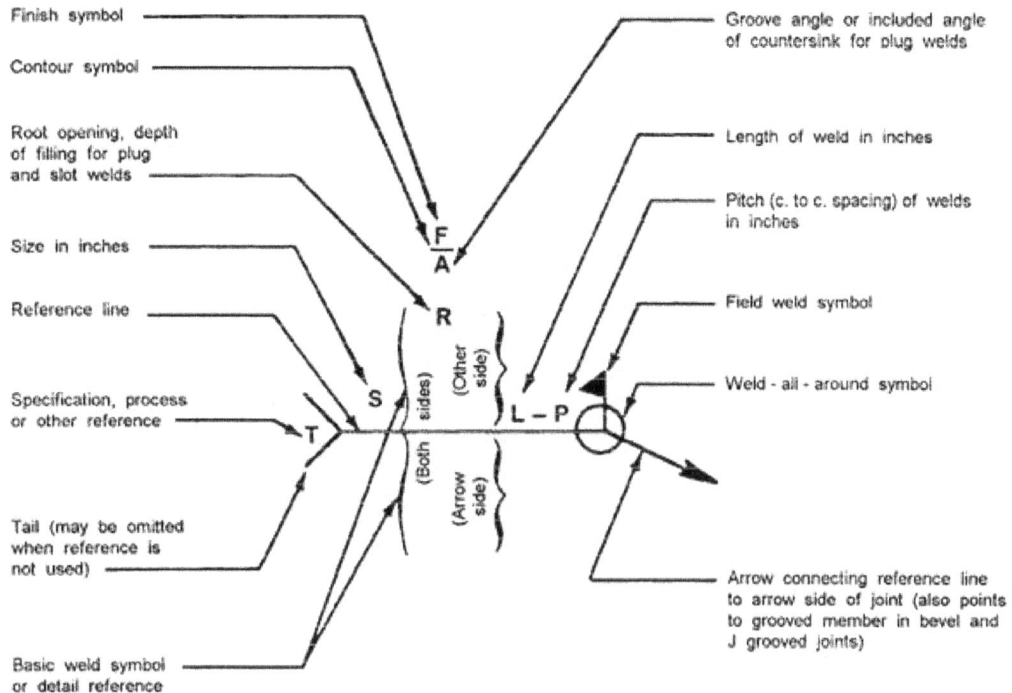

Figure-6. Welding dimension

Till this point, we have learned the basics of weld symbol representations in drawings. So, we are ready to dive into SolidWorks for creating welding representations.

The tools to add weldments are available in the **Weldments** tab of the **Ribbon**; refer to Figure-7. If this tab is not displayed by default, you can add it by right-clicking on any of the tab in the **Ribbon** and then selecting the **Weldments** option from the shortcut menu displayed; refer to Figure-8.

Figure-7. Weldments CommandManager

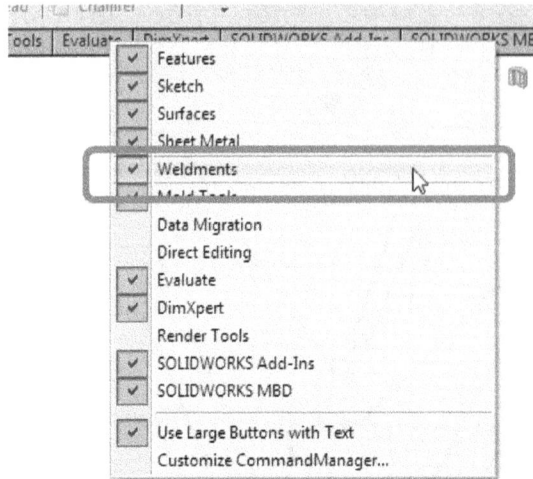

Figure-8. Weldments option in shortcut menu

Most of the tools in this tab have been discussed earlier. The remaining tools are discussed next.

WELDMENT TOOL

The **Weldment** tool is used to start multi-body environment in which you can create parts without merging them into a single body. In this way, we are allowed to join two entities in the modeling environment with the help of weld beads. The **Weldment** tool does not do a function like **Extrude** or **Revolve** tool but it starts **Weldment** environment. Click on this tool to start the environment.

STRUCTURAL MEMBER

The **Structural Member** tool is used to create structural member for welding, like angle, C channel, pipe, and so on. The procedure to use this tool is given next.

- Click on the **Structural Member** tool from the **Ribbon**. The **Structural Member PropertyManager** will be displayed; refer to Figure-9.

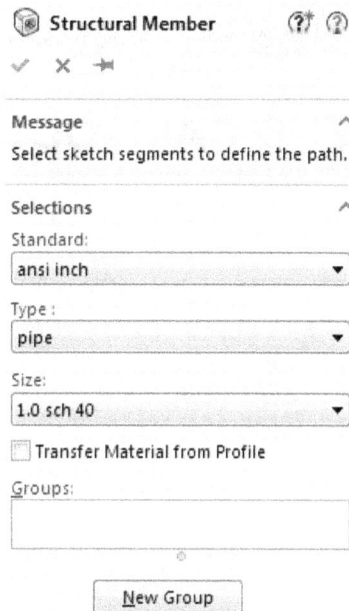

Figure-9. Structural Member PropertyManager

- Select the desired standard, type and size of structural member from the **PropertyManager**.
- One by one click on the line member of the sketch. Preview of the structural member will be displayed; refer to Figure-10.

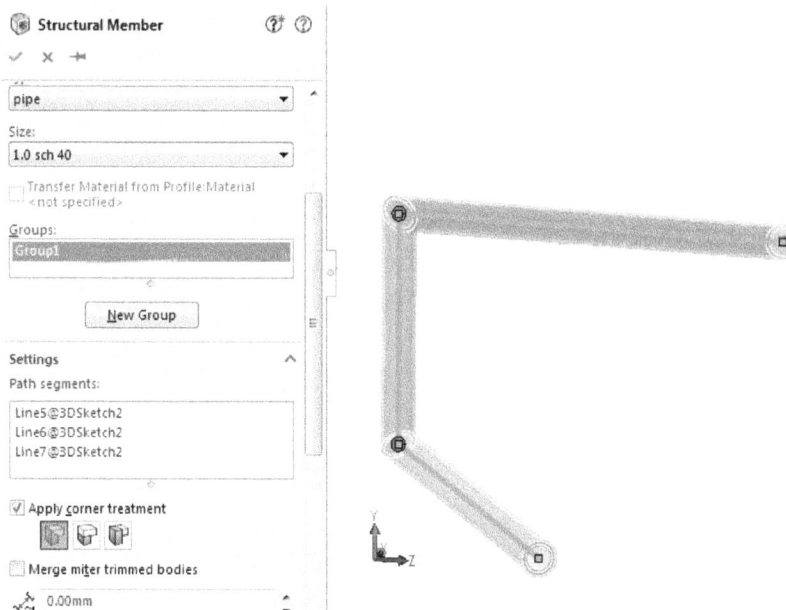

Figure-10. Preview of the structure

- Set the desired parameters and click on the **OK** button from the **PropertyManager** to create the structure.

END CAP TOOL

The **End Cap** tool is used to close the ends of an open structural member by using a lid. The procedure to use this tool is given next.

- Click on the **End Cap** tool from the **Ribbon**. The **End Cap PropertyManager** will be displayed; refer to Figure-11.

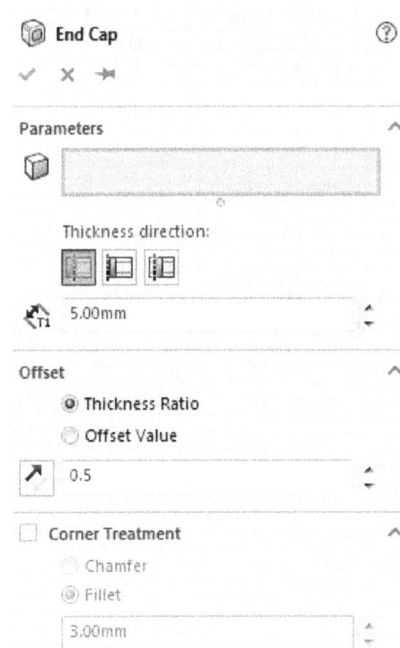

Figure-11. End Cap PropertyManager

- Select the end face of the structural member. Preview of the end cap will be displayed; refer to Figure-12.

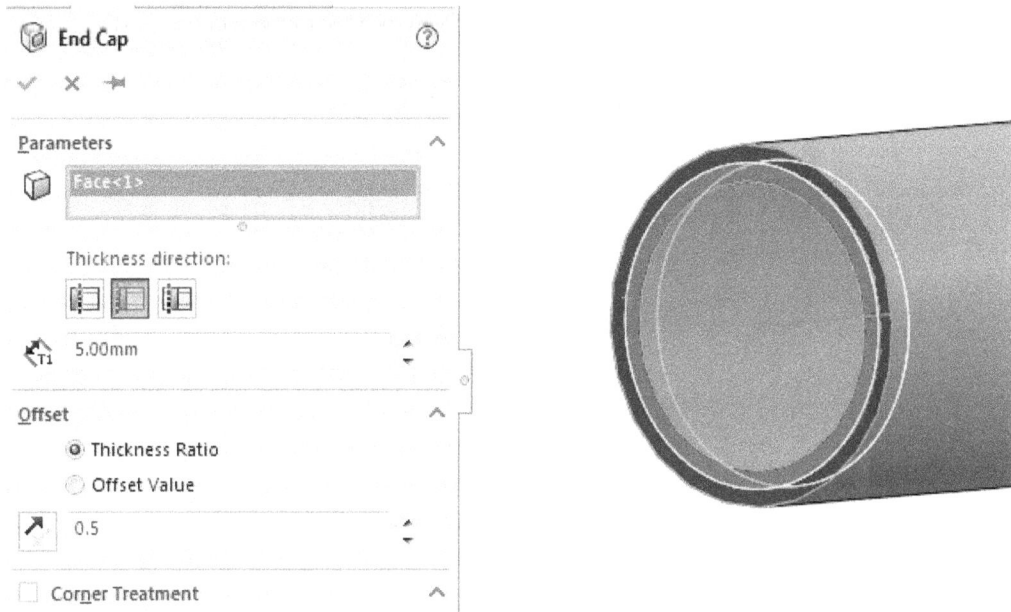

Figure-12. Preview of the end cap

- Select the desired thickness direction using the three buttons in the **Thickness direction:** area of the **Parameters** rollout.
- Specify the desired thickness in the **Thickness** edit box.
- Specify the other required parameters and then click on the **OK** button from the **PropertyManager** to create the end cap.

WELD BEAD

The **Weld Bead** tool is used to apply the desired type of welding bead on the selected edges. The procedure to use this tool is given next.

- Click on the **Weld Bead** tool from the **Ribbon**. The **Weld Bead PropertyManager** will be displayed; refer to Figure-13.

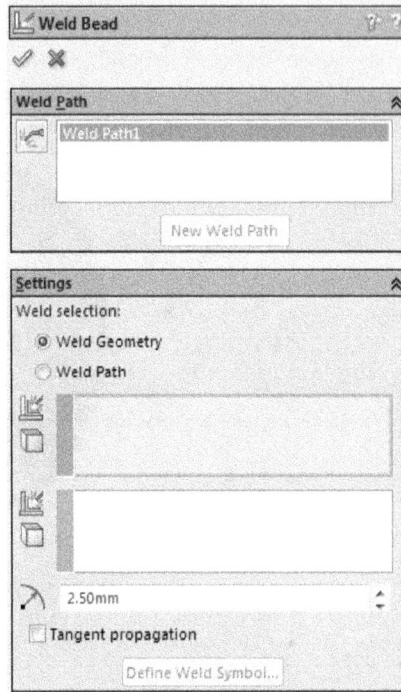

Figure-13. Weld Bead PropertyManager

- If **Weld Geometry** radio button is selected in the **Weld selection** area of the **Settings** rollout then select the face sets of the two joining parts; refer to Figure-14.

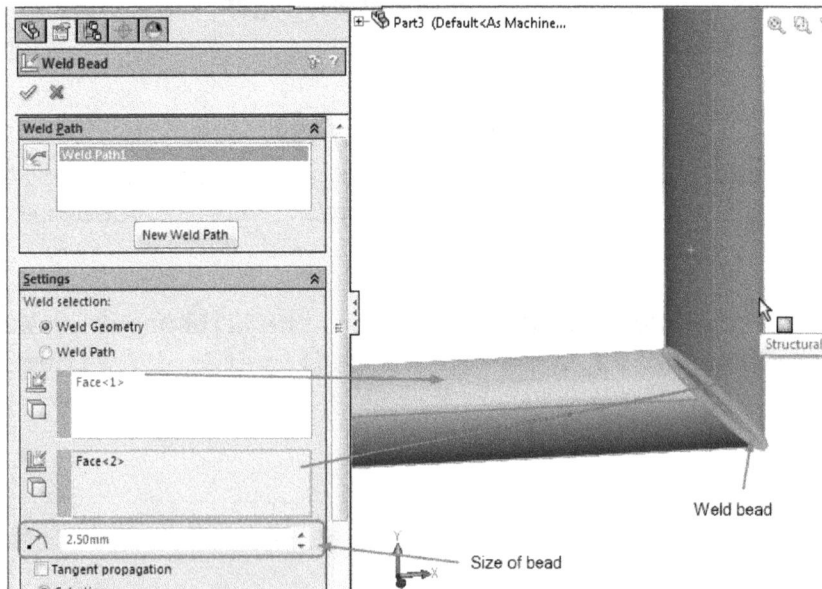

Figure-14. Weld bead using faces

- If you select the **Weld Path** radio button from the **Weld selection** area of the **Settings** rollout then you need to select the edge/edges along which you want to represent welding bead; refer to Figure-15.

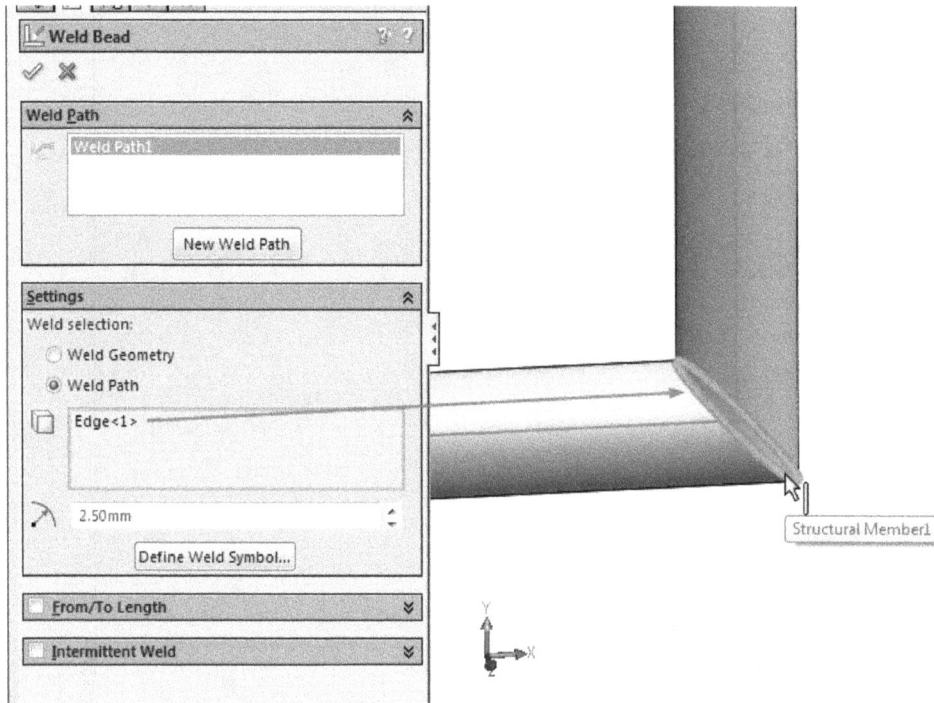

Figure-15. Weld bead using edge

- After creating every closed bead, you need to start a new weld path by clicking on the **New Weld Path** tool from the **Weld Path** rollout in the **PropertyManager**.
- To display the weld symbol on the model, click on the **Define Weld Symbol** button from the **Settings** rollout. A dialog box will be displayed; refer to Figure-16.

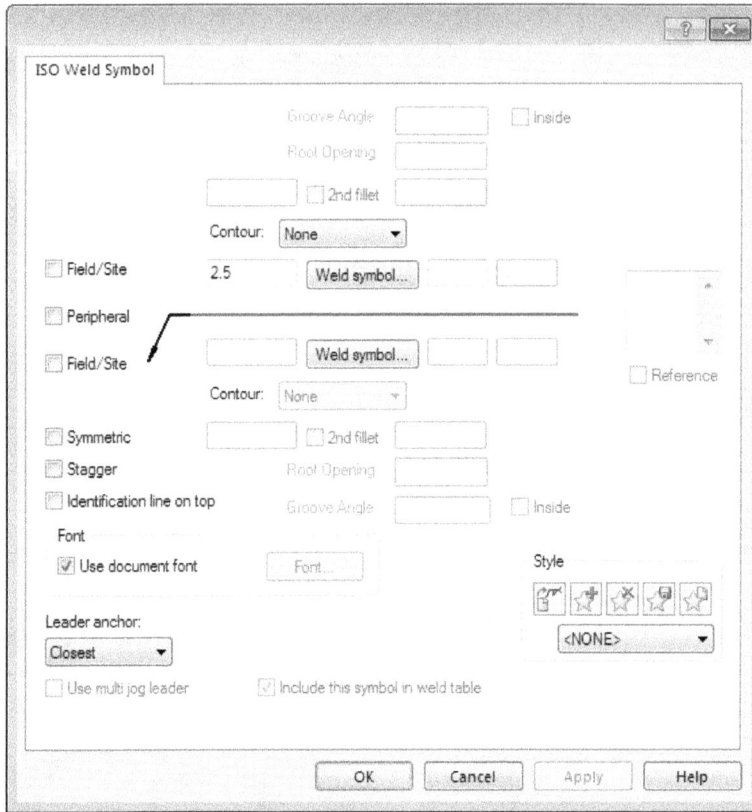

Figure-16. Weld Symbol dialog box

- Now, this is the dialog box in which we need to feed all the information regarding the weld bead. You can go back to the starting of this chapter and specify the dimensions for weld bead.
- Click on the **Weld Symbol** button in the dialog box and select the desired welding symbol; refer to Figure-17. Click on the **More Symbols** option from the flyout if desired symbol is not listed.

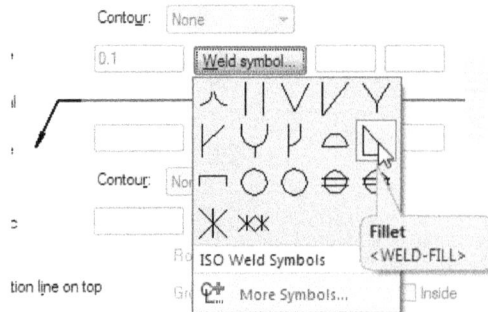

Figure-17. Weld symbols

- After specifying the desired dimension, click on the **OK** button from the dialog box.
- To set the desired length of weld bead, select the **From/To Length** check box and specify the desired length of bead; refer to Figure-18.

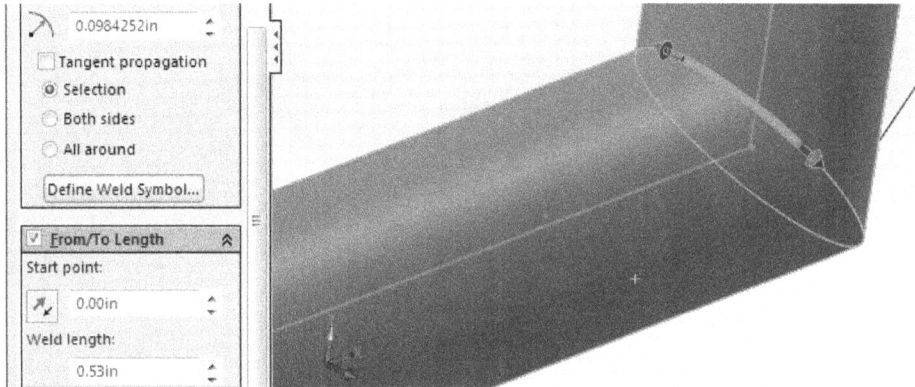

Figure-18. Setting length of weld bead

- Similarly, you can select the Intermittent Weld check box and specify the intermediate gap for welding bead; refer to Figure-19.

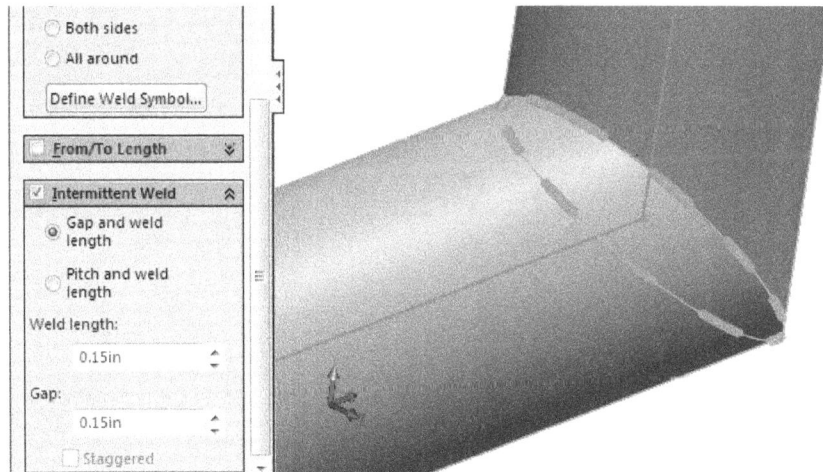

Figure-19. Intermittent weld

- After specifying the desired parameters, click on the **OK** button from the **PropertyManager** to create the bead. The bead will be added in the **Weld Folder** in **FeatureManager Design Tree**; refer to Figure-20.

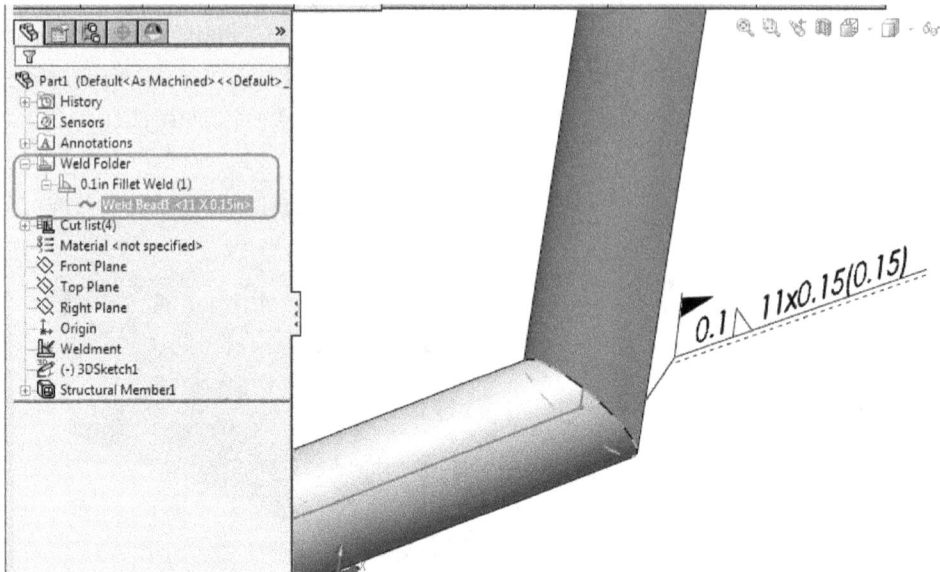

Figure-20. Weld Folder

INSERTING WELDING DATA IN DRAWING

There is no use of assigning welding symbols unless the manufacturer/fabricator does not get them in his drawing. The procedure to insert the welding data in the drawing is given next.

- After creating the welding model, click on the **Make Drawing from Part** option from the **File** menu; refer to Figure-21. The drawing environment will be displayed.

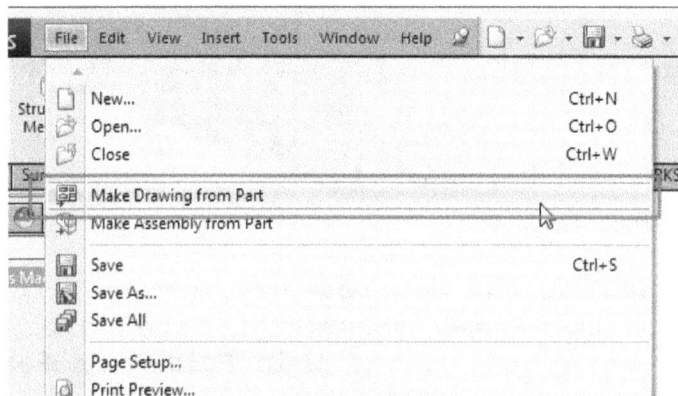

Figure-21. Make Drawing from Part option

- Select the desired sheet size and create the basic views; refer to Figure-22.

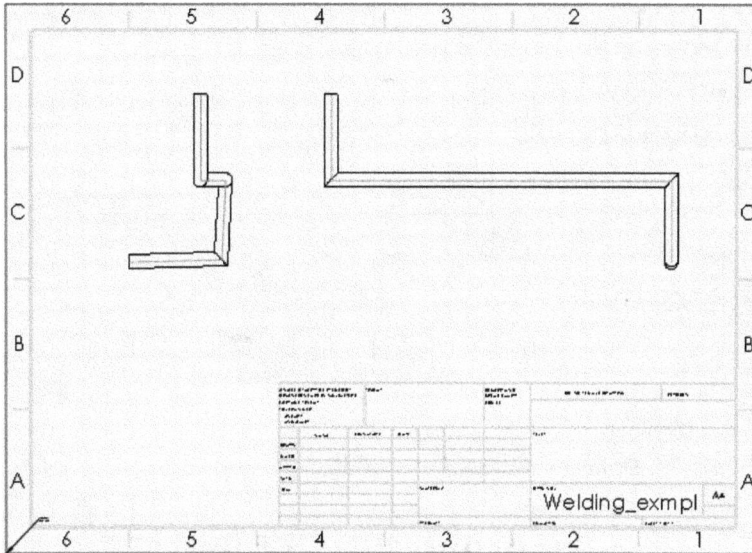

Figure-22. Drawing from part

- Select the view/views in which you want to display the annotations and click on the **Model Items** button from the **Annotations** tab in the **Ribbon**. The **Model Items PropertyManager** will be displayed.
- Select the **Weld Symbols** button from the **Annotations** rollout in the **PropertyManager**; refer to Figure-23.

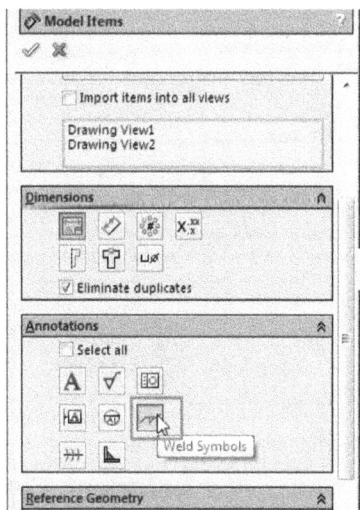

Figure-23. Weld Symbols in Model Items PropertyManager

- Click on the **OK** button from the **PropertyManager**. The welding symbols will be assigned in the drawing; refer to Figure-24.

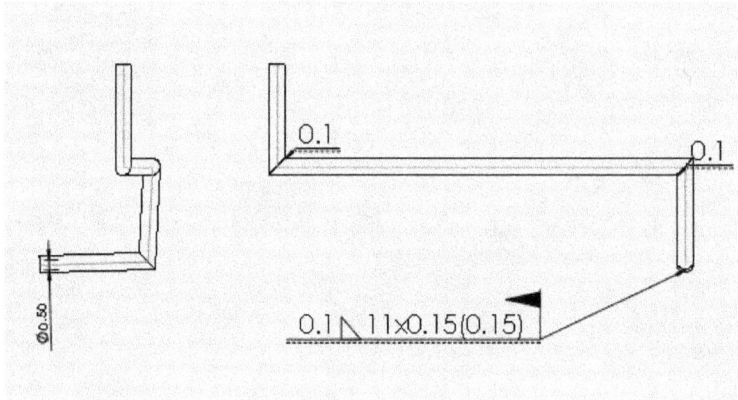

Figure-24. Drawing with welding symbols

Inserting the Cut list

Cut list is a kind of part list similar to bill of material. The cut list is used to identify the structural members that are being joined by using the welding bead. The procedure to insert the cut list is given next.

- Click on the **Weldment Cut List** option from the **Tables** drop-down in the **Annotation** tab of the **Ribbon**; refer to Figure-25. You are asked to select a view for which you want to create the cut list.

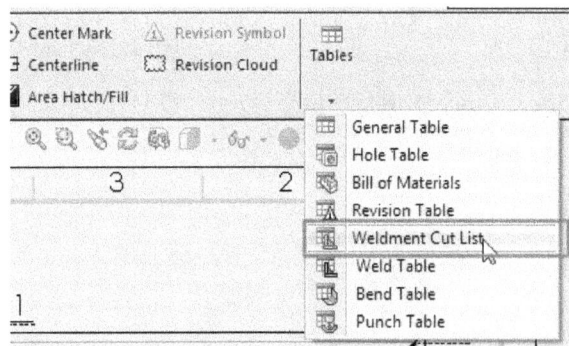

Figure-25. Weldment cut list option

- Select the desired view. The **Weldment Cut List PropertyManager** will be displayed; refer to Figure-26.

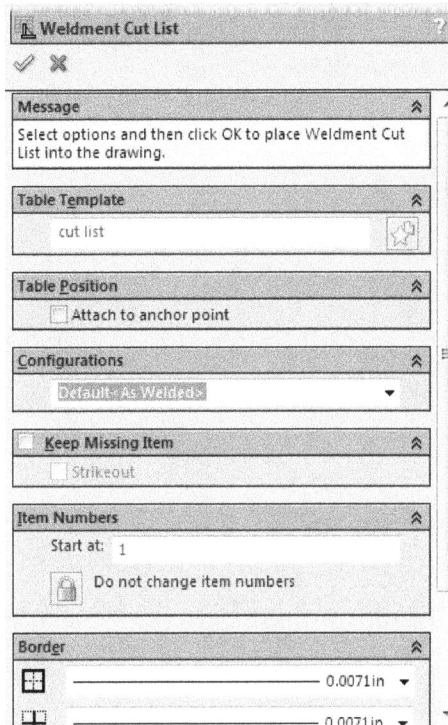

Figure-26. Weldment Cut List PropertyManager

- Specify the desired options, if any. Next, click on the **OK** button from the **PropertyManager**. The cut list will get attached to the cursor; refer to Figure-27.

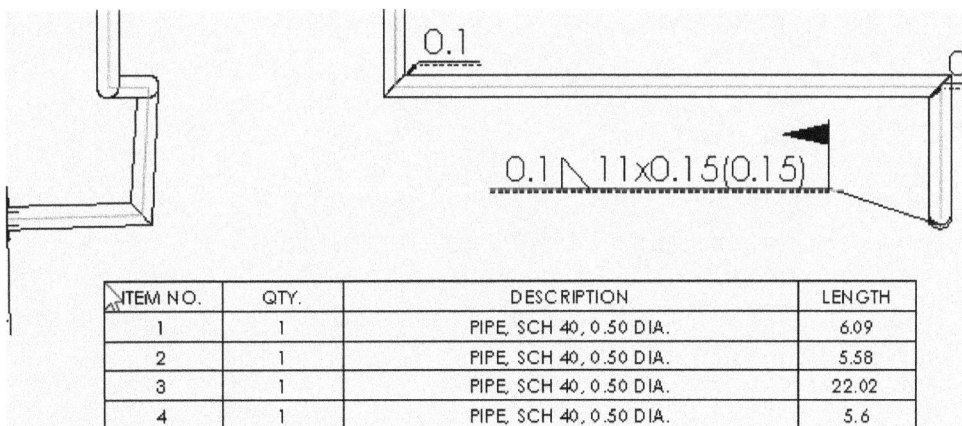

ITEM NO.	QTY.	DESCRIPTION	LENGTH
1	1	PIPE, SCH 40, 0.50 DIA.	6.09
2	1	PIPE, SCH 40, 0.50 DIA.	5.58
3	1	PIPE, SCH 40, 0.50 DIA.	22.02
4	1	PIPE, SCH 40, 0.50 DIA.	5.6

Figure-27. Cut list attached to the cursor

- Click at the desired location to place the cut list. Similarly, you can place the Weld table to specify the welding length and parameters.

SELF ASSESSMENT

Q1. Draw the symbol of Square butt/groove weld.

Q2. Draw the symbol of Single bevel butt weld with broad root face.

Q3. Draw the symbol of Fillet weld.

Q4. Draw the symbol of Plug weld.

Q5. Draw the symbol of Single-J butt/groove weld.

Q6. Which of the following figure shows correct annotation of welding symbol?

a.

b.

c. Both of the above.
d. None of the above.

Q7. Which of the following is not a type of structural member in Structural Member PropertyManager?

a. pipe
b. round bar
c. c channel
d. s section.

FOR STUDENT NOTES

For Student Notes

Index

T

1

www.ingramcontent.com/pod-product-compliance
Lightning Source LLC
Chambersburg PA
CBHW081105050426
R18088300001B/R180883PG42334CBX00001B/1